Intelligent Systems Reference Library

Volume 189

The aim of this series is to publish a Reference Library, including novel advances and developments in all aspects of Intelligent Systems in an easily accessible and well structured form. The series includes reference works, handbooks, compendia, textbooks, well-structured monographs, dictionaries, and encyclopedias. It contains well integrated knowledge and current information in the field of Intelligent Systems. The series covers the theory, applications, and design methods of Intelligent Systems. Virtually all disciplines such as engineering, computer science, avionics, business, e-commerce, environment, healthcare, physics and life science are included. The list of topics spans all the areas of modern intelligent systems such as: Ambient intelligence, Computational intelligence, Social intelligence, Computational neuroscience, Artificial life, Virtual society, Cognitive systems, DNA and immunity-based systems, e-Learning and teaching, Human-centred computing and Machine ethics, Intelligent control, Intelligent data analysis, Knowledge-based paradigms, Knowledge management, Intelligent agents, Intelligent decision making, Intelligent network security, Interactive entertainment, Learning paradigms, Recommender systems, Robotics and Mechatronics including human-machine teaming, Self-organizing and adaptive systems, Soft computing including Neural systems, Fuzzy systems, Evolutionary computing and the Fusion of these paradigms, Perception and Vision, Web intelligence and Multimedia.

** Indexing: The books of this series are submitted to ISI Web of Science, SCOPUS, DBLP and Springerlink.

More information about this series at http://www.springer.com/series/8578

Gloria Phillips-Wren · Anna Esposito ·
Lakhmi C. Jain
Editors

Advances in Data Science: Methodologies and Applications

Springer

Editors
Gloria Phillips-Wren
Sellinger School of Business
and Management
Loyola University Maryland
Baltimore, MD, USA

Lakhmi C. Jain
University of Technology Sydney
Broadway, Australia

Liverpool Hope University
Liverpool, UK

KES International
Shoreham-by-Sea, UK

Anna Esposito
Dipartimento di Psicologia
Università della Campania
"Luigi Vanvitelli", and IIASS
Caserta, Italy

ISSN 1868-4394 ISSN 1868-4408 (electronic)
Intelligent Systems Reference Library
ISBN 978-3-030-51869-1 ISBN 978-3-030-51870-7 (eBook)
https://doi.org/10.1007/978-3-030-51870-7

This Springer imprint is published by the registered company Springer Nature Switzerland AG
The registered company address is: Gewerbestrasse 11, 6330 Cham, Switzerland

Preface

The tremendous advances in inexpensive computing power and intelligent techniques have opened many opportunities for managing and investigating data in virtually every field including engineering, science, healthcare, business, and so on. A number of paradigms and applications have been proposed and used by researchers in recent years as this book attests, and the scope of data science is expected to grow over the next decade. These future research achievements will solve old challenges and create new opportunities for growth and development.

The research presented in this book are interdisciplinary and cover themes embracing emotions, artificial intelligence, robotics applications, sentiment analysis, smart city problems, assistive technologies, speech melody, and fall and abnormal behavior detection.

This book provides a vision on how technologies are entering into ambient living places and how methodologies and applications are changing to involve massive data analysis of human behavior.

The book is directed to researchers, practitioners, professors, and students interested in recent advances in *methodologies and applications* of data science. We believe that this book can also serve as a reference to relate different applications using a similar methodological approach.

Thank are due to the chapter contributors and reviewers for sharing their deep expertise and research progress in this exciting field.

The assistance provided by Springer-Verlag is gratefully acknowledged.

Baltimore, Maryland, USA — Gloria Phillips-Wren
Caserta, Italy — Anna Esposito
Sydney, Australia/Liverpool, UK/Shoreham-by-Sea, UK — Lakhmi C. Jain

Contents

About the Editors

Gloria Phillips-Wren is Full Professor in the Department of Information Systems, Law and Operations Management at Loyola University Maryland. She is Co-editor-in-chief of *Intelligent Decision Technologies International Journal (IDT)*, Associate Editor of the *Journal of Decision Systems (JDS)* Past Chair of SIGDSA (formerly SIGDSS) under the auspices of the Association of Information Systems, a member of the SIGDSA Board, Secretary of IFIP WG8.3 DSS, and leader of a focus group for KES International. She received a Ph.D. from the University of Maryland Baltimore County and holds MS and MBA degrees. Her research interests and publications are in decision making and support, data analytics, business intelligence, and intelligent systems. Her publications have appeared in *Communications of the AIS, Omega*, *European Journal of Operations Research*, *Information Technology & People, Big Data*, and *Journal of Network and Computer Applications*, among others. She has published over 150 articles and 14 books. She can be reached at: gwren@loyola.edu.

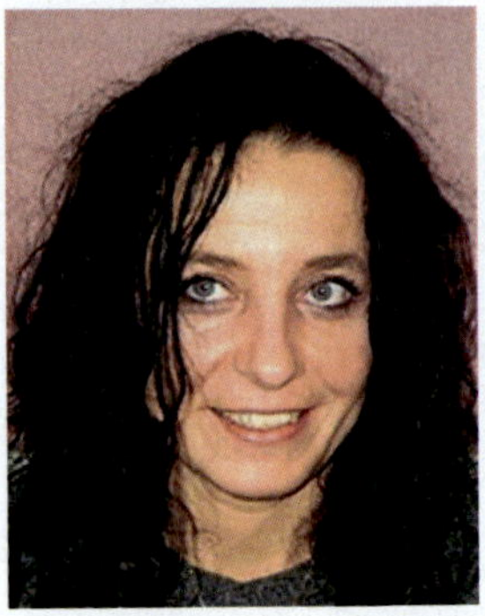

Anna Esposito received her "Laurea Degree" *summa cum laude* in Information Technology and Computer Science from the Università di Salerno with a thesis published on **Complex System**, 6(6), 507–517, 1992), and Ph.D. Degree in Applied Mathematics and Computer Science from Università di Napoli "Federico II". Her Ph.D. thesis published on **Phonetica**, 59(4), 197–231, 2002, was developed at MIT (1993–1995), Research Laboratory of Electronics (Cambridge, USA). Anna has been a Post Doc at the IIASS, and Assistant Professor at Università di Salerno (Italy), department of Physics, where she taught Cybernetics, Neural Networks, and Speech Processing (1996–2000). From 2000 to 2002, she held a Research Professor position at Wright State University, Department of Computer Science and Engineering, OH, USA. From 2003, Anna is Associate Professor in Computer Science at Università della Campania "Luigi Vanvitelli" (UVA). In 2017, she has been awarded of the full professorship title. Anna teach Cognitive and Algorithmic Issues of Multimodal Communication, Social Networks Dynamics, Cognitive Economy, and Decision Making. She authored 240+ peer reviewed publications and edited/co-edited 30+ international books. Anna is the Director of the Behaving Cognitive Systems laboratory (BeCogSys), at UVA. Currently, the lab is participating to the H2020 funded projects: (a) **Empathic**, www.empathic-project.eu/, (b) **Menhir**, menhir-project.eu/ and the national funded projects, (c) **SIROBOTICS**, https://www.istitutomarino.it/project/si-robotics-social-robotics-for-active-and-healthy-ageing/, and (d) **ANDROIDS**, https://www.psicologia.unicampania.it/research/projects.

Lakhmi C. Jain, Ph.D., ME, BE(Hons) Fellow (Engineers Australia) is with the University of Technology Sydney, Australia, and Liverpool Hope University, UK.

Professor Jain founded the KES International for providing professional community the opportunities for publications, knowledge exchange, cooperation, and teaming. Involving around 5,000 researchers drawn from universities and companies world-wide, KES facilitates international cooperation and generate synergy in teaching and research. KES regularly provides networking opportunities for professional community through one of the largest conferences of its kind in the area of KES. www.kesinternational.org.

Chapter 1
Introduction to Big Data and Data Science: Methods and Applications

Gloria Phillips-Wren, Anna Esposito, and Lakhmi C. Jain

Abstract Big data and data science are transforming our world today in ways we could not have imagined at the beginning of the twenty-first century. The accompanying wave of innovation has sparked advances in healthcare, engineering, business, science, and human perception, among others. In this chapter we discuss big data and data science to establish a context for the state-of-the-art technologies and applications in this book. In addition, to provide a starting point for new researchers, we present an overview of big data management and analytics methods. Finally, we suggest opportunities for future research.

Keywords Big data · Data science · Analytics methods

1.1 Introduction

Big data and data science are transforming our world today in ways we could not have imagined at the beginning of the twenty-first century. Although the underlying enabling technologies were present in 2000—cloud computing, data storage,

---

G. Phillips-Wren (✉)
Sellinger School of Business and Management, Department of Information Systems, Law and Operations Management, Loyola University Maryland, 4501 N. Charles Street, Baltimore, MD, USA
e-mail: gwren@loyola.edu

A. Esposito
Department of Psychology, Università degli Studi della Campania "Luigi Vanvitelli", and IIASS, Caserta, Italy
e-mail: iiass.annaesp@tin.it; anna.esposito@unicampania.it

L. C. Jain
University of Technology, Sydney, Australia
e-mail: jainlakhmi@gmail.com; jainlc2002@yahoo.co.uk

Liverpool Hope University, Liverpool, UK

KES International, Selby, UK

G. Phillips-Wren et al. (eds.), *Advances in Data Science: Methodologies and Applications*, Intelligent Systems Reference Library 189,
https://doi.org/10.1007/978-3-030-51870-7_1

internet connectivity, sensors, artificial intelligence, geographic positioning systems (GPS), CPU power, parallel computing, machine learning—it took the acceleration, proliferation and convergence of these technologies to make it possible to envision and achieve massive storage and data analytics at scale. The accompanying wave of innovation has sparked advances in healthcare, engineering, business, science, and human perception, among others. This book offers a snapshot of state-of-the-art technologies and applications in data science that can provide a foundation for future research and development.

'Data science' is a broad term that can be described as "a set of fundamental principles that support and guide the principled extraction of information and knowledge from data" [20], p. 52, to inform decision making. Closely affiliated with data science is 'data mining' that can be defined as the process of extracting knowledge from large datasets by finding patterns, correlations and anomalies. Thus, data mining is often used to develop predictions of the future based on the past as interpreted from the data.

'Big data' make possible more refined predictions and non-obvious patterns due to a larger number of potential variables for prediction and more varied types of data. In general, 'big data' can be defined as having one or more of characteristics of the 3 V's of Volume, Velocity and Variety [19]. Volume refers to the massive amount of data; Velocity refers to the speed of data generation; Variety refers to the many types of data from structured to unstructured. Structured data are organized and can reside within a fixed field, while unstructured data do not have clear organizational patterns. For example, customer order history can be represented in a relational database, while multimedia files such as audio, video, and textual documents do not have formats that can be pre-defined. Semi-structured data such as email fall between these two since there are tags or markers to separate semantic elements. In practice, for example, continual earth satellite imagery is big data with all 3 V's, and it poses unique challenges to data scientists for knowledge extraction.

Besides data and methods to handle data, at least two other ingredients are necessary for data science to yield valuable knowledge. First, after potentially relevant data are collected from various sources, data must be cleaned. Data cleaning or cleansing is the process of detecting, correcting and removing inaccurate and irrelevant data related to the problem to be solved. Sometimes new variables need to be created or data put into a form suitable for analysis. Secondly, the problem must be viewed from a "data-science perspective [of] … structure and principles, which gives the data scientist a framework to systematically treat problems of extracting useful knowledge from data" [20]. Data visualization, domain knowledge for interpretation, creativity, and sound decision making are all part of a data-science perspective. Thus, advances in data science require unique expertise from the authors that we are proud to present in the following pages. The chapters in this book are briefly summarized in Sect. 3 of this article.

However, before proceeding with a description of the chapters, we present an overview of big data management and analytics methods in the following section. The purpose of this section is to provide an overview of algorithms and techniques for data science to help place the chapters in context and to provide a starting point for new researchers who want to participate in this exciting field.

1.2 Big Data Management and Analytics Methods

When considering advances in data science, big data methods require research attention. This is because, currently, big data management (i.e. methods to acquire, store, organize large amount of data) and data analytics (i.e. algorithms devised to analyze and extract intelligence from data) are rapidly emerging tools for contributing to advances in data science. In particular, data analytics are techniques for uncovering meanings from data in order to produce intelligence for decision making. Big data analytics are applied in healthcare, finance, marketing, education, surveillance, and prediction and are used to mine either structured (as spreadsheets or relational databases) or unstructured (as text, images, audio, and video data from internal sources such as cameras—and external sources such as social media) or both types of data.

Big data analytics is a multi-disciplinary domain spanning several disciplines, including psychology, sociology, anthropology, computer science, mathematics, physics, and economics. Uncovering meaning requires complex signal processing and automatic analysis algorithms to enhance the usability of data collected by exploiting the plethora of sensors that can be implemented on the current ICT (Information Communication Technology) devices and the fusion of information derived from multi-modal sources. Data analytics methods should correlate this information, extract knowledge from it, and provide timely comprehensive assessments of relevant daily contextual challenges. To this aim, theoretical fundamentals of intelligent machine learning techniques must be combined with psychological and social theories to enable progress in data analytics to the extent that the automatic intelligence envisaged by these tools augment human understanding and well-being, improving the quality of life of future societies.

Machine learning (ML) is a subset of artificial intelligence (AI) and includes techniques to allow machines the ability to adapt to new settings and detect and extrapolate unseen structures and patterns from noisy data. Recent advances in machine learning techniques have largely contributed to the rise of data analytics by providing intelligent models for data mining.

The most common advanced data analytics methods are **association rule learning analysis**, **classification tree analysis** (CTA), **decision tree algorithms**, **regression analysis**, **genetic algorithms**, and some additional analyses that have become popular with big data such as social media analytics and social network analysis.

1.2.1 Association Rules

Association rule learning analyses include machine learning methodologies exploiting rule-based learning methods to identify relationships among variables in large datasets [1, 17]. This is done by considering the concurrent occurrence of couple or triplets (or more) of selected variables in a specific database under the 'support' and 'confidence' constraints. 'Support' describes the co-occurrence rule associated with the selected variables, and 'confidence' indicates the probability (or the percentage) of correctness for the selected rule in the mined database, i.e. confidence is a measure of the validity or '*interestingness*' of the support rule. Starting from this initial concept, other constraints or measures of interestingness have been introduced [3]. Currently association rules are proposed for mining social media and for social network analysis [6].

1.2.2 Decision Trees

Decision trees are a set of data mining techniques used to identify classes (or categories) and/or predict behaviors from data. These models are based on a tree-like structure, with branches splitting the data into homogeneous and non-overlapping regions and leaves that are terminal nodes where no further splits are possible. The type of mining implemented by decision trees belongs to supervised classes of learning algorithms that decide how splitting is done by exploiting a set of training data for which the target to learn is already known (hence, supervised learning). Once a classification model is built on the training data, the ability to generalize the model (i.e. its accuracy) is assessed on the testing data which were never presented during the training. Decision trees can perform both classification and prediction depending on how they are trained on categorical (i.e., outcomes are discrete categories and therefore the mining techniques are called classification tree analyses) or numerical (i.e., outcomes are numbers, hence the mining techniques are called regression tree analyses) data.

1.2.3 Classification and Regression

Classification tree analysis (CTA) and regression tree analysis techniques are largely used in data mining and algorithms to implement classification and regression. They have been incorporated in widespread data mining software such as SPSS Clementine, SAS Enterprise Miner, and STATISTICA Data Miner [11, 16]. Recently classification tree analysis has been used to model time-to-event (survival) data [13], and regression tree analysis for predicting relationships between animals' body morphological characteristics and their yields (or outcomes of their production) such as meat and milk [12].

1.2.4 Genetic Algorithms

Mining data requires searching for structures in the data that are otherwise unseen, deriving association rules that are otherwise concealed, and assigning unknown patterns to existing data categories. This is done at a very high computational cost since both the size and number of attributes of mined datasets are very large and, consequently, the dimensions of the search space are a combinatorial function of them. As more attributes are included in the search space, the number of training examples is required to increase in order to generate reliable solutions.

Thus, **Genetic algorithms (GA)** have been introduced in data mining to overcome these problems by applying to the dataset to be mined a features selection procedure that reduces the number of attributes to a small set able to significantly arrange the data into distinct categories. In doing so, GAs assign a value of 'goodness' to the solutions generated at each step and a fitness function to determine which solutions will breed to produce a better solution by crossing or mutating the existing ones until an optimal solution is reached. GAs can deal with large search spaces efficiently, with less chance to reach local minima. This is why they have been applied to large number of domains [7, 23].

1.2.5 Sentiment Analysis

Sentiment analysis (emotion and opinion mining) techniques analyze texts in order to extract individuals' sentiments and opinions on organizations, products, health states, and events. Texts are mined at document-level or sentence-level to determine their valence or polarity (positive or negative) or to determine categorical emotional states such as happiness, sadness, or mood disorders such as depression and anxiety. The aim is to help decision making [8] in several application domains such as improving organizations' wealth and know-how [2], increasing customer trustworthiness [22], extracting emotions from texts collected from social media and online reviews [21, 25], and assessing financial news [24]. To do so, several content-based and linguistic text-based methods are exploited such as such as topic modeling [9], natural language processing [4], adaptive aspect-based lexicons [15] and neural networks [18].

1.2.6 Social Network Analysis

Social network analysis techniques are devoted to mine social media contents, e.g. a pool of online platforms that report on specific contents generated by users. Contents can be photos, videos, opinions, bookmarks, and more. Social networks differentiate among those based on their contents and how these contents are shared as acquaintance networks (e.g. college/school students), web networks (e.g. Facebook and LinkedIn, MySpace, etc.), blogs networks (e.g. Blogger, WordPress etc.), supporter networks (e.g. Twitter, Pinterest, etc.), liking association networks (e.g. Instagram, Twitter, etc.), wikis networks (e.g., Wikipedia, Wikihow, etc.), communication and exchanges networks (e.g. emails, WhatsApp, Snapchat, etc.), research networks (e.g. Researchgate, Academia, Dblp, Wikibooks, etc.), social news (e.g. Digg and Reddit, etc.), review networks (e.g. Yelp, TripAdvisor, etc.), question-and-answer networks (e.g. Yahoo! Answers, Ask.com), and spread networks (epidemics, Information, Rumors, etc.).

Social networks are modeled through graphs, where nodes are considered social entities (e.g. users, organizations, products, cells, companies) and connections (called also links or edges or ties) between nodes describe relations or interactions among them. Mining on social networks can be content-based focusing on the data posted or structure-based focusing on uncovering either information on the network structure such as discovering communities [5], or identifying authorities or influential nodes [14], or predicting future links given the current state of the network [10].

1.3 Description of Book Chapters

The research chapters presented in this book are interdisciplinary and include themes embracing emotions, artificial intelligence, robotics applications, sentiment analysis, smart city problems, assistive technologies, speech melody, and fall and abnormal behavior detection. They provide a vision of technologies entering in all the ambient living places. Some of these methodologies and applications focus the analysis of massive data to a human-centered view involving human behavior. Thus, the research described herein is useful for all researchers, practitioners and students interested in living-related technologies and can serve as a reference point for other applications using a similar methodological approach. We, thus, briefly describe the research presented in each chapter.

Chapter 2 by Diraco, Leone and Siciliano investigates the use of big data to assist caregivers to elderly people. One of the problems that caregivers face is the necessity of continuous daily checking of the person. This chapter focuses on the use of data to detect and ultimately to predict abnormal behavior. In this study synthetic data are generated around daily activities, home location where activities take place, and physiological parameters. The authors find that unsupervised deep-learning techniques out-perform traditional supervised/semi-supervised ones, with detection accuracy greater than 96% and prediction lead-time of about 14 days in advance.

Affective computing in the form of emotion recognition techniques and signal modalities is the topic of Chap. 3 by Sharma and Dhall. After an overview of different emotion representations and their limitations, the authors turn to a comparison of databases used in this field. Feature extraction and analysis techniques are presented along with applications of automatic emotion recognition and issues such as privacy and fairness.

Chapter 4 by Siegert and Krüger researches the speaking style that people use when interacting with a technical system such as Alexa and their knowledge of the speech process. The authors perform analysis using the Voice Assistant Conversation Corpus (VACC) and find a set of specific features for device-directed speech. Thus, addressing a technical system with speech is a conscious and regulated individual process in which a person is aware of modification in their speaking style.

Ktistakis, Goodman and Shimizu focus on a methodology for predicting outcomes, the Fuzzy Inference System (FIS), in Chap. 5. The authors present an example FIS, discuss its strengths and shortcomings, and demonstrate how its performance can be improved with the use of Genetic Algorithms. In addition, FIS can be further enhanced by incorporating other methodologies in Artificial Intelligence, particularly Formal Knowledge Representation (FKR) such as a Knowledge Graph (KG) and the Semantic Web. For example, in the Semantic Web KGs are referred to as ontologies and support crisp knowledge and ways to infer new knowledge.

Chapter 6 by Maldonato, Muzii, Continisio and Esposito challenge psychoanalysis with experimental and clinical models using neuroimaging methods to look at questions such as how the brain generates conscious states and whether consciousness involves only a limited area of the brain. The authors go even further to try to demonstrate how neurophysiology itself shows the implausibility of a universal morality.

In Chap. 7, Randazzo, Cirrincione and Pasero illustrate the basic ideas of a family of neural networks for time-varying high dimensional data and demonstrate their performance by means of synthetic and real experiments. The G-EXIN network uses life-long learning through an anisotropic convex polytope that models the shape of the neuron neighborhood and employs a novel kind of edge, called bridge that carries information on the extent of the distribution time change. G-EXIN is then embedded as a basic quantization tool for analysis of data associated with real time pattern recognition.

Electromyography signals (EMG) widely used for monitoring joint movements and muscles contractions is the topic of Chap. 8 by Rescio, Leone, Giampetruzzi and Siciliano. To overcome issues associated with current wearable devices such as expense and skin reactions, a prototype of a new smart sock equipped with reusable stretchable and non-adhesive hybrid polymer electrolytes-based electrodes is discussed. The smart sock can send sEMG data through a low energy wireless transmission connection, and data are analyzed with a machine learning approach in a case study to detect the risk of falling.

Chapter 9 by Marrone introduces the problem of defining in mathematical terms a useful definition of vulnerability for distributed and networked systems such as electrical networks or water supply. This definition is then mapped onto the formalism of Bayesian Networks and demonstrated with a problem associated with smart cities distributed plate car recognition.

Chapter 10 by Griol, Kanagal-Balakrishna and Callejas investigates communication on Twitter where users must find creative ways to express themselves using acronyms, abbreviations, emoticons, unusual spelling, etc. due to the limit on number of characters. They propose a Maximum Entropy classifier that uses an ensemble of feature sets encompassing opinion lexicons, n-grams and word clusters to boost the performance of a sentiment classifier. The authors demonstrate that using several opinion lexicons as feature sets provides a better performance than using just one, at the same time as adding word cluster information enriches the feature space.

Bornschlegl and Hemmje focus on handling Big Data with new techniques for anomaly detection data access on real-world data in Chap. 11. After deriving and qualitatively evaluating a conceptual reference model and service-oriented architecture, two specific industrial Big Data analysis application scenarios involving anomaly detection on car-to-cloud data and predictive maintenance analysis on robotic sensor data, are utilized to demonstrate the practical applicability of the model through proof-of-concept. The techniques empower different end-user stereotypes in the automotive and robotics application domains to gain insight from car-to-cloud as well as from robotic sensor data.

Chapter 12 by Kilingaru, Nedic, Jain, Tweedale and Thatcher investigates Loss of Situation Awareness (SA) in pilots as one of the human factors affecting aviation safety. Although there has been a significant research on SA, one of the major causes of accidents in aviation continues to be a pilot's loss of SA perception error. However, there is no system in place to detect these errors. Monitoring visual attention is one of the best mechanisms to determine a pilot's attention and, hence, perception of a situation. Therefore, this research implements computational models to detect pilot's attentional behavior using ocular data during instrument flight scenario and to classify overall attention behavior during instrument flight scenarios.

Music is the topic of Chap. 13 by Ciaramella, Nardone, Staiano and Vettigli. A framework for processing, classification and clustering of songs on the basis of their emotional content is presented. The main emotional features are extracted after a pre-processing phase where both Sparse Modeling and Independent Component Analysis based methodologies are applied. In addition, a system for music emotion recognition based on Machine Learning and Soft Computing techniques is introduced. A user can submit a target song representing their conceptual emotion and obtain a playlist of audio songs with similar emotional content. Experimental results are presented to show the performance of the framework.

A new data analytics paradigm is presented and applied to energy demand forecasting for smart cities in Chap. 14 by Alamaniotis. The paradigm integrates a group of kernels to exploit the capabilities of deep learning algorithms by utilizing various abstraction levels and subsequently identify patterns of interest in the data. In particular, a deep feedforward neural network is employed with every network node to implement a kernel machine. The architecture is used to predict the energy consumption of groups of residents in smart cities and displays reasonably accurate predictions.

Chapter 15 by Caroppo, Leone and Siciliano considers innovative services to improve quality of life for ageing adults by using facial expression recognition (FER). The authors develop a Convolutional Neural Network (CNN) architecture to automatically recognize facial expressions to reflect the mood, emotions and mental activities of an observed subject. The method is evaluated on two benchmark datasets (FACES and Lifespan) containing expressions of ageing adults and compared with a baseline of two traditional machine learning approaches. Experiments showed that the CNN deep learning approach significantly improves FER for ageing adults compared to the baseline approaches.

1.4 Future Research Opportunities

The tremendous advances in inexpensive computing power and intelligent techniques have opened many opportunities for managing data and investigating data in virtually every field including engineering, science, healthcare, business, and others. Many paradigms and applications have been proposed and used by researchers in recent years as this book attests, and the scope of data science is expected to grow over the next decade. These future research achievements will solve old challenges and create new opportunities for growth and development.

However, one of the most important challenges we face today and for the foreseeable future is 'Security and Privacy'. We want only authorized individuals to have access to our data. The need is growing to develop techniques where threats from cybercriminals such as hackers can be prevented. As we become increasingly dependent on digital technologies, we must prevent cybercriminals from taking control of our systems such as autonomous cars, unmanned air vehicles, business data, banking data, transportation systems, electrical systems, healthcare data, industrial data, and so on. Although researchers are working on various solutions that are adaptable and scalable to secure data and even measure the level of security, there is a long way to go. The challenge to data science researchers is to develop systems that are secure as well as advanced.

1.5 Conclusions

This chapter presented an overview of big data and data science to provide a context for the chapters in this book. To provide a starting point for new researchers, we also provided an overview of big data management and analytics methods. Finally, we pointed out opportunities for future research.

We want to sincerely thank the contributing authors for sharing their deep research expertise and knowledge of data science. We also thank the publishers and editors who helped us achieve this book. We hope that both young and established researchers find inspiration in these pages and, perhaps, connections to a new research stream in the emerging and exciting field of data science.

Acknowledgements The research leading to these results has received funding from the EU H2020 research and innovation program under grant agreement N. 769872 (EMPATHIC) and N. 823907 (MENHIR), the project SIROBOTICS that received funding from Italian MIUR, PNR 2015-2020,

D. D. 1735, 13/07/2017, and the project ANDROIDS funded by the program V: ALERE 2019 Università della Campania "Luigi Vanvitelli", D. R. 906 del 4/10/2019, prot. n. 157264, 17/10/2019.

References

1. Agrawal, R., Imieliński, T., Swami, A.: Mining association rules between sets of items in large databases. ACM SIGMOD Rec. **22**, 207–216 (1993)
2. Chong, A.Y.L., Li, B., Ngai, E.W.T., Ch'ng, E., Lee, F.: Predicting online product sales via online reviews, sentiments, and promotion strategies: a big data architecture and neural network approach. Int. J. Oper. Prod. Manag **36**(4), 358–383 (2016)
3. Cui, B., Mondal, A., Shen, J., Cong, G., Tan, K. L.: On effective e-mail classification via neural networks. In: International Conference on Database and Expert Systems Applications (pp. 85–94). Springer, Berlin, Heidelberg (2005, August)
4. Dang, T., Stasak, B., Huang, Z., Jayawardena, S., Atcheson, M., Hayat, M., Le, P., Sethu, V., Goecke, R., Epps, J.: Investigating word affect features and fusion of probabilistic predictions incorporating uncertainty in AVEC 2017. In: Proceedings of the 7th Annual Workshop on Audio/Visual Emotion Challenge, Mountain View, CA. 27–35, (2017)
5. Epasto, A., Lattanzi, S., Mirrokni, V., Sebe, I.O., Taei, A., Verma, S.: Ego-net community mining applied to friend suggestion. Proc. VLDB Endowment **9**, 324–335 (2015)
6. Erlandsson, F., Bródka, P., Borg, A., Johnson, H.: Finding influential users in social media using association rule learning. Entropy **18**(164), 1–15 (2016). https://doi.org/10.3390/e1805016
7. Espejo, P.G., Ventura, S., Herrera, F.: A survey on the application of genetic programming to classification. IEEE Trans. Syst. Many, and Cybern. Part C: Appl. Rev. **40**(2), 121–144 (2010)
8. Gandomi, A., Haider, M.: Beyond the hype: big data concepts, methods, and analytics. Int. J. Inf. Manage. **35**, 137–144 (2015)
9. Gong, Y., Poellabauer, C.: Topic modeling based on multi-modal depression detection. In: Proceeding of the 7th Annual Workshop on Audio/Visual Emotion Challenge, Mountain View, CA, pp. 69–76, (2017)
10. Güneş, I., Gündüz-Öğüdücü, Ş., Çataltepe, Z.: Link prediction using time series of neighborhood-based node similarity scores. Data Min. Knowl. Disc. **30**, 147–180 (2016)
11. Gupta, B., Rawat, A., Jain, A., Arora, A., Dhami, N.: Analysis of various decision tree algorithms for classification in data mining. Int. J. Comput. Appl. **163**(8), 15–19 (2017)
12. Koc, Y., Eyduran, E., Akbulut, O.: Application of regression tree method for different data from animal science. Pakistan J. Zool. **49**(2), 599–607 (2017)
13. Linden, A., Yarnold, P.R.: Modeling time-to-event (survival) data using classification tree analysis. J Eval. Clin. Pract. **23**(6), 1299–1308 (2017)
14. Liu, C., Wang, J., Zhang, H., Yin, M.: Mapping the hierarchical structure of the global shipping network by weighted ego network analysis. Int. J. Shipping Transp. Logistics **10**, 63–86 (2018)
15. Mowlaei, M.F., Abadeh, M.S., Keshavarz, H.: Aspect-based sentiment analysis using adaptive aspect-based lexicons. Expert Syst. Appl. **148**, 113234 (2020)
16. Nisbet R., Elder J., Miner G.: The three most common data mining software tools. In: Handbook of Statistical Analysis and Data Mining Applications, Chapter 10, pp. 197–234, (2009)
17. Pang-Ning T., Steinbach M., Vipin K.: Association analysis: basic concepts and algorithms. In: Introduction to Data Mining, Chap. 6, Addison-Wesley, pp. 327–414, (2005). ISBN 978-0-321-32136-7
18. Park, S., Lee, J., Kim, K.: Semi-supervised distributed representations of documents for sentiment analysis. Neural Networks **119**, 139–150 (2019)
19. Phillips-Wren G., Iyer L., Kulkarni U., Ariyachandra T.: Business analytics in the context of big data: a roadmap for research. Commun. Assoc. Inf. Syst. **37**, 23 (2015)

20. Provost, F., Fawcett, T.: Data science and its relationship to big data and data-driven decision making. Big Data **1**(1), 51–59 (2013)
21. Rout, J.K., Choo, K.K.R., Dash, A.K., Bakshi, S., Jena, S.K., Williams, K.L.: A model for sentiment and emotion analysis of unstructured social media text. Electron. Commer. Res. **18**(1), 181–199 (2018)
22. Tiefenbacher K., Olbrich S.: Applying big data-driven business work schemes to increase customer intimacy. In: Proceedings of the International Conference on Information Systems, Transforming Society with Digital Innovation, (2017)
23. Tsai, C.-F., Eberleb, W., Chua, C.-Y.: Genetic algorithms in feature and instance selection. Knowl. Based Syst. **39**, 240–247 (2013)
24. Yadava, A., Jhaa, C.K., Sharanb, A., Vaishb, V.: Sentiment analysis of financial news using unsupervised approach. Procedia Comput. Sci. **167**, 589–598 (2020)
25. Zheng, L., Hongwei, W., Song, G.: Sentimental feature selection for sentiment analysis of Chinese online reviews. Int. J. Mach. Learn. Cybernet. **9**(1), 75–84 (2018)

Chapter 2
Towards Abnormal Behavior Detection of Elderly People Using Big Data

Giovanni Diraco, Alessandro Leone, and Pietro Siciliano

Abstract Nowadays, smart living technologies are increasingly used to support older adults so that they can live longer independently with minimal support of caregivers. In this regard, there is a demand for technological solutions able to avoid the caregivers' continuous, daily check of the care recipient. In the age of big data, sensor data collected by smart-living environments are constantly increasing in the dimensions of volume, velocity and variety, enabling continuous monitoring of the elderly with the aim to notify the caregivers of gradual behavioral changes and/or detectable anomalies (e.g., illnesses, wanderings, etc.). The aim of this study is to compare the main state-of-the-art approaches for abnormal behavior detection based on change prediction, suitable to deal with big data. Some of the main challenges deal with the lack of "real" data for model training, and the lack of regularity in the everyday life of the care recipient. At this purpose, specific synthetic data are generated, including activities of daily living, home locations in which such activities take place, as well as physiological parameters. All techniques are evaluated in terms of abnormality-detection performance and lead-time of prediction, using the generated datasets with various kinds of perturbation. The achieved results show that unsupervised deep-learning techniques outperform traditional supervised/semi-supervised ones, with detection accuracy greater than 96% and prediction lead-time of about 14 days in advance.

2.1 Introduction

Nowadays available sensing and assisted living technologies, installed in smart-living environments, are able to collect huge amounts of data by days, months and even years, yielding meaningful information useful for early detection of changes in behavioral and/or physical state that, if left undetected, may be a high risk for frail subjects (e.g., elderly or disabled people) whose health conditions are amenable to change.

G. Diraco (✉) · A. Leone · P. Siciliano
CNR-IMM, Palazzina CNR a/3 - via Monteroni, 73100 Lecce, Italy
e-mail: giovanni.diraco@cnr.it

G. Phillips-Wren et al. (eds.), *Advances in Data Science: Methodologies and Applications*, Intelligent Systems Reference Library 189,
https://doi.org/10.1007/978-3-030-51870-7_2

Early detection, indeed, makes it possible to alert relatives, caregivers, or health-care personnel in advance when significant changes or anomalies are detected, and above all before that critical levels are reached. The "big" data collected from smart homes, therefore, offer a significant opportunity to assist people for early recognition of symptoms that might cause more serious disorders, and so in preventing chronic diseases. The huge amounts of data collected by different devices require automated analysis, and thus it is of great interest to investigate and develop automatic systems for detecting abnormal activities and behaviors in the context of elderly monitoring [1] and smart living [2] applications.

Moreover, the long-term health monitoring and assessment can benefit from knowledge held in long-term time series of daily activities and behaviors as well as physiological parameters [3]. From the big data perspective, the main challenge is to process and automatically interpret—obtaining quality information—the data generated, at high velocity (i.e., high sample rate) and volume (i.e., long-term datasets), by a great variety of devices and sensors (i.e., structural heterogeneity of datasets), becoming more common with the rapid advance of both wearable and ambient sensing technologies [4].

A lot of research has been done in the general area of human behavior understanding, and more specifically in the area of daily activity/behavior recognition and classification as normal or abnormal [5, 6]. However, very little work is reported in the literature regarding the evaluation of machine learning (ML) techniques suitable for data analytics in the context of long-term elderly monitoring in smart living environments. The purpose of this paper is to conduct a preliminary study of the most representative machine/deep learning techniques, by comparing them in detecting abnormal behaviors and change prediction (CP).

The rest of this paper is organized as follows. Section 2.2 contains related works, some background and state-of-the-art in abnormal activity and behavior detection and CP, with special attention paid to elderly monitoring through big data collection and analysis. Section 2.3 describes materials and methods that have been used in this study, providing an overview of the system architecture, long-term data generation and compared ML techniques. The findings and related discussion are presented in Sect. 2.4. Finally, Sect. 2.5 draws some conclusions and final remarks.

2.2 Related Works and Background

Today's available sensing technologies enable long-term continuous monitoring of activities of daily living (ADLs) and physiological parameters (e.g., heart rate, breathing, etc.) in the home environment. At this purpose, both wearable and ambient sensing can be used, either alone or combined, to form multi-sensor systems [7]. In practice, wearable motion detectors incorporate low-cost accelerometers, gyroscopes and compasses, whereas detectors of physiological parameters are based on some kind of skin-contact biosensors (e.g., heart and respiration rates, blood pressure, electrocardiography, etc.) [8]. These sensors need to be attached to a wireless wearable node, carried or worn by the user, needed to process raw data and to communicate detected events with a central base station. Although wearable devices have the advantage of being usable "on the move" and their detection performance is generally good (i.e., signal-to-noise ratio sufficiently high), nonetheless their usage is limited by battery life time (shortened by the intensive use of the wireless communication and on-board processing, both high energy-demanding tasks) [9], by the inconvenience of having to remember to wear a device and by the discomfort of the device itself [10].

Ambient sensing devices, on the other hand, are not intrusive in terms of body obstruction, since they require the installation of sensors around the home environment. Such solutions disappear into the environment, and so are generally well-accepted by end-users [10]. However, the detection performance depends on the number and careful positioning of ambient sensors, whose installation may require modification or redesign of the entire environment. Commonly used ambient sensors are simple switches, pressure and vibration sensors, embedded into carpets and flooring, particularly useful for detecting abnormal activities like falls, since elderly people are directly in contact with the floor surface during the execution of ADLs [11]. Ultra-wideband (UWB) radar is a novel promising, unobtrusive and privacy-preserving ambient-sensing technology that allows to overcome the limitations of vision-based sensing (e.g., visual occlusions, privacy loss, etc.) [12], enabling remote detection (also in through-wall scenarios) of body movements (e.g., in fall detection) [13], physiological parameters [14], or even both simultaneously [15].

As mentioned so far, a multi-sensor system for smart-home elderly monitoring needs to cope with complex and heterogeneous sources of information offered by big data at different levels of abstraction. At this purpose, data fusion or aggregation strategies can be categorized into competitive, complementary, and cooperative [16]. The competitive fusion involves the usage of multiple similar or equivalent sensors, in order to obtain redundancy. In smart-home monitoring, identical sensor nodes are typically used to extend the operative range (i.e., radio signals) or to overcome structural limitations (i.e., visual occlusions). In complementary fusion, different aspects of the same phenomena (i.e., daily activities performed by an elderly person) are captured by different sensors, thus improving the detection accuracy and providing high-level information through analysis of heterogeneous cues. The cooperative fusion, finally, is needed when the whole information cannot be obtained by using any sensor alone. However, in order to detect behavioral changes and abnormalities using a multi-sensor system, it is more appropriate to have an algorithmic framework able to deal with heterogeneous sensors by means of a suitable abstraction layer [17], instead having to design a data fusion layer developed for specific sensors.

The algorithmic techniques for detecting abnormal behaviors and related changes can be roughly categorized into three main categories: supervised, semi-supervised, and unsupervised approaches. In the supervised case, abnormalities are detected by using a binary classifier in which both normal and abnormal behavioral cues (e.g., sequences of activities) are labelled and used for training. The problem with this approach is that abnormal behaviors are extremely rare in practice, and so they must be simulated or synthetically generated in order to train models. Support vector machine (SVM) [18] and hidden Markov model (HMM) [19] are typical (non-parametric) supervised techniques used in abnormality detection systems. In the semi-supervised case, only one kind of label is used to train a one-class classifier. The advantage here is that only normal behavioral cues, that can be observed during the execution of common ADLs, are enough for training. A typically used semi-supervised classifier is the one-class SVM (OC-SVM) [20]. The last, but not least important, category includes the unsupervised classifiers, whose training phase does not need labelled data at all (i.e., neither normal not abnormal cues). The main advantage, in this case, is the easy adaptability to different environmental conditions as well as to users' physical characteristic and habits [21]. Unfortunately, however, unsupervised techniques to be fully operational require a large amount of data, that are not always available when the system is operating for the first time. Thus, a sufficiently long calibration period is preliminary required before the system can be effectively used.

Classical ML methods discussed so far often have to deal with the problem of learning a probability distribution from a set of samples, which generally means to learn a probability density that maximize the likelihood on given data. However, such density does not always exist, as happens when data lie on low-dimensional manifolds, e.g., in the case of highly unstructured data obtained from heterogeneous sources. Under such point of view, conversely, DL methods are more effective because they follow an alternative approach. Instead of attempting to estimate a density, which may not exist, they define a parametric function representing some kind of deep neural network (DNN) able to generate samples. Thus by (hyper-)parameter tuning, generated samples can be made closer to data samples taken from the original data distribution. In such a way, volume, variety and velocity of big data can be effectively exploited to improve detections [22]. In fact, the usage of massive amount of data (volume) is one of the greater advantage of DNNs, which can be also adapted to deal with data abstraction in various different formats (variety) coming from sensors spread around a smart home environment. Moreover, clusters of graphic processing unit (GPU) servers can be used for massive data processing, even in real-time (velocity). However, the application of DL techniques for the purpose of anomaly (abnormal behavior) detection is still in its infancy [23]. Convolutional Neural Network (CNN), that is the current state-of-the-art in object recognition from images [24], exhibits very high feature learning performance but it falls into the first category of supervised techniques. A more interesting DL technique for abnormal activity recognition is represented by Auto-Encoders (AEs), and in particular the Stacked Auto-Encoders (SAEs) [25], which can be subsumed in the semi-supervised techniques when only normal labels are used for training. However, SAEs are basically unsupervised feature learning networks, and thus they can be also exploited for unsupervised anomaly detection. The main limitation of AEs is its requirement of 1D input data, making them essentially unable to capture 2D structure in images. This issue is overcome by the Convolutional Auto-Encoder (CAE) architecture [26], which combines the advantages of CNNs and AEs, besides being suitable for deep clustering tasks [27] and, thus, making it a valuable technique for unsupervised abnormal behavior detection (ABD).

In [28] the main supervised, semi-supervised and unsupervised approaches for anomaly detection were investigated, comparing both traditional ML and DL techniques. The authors demonstrated the superiority of unsupervised approaches, in general, and of DL ones in particular. However, since that preliminary study considered simple synthetic datasets, further investigations are required to accurately evaluate the performance of the most promising traditional and deep learning methods under larger datasets (i.e., big data in long-term monitoring) including more variability in data.

2.3 Materials and Methods

The present investigation is an extension of the preliminary study [28] that compared traditional ML and DL techniques on both abnormality detections and CPs. For each category of learning approach, i.e., supervised, semi-supervised and unsupervised, one ML-based and one DL-based technique were evaluated and compared in terms of detection accuracy and prediction lead-time at the varying of both normal ADLs (N-ADLs) and abnormal ADLs (A-ADLs). All investigated ML-DL techniques are summarized in Table 2.1. At that purpose, a synthetic dataset was generated by referring to common ADLs and taking into account how older people perform such activities at their home environment, following instructions and suggestions provided by consultant geriatricians and existing researches [19, 29]. The synthetic dataset included six basic ADLs, four home locations in which these activities usually take place, and five levels of basic physiological parameters associated with the execution of each ADL.

As an extension of the previous study [28], the objective of this investigation is to evaluate more deeply the techniques reported in Table 2.1 by considering six additional abnormal datasets, instead of only one, obtained in presence of the following changes:

Table 2.1 ML and DL techniques compared in this study

Category	Type	Technique
Supervised	Machine learning	Support vector machine (SVM)
Supervised	Deep learning	Convolutional neural network (CNN)
Semi-supervised	Machine learning	One-class support vector machine (OC-SVM)
Semi-supervised	Deep learning	Stacked auto-encoders (SAE)
Unsupervised	Machine learning	K-means clustering (KM)
Unsupervised	Deep learning	Deep clustering (DC)

- **[St] Starting time of activity**. This is a change in the starting time of an activity, e.g., having breakfast at 9 a.m. instead of 7 a.m. as usual.
- **[Du] Duration of activity**. This change refers to the duration of an activity, e.g., resting for 3 h in the afternoon, instead of 1 h as usual.
- **[Di] Disappearing of activity**. In this case, after the change, one activity is no more performed by the user, e.g., having physical exercises in the afternoon.
- **[Sw] Swap of two activities**. After the change, two activities are performed in reverse order, e.g., resting and then housekeeping instead of housekeeping and resting.
- **[Lo] Location of activity**. One activity usually performed in a home location (e.g., having breakfast in kitchen), after the change is performed in a different location (e.g., having breakfast in bed).
- **[Hr] Heart-rate during activity**. This is a change in heart-rate during an activity, e.g., changing from low to high heart-rate during the resting activity in the afternoon.

Without loss of generality, generated datasets included as physiological parameter only the heart-rate (HR), since heart and respiration rates are both associated with the performed activity. The discrete values assumed by ADLs, locations and heart-rate values included in the generated datasets are reported in Table 2.2.

Furthermore, in this study, both normal and abnormal long-term datasets (i.e., lasting one year each) are realistically generated by suggesting a new probabilistic model based on HMM and Gaussian process. Finally, the evaluation metrics used in this study include, besides the accuracy (the only one considered in the previous study [28]), also the precision, sensitivity, specificity and F1-score:

$$\text{accuracy} = \frac{\text{TP} + \text{TN}}{\text{TP} + \text{TN} + \text{FP} + \text{FN}}, \tag{2.1}$$

$$\text{precision} = \frac{\text{TP}}{\text{TP} + \text{FP}}, \tag{2.2}$$

Table 2.2 Activities, home locations and heart-rate values, used to generate the long-term datasets

Activity of daily living (ADL)	Home location (LOC)	Heart-rate level (HRL)
Eating (AE) Housekeeping (AH) Physical exercise (AP) Resting (AR) Sleeping (AS) Toileting (AT)	Bedroom (BR) Kitchen (KI) Living room (LR) Toilet (TO)	Very low (VL) [<50 beats/min] Low (LO) [65–80 beats/min] Medium (ME) [80–95 beats/min] High (HI) [95–110 beats/min] Very high (VH) [>110 beats/min]

$$\text{sensitivity} = \frac{\text{TP}}{\text{TP} + \text{FN}}, \tag{2.3}$$

$$\textbf{specificity} = \frac{\text{TN}}{\text{TN} + \text{FP}}, \tag{2.4}$$

$$\text{F1 - score} = 2 * \frac{\text{TP}}{2 * \text{TP} + \text{FP} + \text{FN}}, \tag{2.5}$$

where TP is the number of true positives, FP is the number of false positives, TN is the number of true negatives, and FN is the number of false negatives.

In the following of this section, details concerning data generation, supervised, semi-supervised, unsupervised ML/DL techniques for ABD and CP are presented.

2.3.1 Data Generation

In this study, the normal daily behavior has been modelled by using a HMM with three hidden states, Tired (T), Hungry (H), Energized (E), as depicted in Fig. 2.1, representing the user's physical state bearing diverse ADLs. Each arrow of the graph reported in Fig. 2.1 is associated with a probability parameter, which determines the probability that one state π_i follows another state π_{i-1}, i.e., the transition probability:

$$a_{qr} = P(\pi_i = q | \pi_{i-1} = r), \tag{2.6}$$

where q, r $\in$ {T, H, E}. The HMM output is a sequence of triples (a, b, c) $\in$ ADL $\times$ LOC $\times$ HRL, where ADL $=$ {AE, AH, AP, AR, AS, AT}, LOC $=$

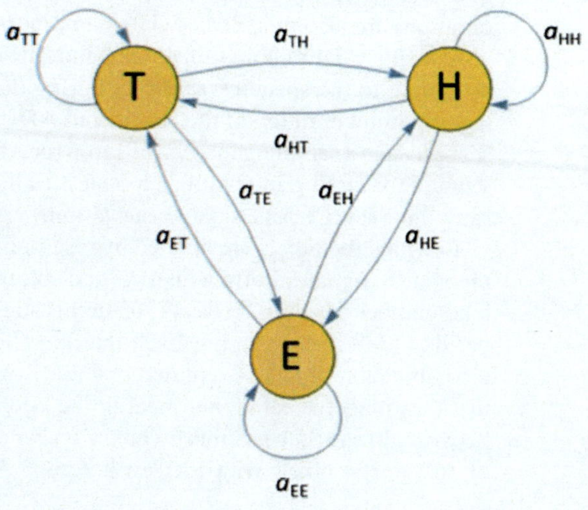

Fig. 2.1 State diagram of the HMM model used to generate long-term activity data

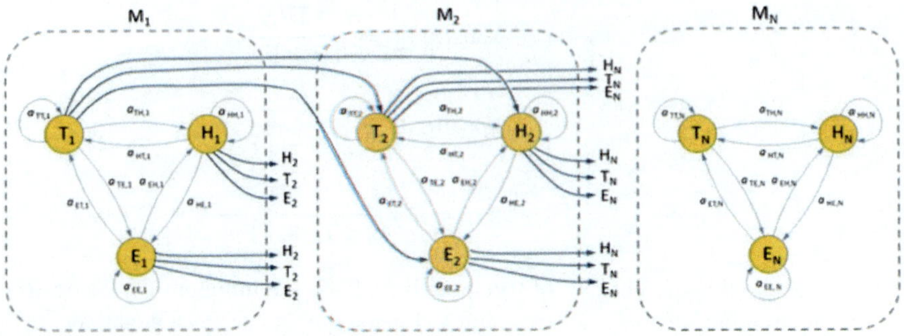

Fig. 2.2 State diagram of the suggested hierarchical HMM, able to model the temporal dependency of daily activities

{BR, KI, LR, TO}, and HRL = {VL, LO, ME, HI, VI} represent, respectively, all possible ADLs, home locations and HR levels (see Table 2.2). In general, a state can produce a triple from a distribution over all possible triples. Hence, the probability that the triple (a, b, c)

is seen when the system is in state k, i.e., the so-called emission probability, is defined as follows:

$$e_k(a, b, c) = P(x_i = (a, b, c)|\pi_i = k). \tag{2.7}$$

Since HMM does not represent the temporal dependency of activity states, a hierarchical approach is proposed here by subdividing the day into N time intervals, and modeling the activities in each time interval with a dedicate HMM sub-model, namely M_1, M_2, ..., M_N, as depicted in Fig. 2.2. For each sub-model M_i, thus, the first state being activated starts at a time T_i modeled as a Gaussian process, while the other states within the same sub-model M_i start in consecutive time slots whose durations are also modeled as Gaussian processes.

Usually, ADLs, home locations, and HR levels are sampled at different rates according to the specific variability during the entire day time. For example, since the minimum duration of the considered ADLs is of about 10 min, it does not make sense to take a sampling interval of 1 min for ADLs. However, for uniformity reasons, a unique sampling interval is adopted for all measurements. In this study, the HR sampling rate (i.e., one sample each 5 min) is selected as reference

to which the others are aligned by resampling them. Then, the generated data are prepared in a matrix form with rows and columns corresponding, respectively, to the total number of observed days (365 in this study) and to the total number of samples per day (288 in this study). Each matrix cell holds a numeric value that indicates a combination of values reported in Table 2.2, for example AE_KI_ME, indicating that the subject is eating her meal in the kitchen and her HR level is medium (i.e., between 80 and 95 beats/min). Thus, a 1-year dataset can be represented by an image of 365 × 288 pixels with 120 levels (i.e., 6 ADLs, 4 locations, and 5 h levels), of which an example is shown in Fig. 2.3. Alternatively, for a better understanding, a dataset can be represented by using three different images of 365 × 288 pixels, one for ADLs (with only 6 levels), one for locations (with only 4 levels), and one for HR levels (with only 5 levels), as shown in Fig. 2.4.

To assess the ability of ML and DL techniques (reported in Table 2.1) to detect behavioral abnormalities and changes, model parameters (i.e., transition probabilities, emission probabilities, starting times and durations) were randomly perturbed in order to generate various kind of abnormal datasets. Without loss of generality, each abnormal dataset includes only one of the abovementioned changes (i.e., St, Du, Di, Sw, Lo, Hr) at a time. At this end, the perturbation is gradually applied between the days 90th and 180th, by randomly interpolating two sets of model parameters, normal and abnormal, respectively. Thus, an abnormal dataset consists of three parts. The first one, ranging from day 1st to day 90th, is referred to normal behavior. The second period, from day 90th to 180th, is characterized by gradually changes, becoming progressively more accentuated. Finally, the third period, starting

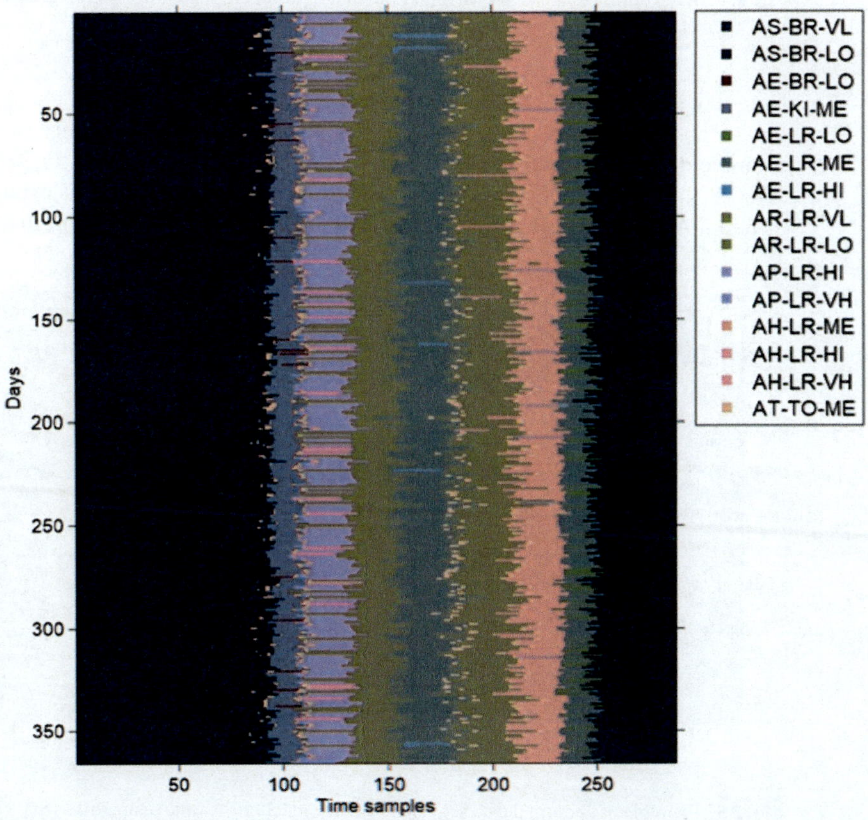

Fig. 2.3 Example of normal dataset, represented as an image of 365 × 288 pixels and 120 levels

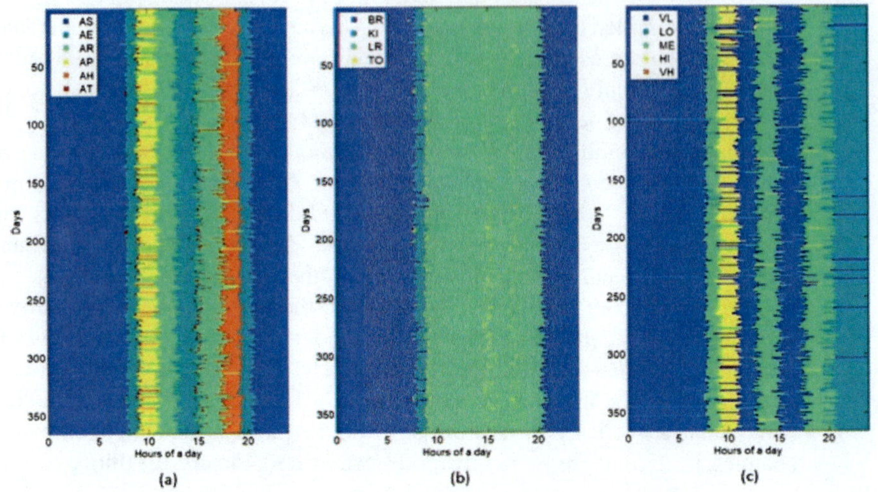

Fig. 2.4 Same normal dataset shown in Fig. 2.3 but represented with different images for **a** ADLs, **b** LOCs and **c** HRLs

from day 180th, is very different from the initial normal period, the change rate is low or absent and the subject's behavior moves into another stability period. An example dataset for each kind of change is reported in figures from Figs. 2.5, 2.6, 2.7, 2.8, 2.9 and 2.10. The detection performance of each technique is evaluated for different A-ADL levels (i.e., percentages of abnormal activities present in a dataset)

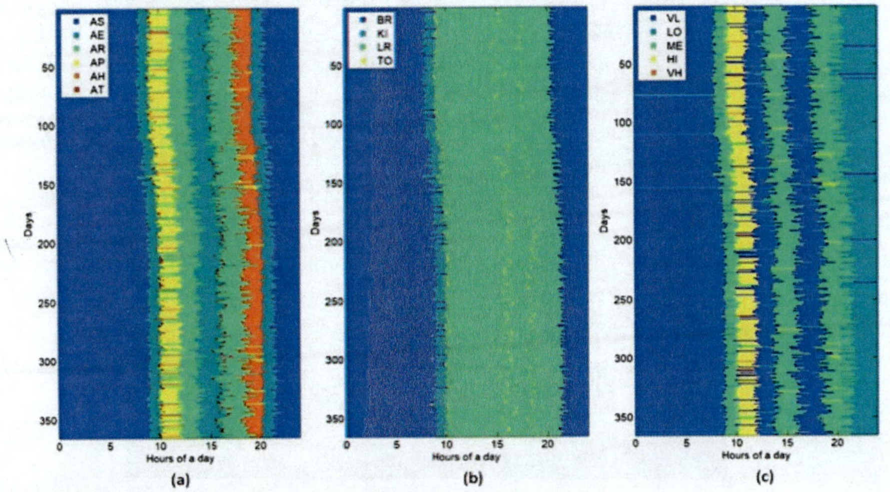

Fig. 2.5 Example of abnormal data set, due to change in "Starting time of activity" (St). The change gradually takes place from the 90th day on

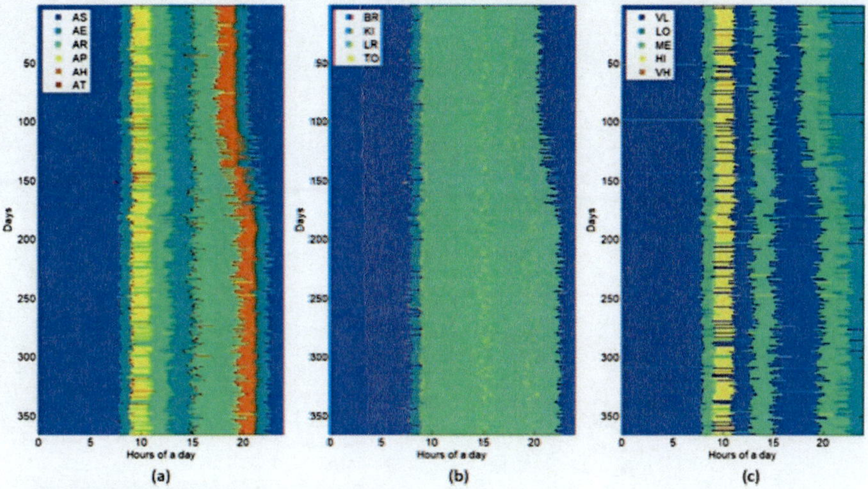

Fig. 2.6 Example of abnormal data set, due to change in "Duration of activity" (Du). The change gradually takes place from the 90th day on

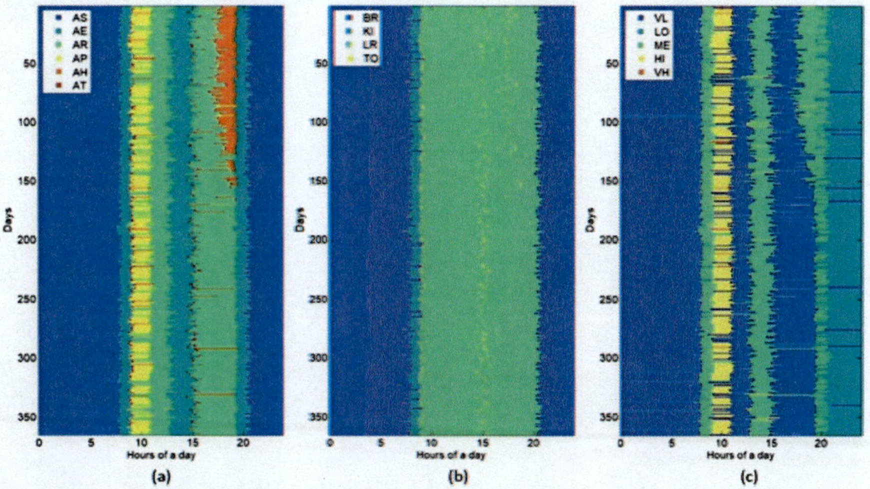

Fig. 2.7 Example of abnormal data set, due to change in "Disappearing of activity" (Di). The change gradually takes place from the 90th day on

as well as different prediction lead-time, which is, the maximum number of days in advance such that the abnormality can be detected with a certain accuracy. Furthermore, in order to better appreciate differences among the three types of detection techniques (i.e., supervised, semi-supervised and unsupervised), beside the A-ADL

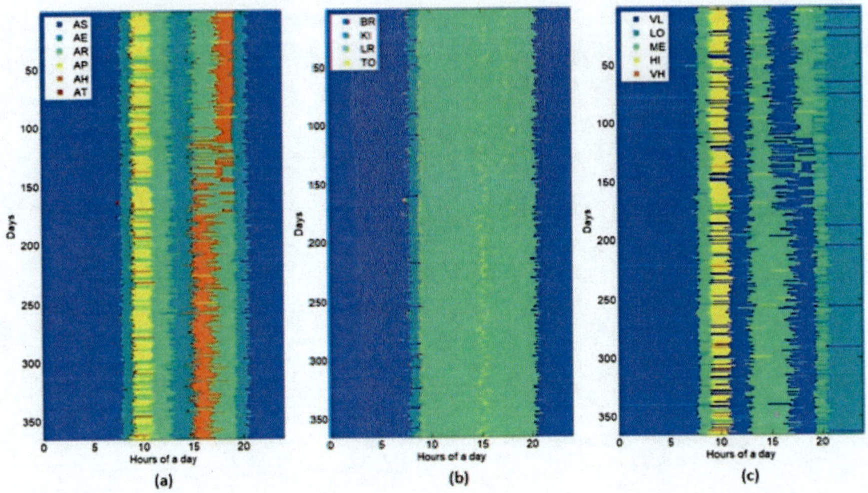

Fig. 2.8 Example of abnormal data set, due to "Swap of two activities" (Sw). The change gradually takes place from the 90th day on

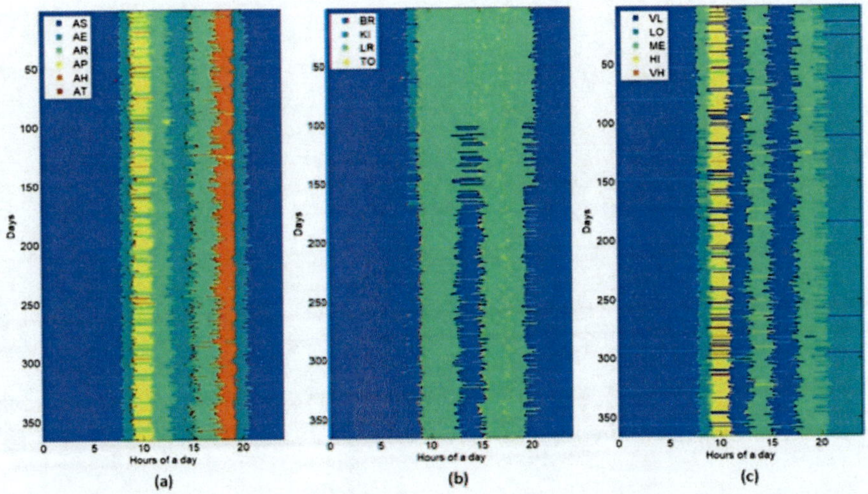

Fig. 2.9 Example of abnormal data set, due to change in "Location of activity" (Lo). The change gradually takes place from the 90th day on

also N-ADL changing is considered, that is, to take into account the potential overlapping of more ADLs in the same sampling interval as well as the occurrence of ALDs never observed before.

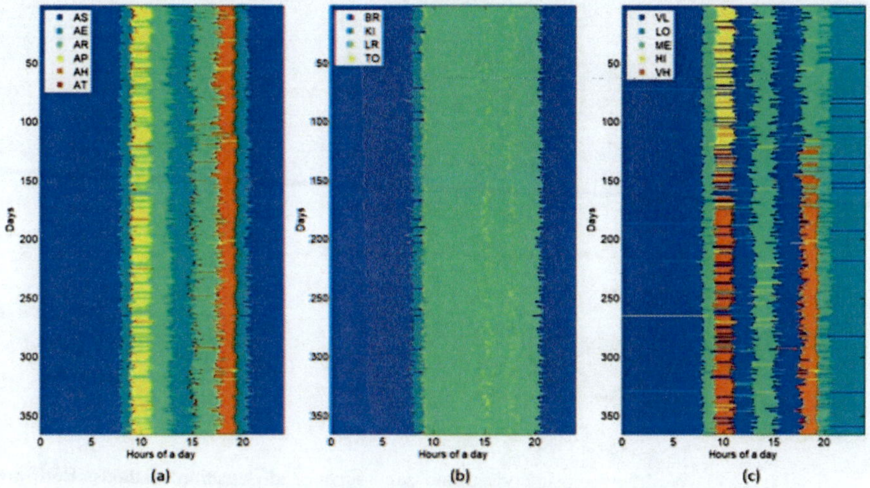

Fig. 2.10 Example of abnormal data set, due to change in "Heart-rate during activity" (Hr). The change gradually takes place from the 90th day on

2.3.2 Learning Techniques for Abnormal Behavior Detection

The problem of ABD can be addressed by means of several learning techniques. Fundamentally, the technique to be used depends on the label availability, so that it is possible to distinguish between the three main typologies of (1) supervised detection, (2) semi-supervised detection and (3) unsupervised detection, as is discussed in this subsection.

2.3.2.1 Supervised Detection

Supervised detection is based on learning techniques (i.e., classifiers) requiring fully labelled data for training. This means that both positive samples (i.e., abnormal behaviors) and negative samples (i.e., normal behaviors) must be observed and labelled during the training phase. However, the two label classes are typically strongly unbalanced, since abnormal events are extremely rare in contrast to normal patterns that instead are abundant. As a consequence, not all classification techniques are equally effective for this situation. In practice, some algorithms are not able to deal with unbalanced data [30], whereas others are more suitable thanks to their high generalization capability, such as SVM [31] and Artificial Neural Networks (ANNs) [32], especially those with many layers like CNNs, which have reached impressive performances in detection of abnormal behavior from videos [33]. The workflow of supervised detection is pictorially depicted in Fig. 2.11.

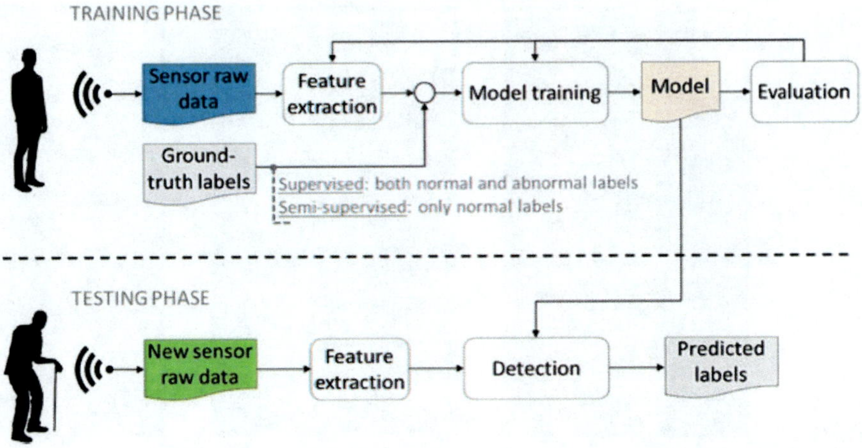

Fig. 2.11 Workflow of supervised and semi-supervised detection methods. Both normal and abnormal labels are needed in the supervised training phase, whereas only normal labels are required in the semi-supervised training

2.3.2.2 Semi-supervised Detection

In real-world applications, the supervised detection workflow described above is not very relevant due to the assumption of fully labelled data, on the basis of which abnormalities are known and labeled correctly. Instead, when dealing with elderly monitoring, abnormalities are not known in advance and cannot be purposely performed just to train detection algorithms (e.g., think, for instance, to falls in the elderly which involve environmental hazards in the home). Semi-supervised detection also uses a similar workflow of that shown in Fig. 2.11 based on training and test data, but training data only involve normal labels without the need to label abnormal patterns. Semi-supervised detection is usually achieved by introducing the concept of one-class classification, whose state-of-the-art implementations—as experimented in this study—are OC-SVM [20] and EAs [25], within ML and DL fields, respectively.

DL techniques learn features in a hierarchical way: high-level features are derived from low-level ones by using layer-wise pre-training, in such a way structures of ever higher level are represented in higher layers of the network. After pre-training, a semi-supervised training provides a fine-tuning adjustment of the network via gradient descent optimization. Thanks to that greedy layer-wise pre-training followed by semi-supervised fine-tuning [34], features can be automatically learned from large datasets containing only one-class label, associated with normal behavior patterns.

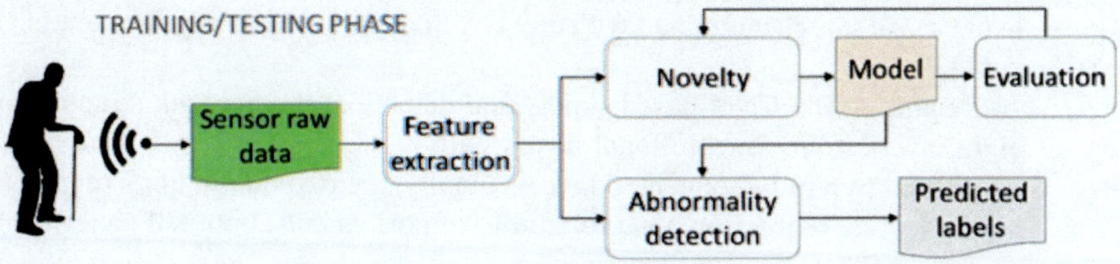

Fig. 2.12 Workflow of unsupervised detection methods

2.3.2.3 Unsupervised Detection

The most flexible workflow is that of unsupervised detection. It does not require that abnormalities are known in advance but, conversely, they can occur during the testing phase and are modelled as novelties with respect to normal (usual) observations. Then, there is no distinction between training and testing phases, as shown in Fig. 2.12. The main idea here is that extracted patterns (i.e., features) are scored solely on the basis of their intrinsic properties. Basically, in order to decide what is normal and is not, unsupervised detection is based on appropriate metrics of either distance or density.

Clustering techniques can be applied in unsupervised detection. In particular, K-means is one of the simples unsupervised algorithms that address the clustering problem by grouping data based on their similar features into K disjoint clusters. However, K-means is affected by some shortcomings: (1) sensitivity to noise and outliers, (2) initial cluster centroids (seeds) are unknown (randomly selected), (3) there is no criterion for determining the number of clusters. The Weighted K-Means [35], also adopted in this study, provides a viable way to approach clustering of noisy data. While the last two problems are addressed by implementing the intelligent K-means suggested by [36], in which the K-means algorithm is initialized by using the so-called anomalous clusters, extracted before running the K-means itself.

2.3.3 *Experimental Setting*

For the experimental purpose, 9000 datasets were generated, i.e., 1500 random instances for each of the six abnormalities shown from Figs. 2.5, 2.6, 2.7, 2.8, 2.9 and 2.10. Each dataset represented a 1-year data collection, as a matrix (image) of 365 rows (days) and 288 columns (samples lasting 5 min each), for a total amount of 105,120 values (pixels) through 120 levels. The feature extraction process was carried out by considering a 50%-overlapping sliding window lasting 25 days, then leading to a feature space of dimension $D = 7200$.

In both supervised and semi-supervised settings, regarding SVM classifier a radial basis function (RBF) kernel was used. The kernel scale was automatically selected using a grid search combined with cross-validation on randomly subsampled training data [37].

Regarding the CNN-based supervised detection, the network structure included eight layers: four convolutional layers with kernel size of 3 × 3, two subsampling layers and two fully connected layers. Finally, the two output units represented, via binary logical regression, the probability of normal and abnormal pattern behaviors.

The SAE network was structured in four hidden layers, the sliding-window feature vectors were given as input to the first layer, which thus included 7200 units (i.e., corresponding to feature space dimension D). The second hidden layer was of 900 units, corresponding to a compression factor of 8 times. The following two hidden layers were of 180 and 60 units, respectively, with compression factors of 5 and 3 times.

In supervised detection settings, the six abnormal datasets were joined in order to perform a 6-fold cross-validation scheme. In semi-supervised detection settings, instead, only normal data from the same dataset were used for training, while testing was carried out using data from day 90 onwards.

Regarding the CAE structure in the DC approach, the encoder included three convolutional layers with kernel size of five, five and three, respectively, followed by a fully connected layer. The decoder structure was a mirror of the encoder one.

All experiments were performed on an Intel i7 3.5 GHz workstation with 16 GB DDR3 and equipped with GPU NVidia Titan X using Keras [38] with Theano [39] toolkit for DL approaches, and Matlab [40] for ML approaches.

2.4 Results and Discussion

This section reports the experimental results in terms of detection accuracy, precision, sensitivity, specificity, F1-score and lead-time of prediction related to all techniques summarized in Table 2.1, and achieved processing the datasets generated by considering six change types (i.e., St, Du, Di, Sw, Lo, Hr) as previously described. The achieved results are reported from Tables 2.3, 2.4, 2.5, 2.6, 2.7 and 2.8, respectively,

Table 2.3 Detection accuracy of all compared techniques

Technique	Accuracy for each change type					
	St	Du	Di	Sw	Lo	Hr
SVM	**0.858**	0.879	0.868	0.888	**0.849**	**0.858**
CNN	0.940	0.959	0.948	0.959	0.910	0.888
OC-SVM	0.910	0.879	0.929	0.940	0.918	0.899
SAE	0.929	0.948	0.970	0.989	0.948	0.940
KM	0.929	0.918	0.940	0.948	0.910	0.888
DC	0.959	*0.978*	*0.970*	0.940	*0.978*	0.959

Table 2.4 Detection precision of all compared techniques

Technique	Precision for each change type					
	St	Du	Di	Sw	Lo	Hr
SVM	**0.951**	0.960	**0.956**	0.961	**0.951**	0.959
CNN	0.985	0.992	0.981	0.989	0.976	0.968
OC-SVM	0.973	0.960	0.981	0.985	0.984	0.972
SAE	0.977	0.989	0.989	0.996	0.985	0.985
KM	0.977	0.977	0.981	0.981	0.969	0.964
DC	0.985	*0.993*	*0.989*	0.981	0.993	*0.989*

Table 2.5 Detection sensitivity of all compared techniques

Technique	Sensitivity for each change type					
	St	Du	Di	Sw	Lo	Hr
SVM	**0.855**	0.876	0.865	0.887	**0.844**	**0.847**
CNN	0.935	0.953	0.949	0.956	0.902	0.880
OC-SVM	0.905	0.876	0.924	0.935	0.905	0.891
SAE	0.927	0.942	0.971	0.989	0.945	0.935
KM	0.927	0.913	0.938	0.949	0.909	0.884
DC	0.960	*0.978*	*0.971*	0.938	*0.978*	0.956

Table 2.6 Detection specificity of all compared techniques

Technique	Specificity for each change type					
	St	Du	Di	Sw	Lo	Hr
SVM	**0.867**	0.889	**0.878**	0.889	**0.867**	0.889
CNN	0.956	0.978	0.944	0.967	0.933	0.911
OC-SVM	0.922	0.889	0.944	0.956	0.956	0.922
SAE	0.933	0.967	0.967	0.989	0.956	0.956
KM	0.933	0.933	0.944	0.944	0.911	0.900
DC	0.956	*0.978*	*0.967*	0.944	*0.978*	*0.967*

Table 2.7 Detection F1-score of all compared techniques

Technique	F1-score for each change type					
	St	Du	Di	Sw	Lo	Hr
SVM	**0.900**	0.916	0.908	0.922	**0.894**	**0.900**
CNN	0.959	0.972	0.965	0.972	0.938	0.922
OC-SVM	0.938	0.916	0.951	0.959	0.943	0.930
SAE	0.951	0.965	0.980	0.993	0.965	0.959
KM	0.951	0.944	0.959	0.965	0.938	0.922
DC	0.972	*0.985*	*0.980*	0.959	*0.985*	0.972

Table 2.8 Lead-time of prediction of all compared techniques

Technique	Lead-time (days) for each change type					
	St	Du	Di	Sw	Lo	Hr
SVM	8	6	11	9	**5**	**3**
CNN	10	8	16	12	6	4
OC-SVM	8	6	10	6	7	**5**
SAE	13	11	19	17	13	11
KM	7	**5**	8	6	5	**3**
DC	*17*	15	*20*	*18*	16	14

for each aforesaid performance metric. As discussed in the previous section, such abnormalities regard both N-ADLs and A-ADLs. The former regard the overlapping of different activities within the same sampling interval or the occurrence of new activities (i.e., sequences not observed before that may lead to misclassification). Instead, the latter take into account six types of change from the usual activity sequence.

From Table 2.3, it is evident that with the change type Sw, there are little differences in detection accuracy, which become more marked with other kind of change such as Lo and Hr. In particular, the supervised techniques exhibit poor detection accuracy with change types as Lo and Hr, while the semi-supervised and unsupervised techniques based on DL maintain good performance also in correspondence of those change types. Similar considerations can be carried out by observing the other performance metrics from Tables 2.4, 2.5, 2.6 and 2.7.

The change types Lo (Fig. 2.9) and Hr (Fig. 2.10) influence only a narrow region of the intensity values. More specifically, only location values (Fig. 2.9b) are interested in Lo-type datasets, and only heart-rate values (Fig. 2.10b) in the Hr case. On the other hand, other change types like Di (Fig. 2.7) or Sw (Fig. 2.8) involve all values, i.e., ADL, LOC and HRL, and so they are simpler to be detected and predicted. However, the ability of DL techniques to capture spatio-temporal local features (i.e., spatio-temporal relations between activities) allowed good performance to be achieved also with change types whose intensity variations were confined in narrow regions.

The lead-times of prediction reported in Table 2.8 were obtained in correspondence of the performance metrics discussed above and reported from Tables 2.3, 2.4, 2.5, 2.6 and 2.7. In other words, such times refer to the average number of days, before the day 180th (since from this day on, the new behavior becomes stable), at which the change can be detected with the performance reported from Tables 2.3, 2.4, 2.5, 2.6 and 2.7. The longer the lead-times of prediction the earlier the change can be predicted. Also in this case, better lead-times were achieved with change types Di and Sw (i.e., characterized by wider regions of intensity variations) and with techniques SAE and DC, since they are able to learn discriminative features more effectively than the traditional ML techniques.

2.5 Conclusions

The contribution of this study is twofold. First, a common data model able to represent and process simultaneously both ADLs, home locations in which such ADLs take place (LOCs) and physiological parameters (HRLs) as image data is presented. Second, the performance of state-of-the-art ML-based and DL-based detection techniques are evaluated by considering big data sets, synthetically generated, including both normal and abnormal behaviors. The achieved results are promising and show the superiority of DL-based techniques in dealing with big data characterized by different kind of data distribution. Future and ongoing activities are focused on the evaluation of prescriptive capabilities of big data analytics aiming to optimize time and resources involved in elderly monitoring applications.

References

1. Gokalp, H., Clarke, M.: Monitoring activities of daily living of the elderly and the potential for its use in telecare and telehealth: a review. Telemedi. e-Health **19**(12), 910–923 (2013)
2. Sharma, R., Nah, F., Sharma, K., Katta, T., Pang, N., Yong, A.: Smart living for elderly: design and human-computer interaction considerations. Lect. Notes Comput. Sci. **9755**, 112–122 (2016)
3. Parisa, R., Mihailidis, A.: A survey on ambient-assisted living tools for older adults. IEEE J. Biomed. Health Informat. **17**(3), 579–590 (2013)
4. Vimarlund, V., Wass, S.: Big data, smart homes and ambient assisted living. Yearbook Medi. Informat. **9**(1), 143–149 (2014)
5. Mabrouk, A.B., Zagrouba, E.: Abnormal behavior recognition for intelligent video surveillance systems: a review. Expert Syst. Appl. **91**, 480–491 (2018)
6. Bakar, U., Ghayvat, H., Hasanm, S.F., Mukhopadhyay, S.C.: Activity and anomaly detection in smart home: a survey. Next Generat. Sens. Syst. **16**, 191–220 (2015)
7. Diraco, G., Leone, A., Siciliano, P., Grassi, M., Malcovati, P.A.: Multi-sensor system for fall detection in ambient assisted living contexts. In: IEEE SENSORNETS, pp. 213–219 (2012)
8. Taraldsen, K., Chastin, S.F.M., Riphagen, I.I., Vereijken, B., Helbostad, J.L.: Physical activity monitoring by use of accelerometer-based body-worn sensors in older adults: a systematic literature review of current knowledge and applications. Maturitas **71**(1), 13–19 (2012)
9. Min, C., Kang, S., Yoo, C., Cha, J., Choi, S., Oh, Y., Song, J.: Exploring current practices for battery use and management of smartwatches. In: Proceedings of the 2015 ACM International Symposium on Wearable Computers, pp. 11–18, September (2015)
10. Stara, V., Zancanaro, M., Di Rosa, M., Rossi, L., Pinnelli, S.: Understanding the interest toward smart home technology: the role of utilitaristic perspective. In: Italian Forum of Ambient Assisted Living, pp. 387–401. Springer, Cham (2018)
11. Droghini, D., Ferretti, D., Principi, E., Squartini, S., Piazza, F.: A combined one-class SVM and template-matching approach for user-aided human fall detection by means of floor acoustic features. In: Computational Intelligence and Neuroscience (2017)
12. Hussmann, S., Ringbeck, T., Hagebeuker, B.: A performance review of 3D TOF vision systems in comparison to stereo vision systems. In: Stereo Vision. InTech (2008)
13. Diraco, G., Leone, A., Siciliano, P.: Radar sensing technology for fall detection under near real-life conditions. In: IET Conference Proceedings, pp. 5–6 (2016)
14. Lazaro, A., Girbau, D., Villarino, R.: Analysis of vital signs monitoring using an IR-UWB radar. Progress Electromag. Res. **100**, 265–284 (2010)

15. Diraco, G., Leone, A., Siciliano, P.: A radar-based smart sensor for unobtrusive elderly monitoring in ambient assisted living applications. Biosensors **7**(4), 55 (2017)
16. Dong, H., Evans, D.: Data-fusion techniques and its application. In: Fourth International Conference on Fuzzy Systems and Knowledge Discovery (FSKD 2007), vol. 2, pp. 442–445. IEEE (2007)
17. Caroppo, A., Diraco, G., Rescio, G., Leone, A., Siciliano, P. (2015). Heterogeneous sensor platform for circadian rhythm analysis. In: Advances IEEE International Workshop on in Sensors and Interfaces (ISIE), 10 August 2015, pp. 187–192 (2015)
18. Miao, Y., Song, J.: Abnormal event detection based on SVM in video surveillance. In: IEEE Workshop on Advance Research and Technology in Industry Applications, pp. 1379–1383 (2014)
19. Forkan, A.R.M., Khalil, I., Tari, Z., Foufou, S., Bouras, A.: A context-aware approach for long-term behavioural change detection and abnormality prediction in ambient assisted living. Pattern Recogn. **48**(3), 628–641 (2015)
20. Hejazi, M., Singh, Y.P.: One-class support vector machines approach to anomaly detection. Appl. Artifi. Intell. **27**(5), 351–366 (2013)
21. Otte, F.J.P., Rosales Saurer, B., Stork, W. (2013). Unsupervised learning in ambient assisted living for pattern and anomaly detection: a survey. In: Communications in Computer and Information Science 413 CCIS, pp. 44–53 (2013)
22. Chen, X.W., Lin, X.: Big data deep learning: challenges and perspectives. IEEE Access **2**, 514–525 (2014)
23. Ribeiro, M., Lazzaretti, A.E., Lopes, H.S.: A study of deep convolutional auto-encoders for anomaly detection in videos. Pattern Recogn. Lett. **105**, 13–22 (2018)
24. Krizhevsky, A., Sutskever, I., Hinton, G.E.: Imagenet classification with deep convolutional neural networks. In: Advances in Neural Information Processing Systems, pp. 1097–1105 (2012)
25. Krizhevsky, A., Hinton, G.E.: Using very deep autoencoders for content-based image retrieval. In: ESANN, April (2011)
26. Masci, J., Meier, U., Cireşan, D., Schmidhuber, J.: Stacked convolutional auto-encoders for hierarchical feature extraction. In: International Conference on Artificial Neural Networks, pp. 52–59. Springer, Berlin, Heidelberg (2011)
27. Guo, X., Liu, X., Zhu, E., Yin, J.: Deep clustering with convolutional autoencoders. In: International Conference on Neural Information Processing, November, pp. 373–382. Springer (2017)
28. Diraco, G., Leone, A., Siciliano, P.: Big data analytics in smart living environments for elderly monitoring. In: Italian Forum of Ambient Assisted Living Proceedings, pp. 301–309. Springer (2018)
29. Cheng, H., Liu, Z., Zhao, Y., Ye, G., Sun, X.: Real world activity summary for senior home monitoring. Multimedia Tools Appl. **70**(1), 177–197 (2014)
30. Almas, A., Farquad, M.A.H., Avala, N.R., Sultana, J.: Enhancing the performance of decision tree: a research study of dealing with unbalanced data. In: Seventh International Conference on Digital Information Management, pp. 7–10. IEEE ICDIM (2012)
31. Hu, W., Liao, Y., Vemuri, V.R.: Robust anomaly detection using support vector machines. In: Proceedings of the International Conference on Machine Learning, pp. 282–289 (2003)
32. Pradhan, M., Pradhan, S.K., Sahu, S.K.: Anomaly detection using artificial neural network. Int. J. Eng. Sci. Emerg. Technol. **2**(1), 29–36 (2012)
33. Sabokrou, M., Fayyaz, M., Fathy, M., Moayed, Z., Klette, R.: Deep-anomaly: fully convolutional neural network for fast anomaly detection in crowded scenes. Comput. Vis. Image Underst. **172**, 88–97 (2018)
34. Erhan, D., Bengio, Y., Courville, A., Manzagol, P.A., Vincent, P., Bengio, S.: Why does unsupervised pre-training help deep learning? J. Mach. Learn. Res., 625–660 (2010)
35. De Amorim, R.C., Mirkin, B.: Minkowski metric, feature weighting and anomalous cluster initializing in K-Means clustering. Pattern Recogn. **45**(3), 1061–1075 (2012)

36. Chiang, M.M.T., Mirkin, B.: Intelligent choice of the number of clusters in k-means clustering: an experimental study with different cluster spreads. J. Classif. **27**(1), 3–40 (2010)
37. Varewyck, M., Martens, J.P.: A practical approach to model selection for support vector machines with a Gaussian kernel. IEEE Trans. Syst. Man Cybernet., Part B (Cybernetics) **41**(2), 330–340 (2011)
38. Chollet, F.: Keras. GitHub repository. https://github.com/fchollet/keras (2015)
39. Bastien, F., Lamblin, P., Pascanu, R., Bergstra, J., Goodfellow, I.J., Bergeron, A., Bouchard, N., Bengio, Y.: Theano: new features and speed improvements. In: Deep Learning and Unsupervised Feature Learning NIPS Workshop (2012)
40. Matlab R2014, The MathWorks, Inc., Natick, MA, USA. https://it.mathworks.com

Chapter 3
A Survey on Automatic Multimodal Emotion Recognition in the Wild

Garima Sharma and Abhinav Dhall

Abstract Affective computing has been an active area of research for the past two decades. One of the major component of affective computing is automatic emotion recognition. This chapter gives a detailed overview of different emotion recognition techniques and the predominantly used signal modalities. The discussion starts with the different emotion representations and their limitations. Given that affective computing is a data-driven research area, a thorough comparison of standard emotion labelled databases is presented. Based on the source of the data, feature extraction and analysis techniques are presented for emotion recognition. Further, applications of automatic emotion recognition are discussed along with current and important issues such as privacy and fairness.

3.1 Introduction to Emotion Recognition

Understanding one's emotional state is a vital step in day to day communication. It is interesting to note that human beings are able to interpret other's emotion with great ease using different cues such as facial movements, speech and gesture. Analyzing emotions help one to understand other's state of mind. Emotional state information is used for intelligent Human Computer/Robot Interaction (HCI/HRI) and for efficient, productive and safe human-centered interfaces. The information about the emotional state of a person can also be used to enhance the learning environment so that students can learn better from their teacher. Such information is also found to be beneficial in surveillance where the overall mood of the group can be detected to prevent any destructive events [47].

G. Sharma (✉)
Human-Centered Artificial Intelligence group, Monash University, Melbourne, Australia
e-mail: garima.sharma1@monash.edu

A. Dhall (✉)
Human-Centered Artificial Intelligence group, Monash University, Melbourne, Australia

Indian Institute of Technology, Ropar, India
e-mail: abhinav.dhall@monash.edu

G. Phillips-Wren et al. (eds.), *Advances in Data Science: Methodologies and Applications*, Intelligent Systems Reference Library 189,
https://doi.org/10.1007/978-3-030-51870-7_3

The term emotion is often used interchangeably with affect. Thoits [133] argued that *affect* is a non-conscious evaluation of an emotional event. Whereas, *emotion* is a culturally biased reaction to a particular affect. Emotion is an ambiguous term as it has different interpretations from different domains like psychology, cognitive science, sociology, etc. Relevant to affective computing, emotion can be explained as a combination of three components: *subjective experience*, which is biased towards a subject; *emotion expressions*, which include all visible cues like facial expressions, speech patterns, posture, body gesture; and *physiological response* which is a reaction of a person's nervous system during an emotion [5, 133].

A basic cue for identifying a person's emotional state is to detect his/her *facial expressions*. There are various psychological theories available which help one to understand a person's emotion by their facial expressions. The introduction of *Facial Action Coding System (FACS)* [44] has helped researchers to understand the relationship between facial muscles and facial expressions. For example, one can distinguish two different types of smiles using this coding system. After years of research in this area, it has become possible to identify facial expressions with greater accuracy. Still, a question arises that, whether only expressions are sufficient to identify emotions? Some people are good at concealing their emotions. It is easier to identify an expression; however, it is more difficult to understand a person's emotion i.e. the state of the mind or what a person is actually feeling.

Along with the facial expressions, we human's also rely on other non-verbal cues such as gestures and verbal cues such as speech. In the affective computing community, along with the analysis of the facial expressions, researchers have also used the speech properties like pitch, volume and physiological signals like Electroencephalogram (EEG) signals, heart rate, blood pressure, pulse rate, flow of words in the written text to understand a person's affect with more accuracy. Hence, the use of different modalities can improve a machine's ability to identify the emotions similar to how human beings perform the task.

The area of affective computing though not very old, has seen a sharp increase in the number of contributing researchers. This impact is due to the interest in developing human centered artificial intelligence, which are in trend these days. Various emotion based challenges are being organized by the researchers, such as Aff-Wild [152], Audio/Visual Emotion Challenge (AVEC) [115], Emotion recognition in the wild (EmotiW) [33], etc. These challenges provide an opportunity for researchers to benchmark their automatic methods against the prior works and each other.

3.2 Emotion Representation Models

The emotional state of a person represents the way a person feels due to the occurrence of various events. Different external actions can lead to a change in the emotional state. For an efficient HCI there is a need for an objective representation of emotion. There exists various models which interpret emotions differently. Some of the models are applicable to audio, visual, textual content and others are limited to only visual data. Some of the widely used emotion representation models are discussed below.

3.2.1 Categorical Emotion Representation

This emotion representation has discrete categories for different emotions. This is based on the theory by Ekman [35], which argues that emotion can be represented in six universal categories. These categories are also known as *basic emotions* which are *happiness, sadness, fear, disgust, surprise and anger*. *Neutral* is added to this to represent the absence of any expression. This discrete representation is the most commonly used representation of emotions as it is easy to categorize any image, video, audio or text to one of these categories.

It is non trivial to draw a clear boundary between two universal emotions as they may be present together in a sample. In general, human beings feel different kinds of emotions, which are a combination of the basic categories like *happily surprised, fearfully surprised*, etc. Hence, 17 categories were defined to include a wide range of emotions [34]. These categories are termed as *compound emotions*. Inspite of adding more categories to represent real life emotions, it is still a challenging task to identify compound emotions as their occurrence depends on the identity and culture. The use of basic and compound emotions depends on the application of the task. In spite of having some limitations, basic emotions are mostly used for tasks to achieve a generalized performance across different modalities of the data. In an interesting recent work, Jack et al. [64] found that there are only four universal emotions, which are common across different cultures, instead of the earlier believed six universal emotions.

3.2.2 Facial Action Coding System

The presence of any expression can also be estimated by the change in the muscle movements as defined by Facial Action Coding System (FACS) [44]. This system defines Action Units (AU) which map the activation of muscles in the face, representing the facial deformations. Originally, 32 such AUs were defined to represent the presence of an expression. Later, the system was extended to include 14 additional action descriptors, which contain the information of head pose, gaze, etc. [36]. The emotion detection system can predict the occurrence of particular AUs as a classification problem or the intensity of AU as a regression problem. AUs such as *inner brow raise, cheek raise, lip puller, nose wrinkle*, etc. provide independent and impulsive actions. Despite of having many benefits of using FACS for emotion detection, there exists a dependency on a trained expert to understand and annotate the data. This requirement makes it complicated to use AUs to represent emotions. It is to be noted that FACS is a visual modailty only based emotion representation.

3.2.3 *Dimensional (Continous) Model*

Dimensional model assumes that each emotional state lies somewhere in a continous dimension rather than being an independent state. The circumplex model [117] is the most popular dimensional model to describe emotions. It represents emotion in terms of continuous value for *valence* and *arousal*. These values represent the changes in the emotion from positive to negative and the intensity of the emotion, respectively. This method provides a suitable logical representation to map each emotion with respect to other emotions. The two dimensions of the model were later extended to include *dominance* (or *potency*). It represents a certain way one emotion can be controlled over others due to different personal or social boundaries [98].

The dimensional model can be successfully used to analyze the emotional state of a person in continuous value and time. The model can be used corresponding to audio and visual data. The value of arousal and valence can also be specified for different keywords to recognize the emotions from the textual data. However, it is still complicated to find the relation between the dimensional model and Ekman's emotion categories [52]. The representation of basic emotions categories doesn't include the complete arousal-valence space. Some psychologists claim that emotional information cannot be represented in just two or three dimensions [52].

3.2.4 *Micro-Expressions*

Apart from understanding facial expressions and AUs for emotion detection, there exists another line of works, which focus on the subtle and brief facial movements present in a video, which are difficult for the humans to recognise. Such facial movements are termed as *micro-expressions* as they last less than approximately 500 ms as compared to normal facial expressions (macro-expressions) which may last for a second [150]. The concept of micro-expressions was introduced by Haggard and Issacs [53] and it gained much success as micro-expression is an involuntary act and it is difficult to voluntarily control them.

3.3 Emotion Recognition Based Databases

Affective computing is a data-driven research area. The performance of an emotion detection model gets effected by the type of data present. Factors such as recording environment, selection of subjects, time duration, emotion elicitation method, imposed constraints, etc. are considered while the creation or selection of database to train a classifier. The amount of illumination, occlusion, camera settings, etc. are the other important factors, which requires consideration. A large number of databases are already present in the literature over these variations which can be used depending on the problem.

Table 3.1 compares some of the commonly used emotion databases with respect to the variations present in them. All the mentioned databases in the table are open sourced and can be used for the academic use. These databases include different modalities i.e. image, audio, video, group, text and physiological signals as specified in the table. Some databases also include the spontaneous expressions, which are used in a several current studies [28, 144].

3.4 Challenges

As the domain of emotion recognition has a high number of possible applications, research is going on to make the process more automatic and applied. Due to the adaptation of benchmarking challenges such as Aff-Wild, AVEC and EmotiW, few obstacles are being successfully addressed. Major challenges are mentioned below:

Data Driven—Currently the success of emotion recognition techniques is partly due to the advancements of different deep neural networks. Due to deep networks, it has become possible to extract complex and discriminative information. However, neural networks require a large amount of data to learn useful representations for any given task. For automatic emotion recognition task, having data corresponding to real world emotions is non trivial; however, one may record the person's facial expressions or speech to some extent, although these expressions may vary for the real and fake emotions.

For many years, posed facial expressions of professional actors have been used to train models. Although, these models perform poorly, when applied on data from real world settings. Currently, many databases exists, which have spontaneous audio-visual emotions. Most of these temporal databases are limited to the size and the number of samples corresponding to each emotion category. It is non-trivial to create a balanced database as it is difficult to induce few emotions like fear, disgust, etc. as compared to happy and angry.

Intra-class Variance—If the data is recorded in different settings for the same subject with or without same stimuli, the emotion elicited may vary due to the prior emotional state of the person and local context. Due to different personalities, different people may show the same expression differently or react differently to the same situation. Hence, the final obtained data may have high intra-class variance which remains a challenge to the classifier.

Culture Bias—All emotion representation models define the occurrence of emotions based on audible or visible cues. The well established categorical model for basic emotions by Ekman has also defined the seven categories as universal. However, many recent studies have shown that the emotion categories depends on the ethnicity and the culture of a person [63]. The way of expressing emotion varies from culture to culture. Sometimes, people use their hand and body gestures to con-

Table 3.1 Comparison of commonly used emotion detection databases. Online readers can access the website of these databases by clicking on the name of the database for more information. Number of samples for text databases is in words. Number of samples in each database is an approximate count

Dataset	No. of samples	No. of subjects	P/ NP	Recording environment	Labels	Modalities	Studies
AffectNet [102]	1 M	400 K	NP	Web	BE ⋄, CoE	I	[144]
EmotionNet [41]	100 K	–	NP	Web	AU, BE, CE	I	[67]
ExpW [157]	91 K	–	NP	Web	BE †	I	[80]
FER-2013 [50]	36 K	–	NP	Web	BE	I	[70]
RAF-DB [78]	29 K	–	NP	Web	BE, CE	I	[47]
GAFF [33]	15 K	-	NP	Web	3 Group emotions	I, G	[47]
HAPPEI [31]	3 K	–	NP	Web	Val (Discrete)	I, G	[47]
AM-FED+ [95]	1 K	416	NP	Unconst.	AU	V	–
BU-3DFE [151]	2.5 K	100	P	Const.	BE + Intensity	V ⁕	[91]
CK+ [89]	593	123	P, NP	Const.	BE	V	[91]
CASME II [149]	247	35	NP	Const.	Micro-AU, BE ⋄	V	[28]
DISFA [94]	100 K	27	NP	Const.	AU	V	[92]
GFT [48]	172 K	96	NP	Const.	AU	V	[38]
ISED [56]	428	50	NP	Const.	BE ⋄	V	[91]
NVIE [142]	–	215	P, NP	Const.	BE	V ⋆	[12]
Oulu-CASIA NIR-VIS [158]	3 K	80	P	Const.	BE	V ⋆	[92]
SAMM [29]	159	32	NP	Const.	Micro-AU, BE	V	[28]
AFEW [32]	1 K	–	NP	Web	BE	A, V	[70], [92]
BAUM-1 [154]	1.5 K	31	P, NP	Const.	BE †	A, V	[108]
Belfast [128]	1 K	60	NP	Const.	CoE	A, V	[62]
eNTERFACE [93]	1.1 K	42	NP	Const.	BE	A, V	[108]
GEMEP [10]	7 K	10	NP	Const	Ar, Val (Discrete)	A, V	[91]
IEMOCAP [16]	7 K	10	P, NP	Const, Unconst	BE †	A, V	[74]
MSP-IMPROV[18]	8 K	12	P	Const.	BE ⋄	A, V	[74]
RAVDESS [84]	7.3 K	24	P	Const.	BE	A, V	[153]
SEMAINE [97]	959	150	NP	Const.	Ar, Val	A, V	[91]
AIBO [13]	13 K	51	NP	Const.	BE †	A	[141]
MSP-PODCAST [85]	84 K	–	NP	Const	CoE, Dom	A	[57]
Affective dictionary [143]	14 K	1.8 K	–	–	CoE	T	[160]
Weibo [79]	16 K	–	NP	Web	BE	T	[147]
Wordnet-Affect [131]	5 K	–	–	–	BE ⋄	T	[130]
AMIGOS [25]	40	40	NP	Const.	CoE	V, G, Ph	[125]

(continued)

Table 3.1 (continued)

Dataset	No. of samples	No. of subjects	P/ NP	Recording environment	Labels	Modalities	Studies
BP4D+ [156]	–	140	NP	Const.	AU (Intesity) †	V ⋆, Ph	[38]
DEAP [71]	120	32	NP	Const.	CoE	V, Ph	[61]
MAHNOB-HCI [129]	–	27	NP	Const.	Ar, Val (Discrete)	A, V, Ph†	[61]
RECOLA [116]	46	46	NP	Const.	CoE†	A, V, Ph†	[115]

I—Image, A—Audio, V—Video, G—Group, T—Text, Ph—Physiological, K—Thousand, M—Million, BE—Basic categorical emotion (6 or 7), CE—Compound emotions, CoE—Continues emotions, Val—valence, Ar—Arousal, P—Posed, NP—Non posed, Const.—Constrained, Unconst—Unconstrained
⋄—Contains less than 6 or 7 basic emotions, ⋆—Also include infra red recordings
†—Contains extra information (than emotions), ⁕—Includes 3-D data

vey their emotions. The research in creating a generic universal emotion recognition system faces the challenge of inclusivity of all ethnicities and cultures.

Data Attributes—Attributes such as head pose, non frontal face, occlusion and illumination effect the data alignment process. The presence of these attributes acts as a noise in the features which can degrade the performance of the model. Also, the real world data may contain some or all of these attributes. Hence, there is a scope of improvement to neutralize the effects of these attributes.

Single Versus Multiple Subjects—A person's behaviour is affected by the presence of other people around them. For such cases, the amount of occlusion increases to a large extent due to the location of the placed camera. Also, the face captured for these settings are usually very small to identify the visible cues in them. There are a wide number of applications which need to analyze a person's behaviour in a group, the most important of which is for surveillance. There are some already proposed methods which can detect multiple subjects in visual data; however, analyzing their collective behaviour still need some progress.

Context and Situation—Emotions of a person can be estimated efficiently by using different types of data such as audio, physiological and visual. However, it is still non-trivial to predict emotion of a person from these information. The effect of the environment may be easily observed in case of emotion analysis. In formal settings (such as a workplace), people may be more cautious, while writing. However, in an informal environment, people tend to use casual or sarcastic words to express themselves. In a recent study, Lee et al. [76] found that contextual information is important as a person's reaction depends on the environment and situation.

Privacy—The privacy issue is now an active topic of discussion in the affective computing community. The learning task in various domains require data, which is collected from various sources. Sometimes the data is used and distributed for academic or commercial purposes, which may directly or indirectly violate the right to privacy of a person. Due to its significance, privacy will also be discussed in the Sect. 3.11.

3.5 Visual Emotion Recognition Methods

Visual content plays a major role in emotion detection as facial expression provide meaningful emotion specific information. To perform the *Facial Expression Recognition (FER)* task, input data may have spatial information in the form of images or spatio-temporal data from videos. Videos have extra advantage in this task as one can use the variation in the expressions across time. Another important application of FER is the identification of micro-expressions, which can be accomplished by using spatio-temporal data. Despite having many advantages, it is computationally expensive to extract features from videos and to process them for emotion detection.

The emotion detection process can be extended from a single person to multiple persons. One can use these methods to understand the behaviour of a group of people by analyzing the expressions of each identity. There are some other factors which need to be considered for a group such as a context, interpersonal distance, etc. which effect the group dynamics.

The emotion recognition process in visual data is effected by the occurrence of deep learning methods. A different set of methods are used in this process prior to the deep learning and after it. However, understanding of traditional processes is important to understand the process. Due to this reason, the methods used before and after the introduction of deep learning techniques are explained in detail.

3.5.1 Data Pre-processing

The input data for any FER task consist of facial images/videos which may have faces in different pose, illumination. One needs to convert the raw input data to a form such that only meaningful information is extracted from them. First, *face detector* is used to detect the location of faces present in the images. Viola Jones technique [139] is a classic example and is one of the most widely used face detector. Face detector locates the face, which needs to be aligned with respect to the input image. *Face alignment* is performed by applying affine transformations to convert non frontal facial image to frontal image. A common technique to perform this operation is to identify the location of the nose, eyes and mouth and then transform an image with respect to these points. To perform these transformations smoothly, more number of points are selected. One of the such methods is Active Appearance Model (AAM) [24], which is a generative technique to deform objects based on their geometric and appearance information. Along with FER, AAMs have been widely used in problems like image segmentation and object tracking. Despite of all these advantages, AAM

Table 3.2 Comparison of open source face detection and analysis libraries

Library	Face detection	Face tracker	Facial landmarks	Head pose	Action units	Studies
Chehra [7]	✓	✓	✓	✓	✗	[125]
Dlib [69]	✓	✓	✓	✓	✗	[70, 135]
KLT [88, 124, 134]	✓	✗	✗	✗	✗	[46]
MTCNN [155]	✓	✗	✓	✗	✗	[47]
NPD Face [81]	✓	✗	✗	✗	✗	[135]
Open Face 2.0 [9]	✓	✓	✓	✓	✓	[43]
Tiny Face [59]	✓	✗	✗	✗	✗	[111]
Viola Jones [139]	✓	✗	✗	✗	✗	[82, 111]

lacks to align images smoothly and in real time. It also produces varied results in case of inconsistent input data. These limitations are overcome in Constrained Local Models (CLM) [119] in which key features are detected by the use of linear filters on the extracted face image. The CLM features are robust to illumination changes and more generic towards unseen data. Some open source libraries used in the data pre-processing step are shown in Table 3.2.

3.5.2 Feature Extraction

In order to use images or videos for any learning task, one needs to identify the appropriate way of *data registration*. Facial data can be registered in different ways depending on the end goal, input data and features to be used to encode the representation of the data [120]. The *full facial image* may be used to extract all the information present in an image. The method is useful in the case when there are small variation in the images across classes and one want to use all the information explicitly present in the input images.

Part based registration methods divide the input image into different parts by focusing on the specific part of the face. These parts can be decided by the additional information like the position of the components of the image [120]. For facial images, the sub parts of an image may constitute of eye, lips, forehead region, etc. This method ensures the consideration of low-level features. Similar to part based methods, *point based* methods are also used to encode low-level information. These methods focus on particular geometric locations [120]. These points can be initialized by the interest

Table 3.3 Comparison of low-level features. Here, Geo. and App. refer to geometric and appearance based features

Feature	AAM [24]	LBP [2]	LPQ [106]	HOG [27]	PHOG [14]	SIFT [87]	Gabor
Geo/App	Geo.	App.	App.	App.	App.	App.	App.
Temporal	–	LBP-TOP [159]	LPQ-TOP [65]	HOG-TOP [21]	–	–	Motion energy [146]
Local/Holistic	Global	Local	Local	Local	Local	Local	Holistic
Studies	[142]	[149, 158]	[28, 30]	[126]	[30, 126]	[126]	[107]

point detector or the facial fiducial points. Point based methods are beneficial to encode the shape related information to maintain the consistency in input images. This can be used for spatial as well as spatio-temporal information (Table 3.3).

In an abstract manner, the face provides three main types of information. The *static* variations, which remains almost constant for an identity like the facial shape or the skin color. The *slower* changes which a face has along a longer time span like wrinkles in the face. The *rapid* changes in a face take place for a short span of time like the small changes in the facial muscle. For emotion detection task, these rapid variations are more focused, whereas the static and slower variations still remain a challenge to tackle with.

Geometric Features—Emotion detection process requires a suitable *data representation* method to encode the non-deformable changes in the face. *Geometric features* represent the shape or the structural information of an image. For a facial image, these features encode the position/location of facial components like eyes, nose, mouth, etc. Hence, geometric features can encode the semantic information present in an image. These features can also be extracted in a holistic or parts based manner. With the development of many facial landmark detectors, it has become an easy task to find the precise location of the parts of the face in real time. The extracted geometric features are invariant to illumination and affine transformations. Further, it is easy to model this representation to create a 3-D model of face to make the process pose invariant also. Although geometric features provide a better representation of the shape of the object, it may lack in representing smaller variations in the facial expressions. The expressions which do not have much change in the AUs can not be represented well by geometric features alone.

Spatial Features—The *spatial features* (appearance based) focus on the texture of the image by using the pixel intensity values of the image. For emotion detection task, change in the expressions in a person's face is encoded in the appearance based features. The representation of the facial information can be performed either in *holistic* way or the *part-wise*. The holistic features focus on the high level information by using the complete image. These features encode the wide variations which take place in the appearance of the object. Appearance features can also be extracted on the parts of the face. Appearance features are extracted from small patches across different keypoints on a facial image.

To represent emotion specific information, it is necessary to capture the subtle changes in the facial muscles by focusing on the fine level details of the image. Based on the type of information, feature descriptors can be classified into three categories: low-level, mid-level and high-level. These features can also be computed in a holistic or part based manner. The *low-level* image descriptor encodes pixel level information like edges, lines, color, interest points, etc. These features are invariant to affine transformations and illumination variation. Commonly used low-level features are: *Local Binary Pattern* (LBP) [2] features extract the texture of the image by counting the pixels greater than the fixed threshold with respect to the neighboring pixels. *Local Phase Quantisation* (LPQ) [106] is widely used to encode blur insensitive image texture. LPQ also counts the number of pixels locally after calculating local fourier transformations.

A certain class of low-level features focuses on the change in the gradients across pixels. *Histogram of Gradients* (HOG) [27] is such a popular method which calculates the change in the gradient magnitude and orientations. A histogram is computed for the orientation of the gradients which specifies the chances of a gradient with particular orientation corresponding to a local patch. The simplicity of HOG is later extended to *Pyramid of Histogram of Gradients* (PHOG) [14]. PHOG captures the distribution of edge orientation over a region to record its local shape. The image region is divided into different resolutions to encode the spatial layout of the image. *Scale Invariant Feature Transform* (SIFT) [86] find the keypoints across different scales and assign orientations to each keypoint. These orientations are assigned on the basis of local gradient directions. The local shape of the face can also be encoded by calculating the histogram of directional variations of the edges. *Local Prominent Directional Pattern (LPDP)* [91] uses this statistical information from a small local neighboring area for a given target pixel. The texture of the input image can also be extracted by using *Gabor filters*. It is a type of bandpass filter which accepts the certain range of frequency and rejects others. The input image is convolved with Gabor filters of different sizes and orientations.

Mid-level features are computed by combining several low-level features for the complete facial image. One of the methods widely used for mid-level representation is *Bag of visual words (BOVW)*. In this method, a vocabulary is created by extracting low-level features from different locations in the image. Features for the new target image is then matched with vocabulary without getting affected by translation or rotation. To find a feature in the vocabulary, spatial pyramid method can be used, in which feature matching is performed at different scales. The use of a spatial pyramid makes the process invariant to scaling. The information learned by low and mid level features can be combined to gain the semantic information which a human can relate to. Such features are known as *high-level* features. An example of high level features for emotion detection task can be a model, which identify the name of the expression (not just the class) or the active AU's as output by using certain features.

Spatio-temporal Features—A large number of computer vision based applications require to extract spatial as well as temporal features from a video. Information can be extracted in two ways across the frames. The first type captures the motion due to the transition from one frame to another (optical flow). The other type of features are dynamic appearance features which capture the change in the appearance of objects across time. The motion based features doesn't encode the identity specific information; however, these features depends on the variation of illumination and head pose. A video can be considered as a stack of frames in 3-dimensional space, each of which has small variation along its depth. A simple and efficient solution to extract spatial as well as temporal features from video is the use of low-level feature descriptors across Three Orthogonal Planes (TOP) of the video. Extraction of features from TOP, is used with various low-level feature descriptors such as LBP-TOP [159], LPQ-TOP [65], HOG-TOP [21], etc. Features are computed along spatial and temporal plane i.e. along *xy*, *xt* and *yt* planes. The concept of Gabor filters is also extended to *Gabor motion energy filters* [146]. These filters are created by adding 1-D temporal filters on frequency tuned Gabor energy filters.

To encode features from a facial region, the representation strategy should be invariant to the illumination settings, the head pose of a person and the alignment of the face at the time of recording. It is more meaningful to extract identity independent information from a face which is a challenge for appearance based feature descriptors as they encode the complete pixel wise information from a image. Also, it is important to note that now learnt features from deep neural networks are also widely used as low-level and high-level features [109].

3.5.3 Pooling Methods

Generally, low-level feature descriptors produce large dimensional feature vectors. It is important to investigate dimensionality reduction techniques. All the low-level feature descriptors for which a histogram is created for a local region, the dimension of feature vector can be reduced by controlling the size of the bin of a local patch. A classic example of such low-level feature descriptor is *Gabor filters*. The use of these large number of filters produces a high dimensional data.

Principle Component Analysis (PCA) is a method which has been widely used to reduce the dimension of features. PCA find the linearly independent dimensions which can represent the data points with minimal loss in an unsupervised manner. Dimensions of feature vector can also be reduced in a supervised manner. *Linear Discriminant Analysis (LDA)* is a popular method used for data classification as well as dimensionality reduction. LDA finds a common subspace in which original features can be represented in *K-1* number of features, where *K* is the number of classes in the data. Thus, features can be classified in reduced subspace by using the less number of features.

3.5.4 Deep Learning

With the recent advancements in computer hardware, it has become possible to compute a large number of computations in a fraction of seconds. Growth in Graphical Processing Unit (GPU) has resulted in the ease to use deep learning based methods in any domain including computer vision. The readers are pointed to Goodfellow et al. [49] for the details of deep learning concepts. The use of Convolutional Neural Networks (CNN) has achieved an efficient performance in the emotion detection task. Introduction of CNN has made it easy to extract the features from the input data. Earlier, the choice of handcrafted features used to depend on the input data which affects the performance of FER explicitly.

CNN directly converts the input data to a set of relevant features which can be used for the prediction. Also, one can directly use the complete facial data and let the deep learning model decide the relevant features for the FER task. The deep learning based techniques require a large amount of input data to achieve an efficient performance. The requirement is fulfilled by many researchers who have contributed large databases to the affective computing community as explained in Sect. 3.3.

3.5.4.1 Data Pre-processing

CNN learn different filters corresponding to the given input image. Hence, all the input data must be in the same format such that filters can learn the generalized representation on all the training data. Different face detector libraries are available nowadays which can be used with deep learning based methods to detect a face, landmarks or fiducial points, head pose, etc. in real time. Some libraries even produce aligned and frontal faces as their output [9]. Among all the open source face detection libraries shown in Table 3.2, Dlib, Multi-task Cascaded Convolutional Networks (MTCNN), OpenCV and Openface are widely used with deep learning methods. Also, as neural networks require a large amount of data, data augmentation techniques are used to produce extra data. Such techniques apply transformations like translation, scaling, rotation, addition of noise, etc. and help to reduce the over-fitting.

Data augmentation techniques are also required when the data is not class-wise balanced, which is a common situation while dealing with real world spontaneous FER system. Several studies show that new minority class data can be sampled from class-wise distribution in a higher dimension [83]. Recently proposed networks like Generative Adversarial Networks (GAN) are able to produce identity independent data in a high resolution [137]. All these methods have helped researchers to overcome the high data requirement of deep learning networks.

3.5.4.2 Feature Extraction

Deep learning based methods extract features from input data by capturing high-level and low-level information from a series of filters. A large number of filters vary in their size and learn information ranging from edges, shapes to the identity of the person. These networks have convolution layer which learn filters on a complete 2-D image by a convolution operation. It learns shared weights and ignores small noise produced from the data registration process. Learned filters are invariant to illumination and translation. A network can have multiple convolution layers each of which can have the different number of filters. Filters at initial layers learn high level information whereas higher convolution filters focus on learning low-level information.

Spatial Features—The full input image or part of the image can be used as input to CNN. It converts the input data to a feature representation by learning different filters. These features can be further used to learn the model. Various deep learning based networks like Alexnet [73], Resnet [58], DensNet [60], VGG [127], Capsule network [118], etc. exists, each of which have convolutions and fully connected layers in different combinations to learn better representation. The autoencoder based networks are also used to learn the representations by regenerating the input image from the learned embeddings [144]. The comparison of different widely used such networks is also discussed in Li et al. [77].

Spatio-temporal Features—Currently, several deep learning based modules are available, which are able to encode the change in the frames corresponding to the appearance of objects across time. The videos can also be represented in the form of 3-D data. Hence, 3-D convolution operation may be used to learn the filters. However, feature extraction using 3-D convolution is a complex task. First, frames need to be identified such that selected frames have uniform variation for the expression. Also, 3-D convolution requires a large amount of memory due to the large number of calculations associated with it.

Variations along the temporal space can also be encoded by using Long Short Term Memory (LSTM) and Gated Recurrent Network (GRU) [23]. These methods learn the temporal variations for the given set of sequence vectors. Several variations of LSTM like ConvLSTM [148] and bidirectional-LSTM [51] also exists to learn better representation of a video.

3.5.4.3 Pooling Methods

The success of deep neural networks lies in the use of deep networks which include the large number of filters. The filters are responsible to encode all the information present in input data. However, large number of filters also increase the computations involved in the process. To reduce the size of filters, pooling operations are performed which consist of *max pooling*, *min pooling* and *average pooling*. These operations reduce the size of features by finding maximum, minimum and average feature values, respectively. These operations are also found to be useful for discarding information while learning, which is essential to reduce overfitting.

3.6 Speech Based Emotion Recognition Methods

According to *7-38-53 rule* by Mehrabian et al. [99], 7% of any communication depends on verbal content, 38% depends on the tone of the voice and 53% on the body language of a person. Hence, the acoustic features like pitch (fundamental frequency), timing (speech rate, voiced, unvoiced, sentence duration, etc.), voice quality, etc. can be utilized to detect the emotional state of a person. However, it is still a challenge to identify the significance of different speaking styles and rates and their impact on emotions. The features are extracted from audio signals by focusing on the different attributes of speech. Murray et al. [104] identified that quality of voice, timing and the pitch contour are mostly affected by the emotion of the speaker. The acoustic features can be categorized as *continuous*, *qualitative*, *spectral* and *TEO-based* features [37].

Continuous or *prosodic* features contribute more to the emotion detection task as they focus on the cues like tone, stress, words, pause in between words, etc. These features include pitch related features, formant frequencies, timing features, voice quality and articulation parameters [75]. McGilloway et al. [96] provided 32 acoustic features by using their Automatic Statistical Summary of Elementary Speech Structures (ASSESS) system most of which are related to prosodic features. Some of these features are tune duration, mean intensity, inter quartile range, energy intensity contour, etc. The widespread study of Murray et al. [104] also provided the effect of 5 basic emotions on the different aspects of speech, most of which are prosodic features. Sebe et al. [122] also used the logarithm of energy, syllable rate and pitch as prosody features. All of these prosodic features focuses on global level features by extracting utterance level statistics from the speech. However, the features can't encode the small dynamic variations along the utterance [17]. It becomes a challenge to identify the emotion of the person in the presence of the two emotions together in the speech from these set of features. This limitation is overcome by focusing on the changes in segment-level.

Qualitative features emphasis on the voice quality for the perceived emotion. These type of features can be categorized in voice level features, voice pitch based, phrase, word, phoneme, feature boundaries and temporal structures [26]. Voice level features consider the amplitude and the duration of the speech. Boundary detection for phrases, words is useful to understand the semantics for connected speech. A simple way to detect the boundary is by identifying the pauses in between the words. Temporal structure measure the voice pitch in terms of rises, falls and level stretches. Jitter and shimmer are also the commonly used features, which encode the frequency and amplitude of the vocal fold vibrations [8]. Many a time attributes like breathy, harsh and tense are also used to define the quality of a voice [26].

Spectral features are used to extract short speech signals. These features can be extracted from the speech signals directly or by applying filters to get better distribution over the audible frequency range. Many studies also include these features as quality features. *Linear Predictive Cepstral Coefficients* (LPCC) [110] is one such kind of feature used to represent the spectral envelope of the speech. The linear predicitve analysis represents the speech signals as an approximation of the linear combination of past speech signals. The method is used to extract accurate speech parameters and is faster to compute.

Mel Frequency Cepstral Coefficients (MFCC) is a popular spectral based method used to represent sound in many speech domains like music modelling, speaker identification, voice, etc. It represents the short term spectrum of sound waves. MFCC features approximate the human's auditory system where the pitch is perceived in non linear manner. In MFCC, the frequency band used is of equal space in mel scale (scale providing the mappings of actual frequency and perceived audio pitch). The speech signal is passed through a number of mel filters. Several implementations of MFCC exist which depends on the type of approximation to the nonlinear pitch, design and the compression method used for the filter banks [45]. *Log Frequency Power Coefficients (LFPC)* also approximates the human's auditory system by calculating the logarithmic filtering of the signal. LFPC can encode the fundamental frequency from the signal in a better way as compared to MFCC for emotion detection [105]. Many different variations of spectral features are also proposed by modifying these set of features [145]. The linear predictor coefficients were also extended to cepstral based *One Sided Autocorrelation Linear Predictor Coefficients (OSALPCC)* [15].

TEO based features are used to find the stress in the speech. The concept is based on the Teager-energy-operator (TEO) study done by Teager [132] and Kaiser [68]. The studies show that the non linear air flow in the human's vocal system produces speech and one need to detect this energy to listen to it. The TEO is used to successfully analyze the pitch contour for the detection of neutral, loud, angry, Lombard effect, and clear [19]. Zhou et al. [161] proposed three non linear TEO based features namely TEO-decomposed FM variation (TEO-FM-Var), normalized TEO autocorrelation envelope area (TEO-Auto-Env), and critical band based TEO autocorrelation envelope area (TEO-CB-Auto-Env). These features are proposed by discarding the word level dependency of the stress. The focus of these features is to find the correlation between nonlinear excitation attributes of the stress.

To define a common standard set of features for audio signals, Eyben et al. [39] proposed a *Geneva Minimalistic Acoustic Parameter Set (GeMAPS)*. The authors performed an extensive interdisciplinary research to define a common standard that can be used to benchmark the auditory based research. GeMAPS defines two set of parameters on the basis of their ability to capture the various physiological changes in affect related processes, their theoretical significance and the relevance found in the past literature. One is minmalistic parameter set which contains 18 low-level descriptors based on prosodic, excitation, vocal tract and spectral features. The other is an extended parameter set containing 7 low-level descriptors including cepstral and frequency related parameters. Similar to GeMAPS, *Computational Para-linguistics Challenge (COMPARE)* parameter set is widely used in INTERSPEECH challenges [121]. COMPARE defines 6,373 audio features among which 65 are acoustic low-level descriptors based on energy, spectral and voicing related information. The GeMAPS and COMPARE are highly used in the recent studies of emotion detection from speech [136].

The success of the bag of words method has motivated researchers to extend it for speech also. Bag of audio words [66] and bag of context-aware words [55] are such proposed methods where a codebook is created of audio words. The context information is added to this method by generating features from the complete segment to obtain much higher level representation.

There exist a large number of toolkits which can be used to extract the features from the speech signal. Some of these toolkits are aubio, Maaate, YAAFE, etc. The in detailed comparison of such toolkits is already provided by Moffat et al. [101]. The another popular library is *OpenSMILE* [40], where SMILE stands for Speech and Music Interpretation by Large-space Extraction. It is used to extract audio features and to recognize the pattern present in the audio in real time. It provides low-level descriptors including FFT, cepstral, pitch, quality, spectral, tonal, etc. The toolkit also extracts various functionals like mean, moment, regression, DCT, zero crossings, etc.

Similar to the feature extractor process in images, features can be extracted by dividing the speech into multiple intervals or using the complete timestamp. This difference in the extraction process produces global versus local features. The selection of feature extraction depends on the classification problem. The extracted audio features are used to learn for the presence of a given emotion. Earlier SVM, HMM kind of models has been used to accomplish this task which is now replaced by different kind of neural networks. Different types of neural networks like LSTM, GRU, etc. are also used to learn the change in the sequence. The audio signals can be used along with the visual information to achieve better performance of the emotion detection model. For such cases, information from two modalities can be fused in different types [54]. The fusion methods are discussed in Sect. 3.9.

3.7 Text Based Emotion Recognition Methods

The recent trends in social media have provided opportunities to analyse data from the text modality. Users upload a large number of posts and Tweets to describe their thoughts, which can be used to detect the emotional state of the users. This problem is largely explored in the field of sentiment analysis [20]. Analysis of emotion differs from that of sentiment analysis as an emotion defines the state of the feeling whereas a sentiment shows the opinion or the judgment produced from a particular feeling [103]. Emotions occur in the pre-conscious state; however, the sentiments result due to the occurrence of emotions in a conscious state.

To interpret the syntactic and semantic meaning of a given text, the data is converted into a vector form. There are different methods to compute these representations [72]. Keyword or lexicon based methods use a predefined dictionary, which contains an affective label corresponding to a given keyword. The labels follow either the dimensional or the categorical model for emotion representation. Few examples of such dictionaries are already shown in Table 3.1. The dictionary can be created manually or automatically such as WordNet-Affect dictionary [131]. Creating such a dictionary requires prior knowledge of linguistics. Further, the annotations can get effected by the ambiguity of words and the context associated with them.

The category of emotion can also be predicted by using a learning based method. In this category, a trained classifier takes the segment wise input data in a sliding window manner. As the output is produced by considering a part of the data at a given time, contextual information may be lost during the process. The input data can also be converted to word embeddings based on their semantic information. To create such embedding, each word is converted into a vector in the latent space such that two semantically similar words remain closer in the latent space. Word2Vec [100] is one such widely used model used to compute the word embeddings from the text. These embeddings are known to capture the low-level semantic information as well as contextual information present in the data. Asghar et al. [6] used a 3-D affective space to accomplish this task. These methods can be used to interpret the text either in a supervised or an unsupervised manner to find a target emotion. The comparison of such methods is presented by Kratzwald et al. [123].

The word embeddings represent the data in the latent space, which can be of high dimensional. Techniques like latent semantic analysis, probabilistic latent semantic analysis or non negative matrix factorization can be applied to obtain a compact representation [123]. Learning algorithms such as Recurrent Neural Networks (RNN) etc. are used to learn the sequence of the data. Several studies are also conducted by applying transfer learning, which uses a model trained on a different domain to predict the target domain after fine tuning [72].

3.8 Physiological Signals Based Emotion Recognition Methods

The modalities discussed so far focus on the audible or visible cues, which human express in response to a particular situation or action. It is known that some people can conceal their emotions better than others. Attributes like micro-expressions try to bridge this gap of perceived emotion/affect to the actual one felt by the person.

It will be useful for a large set of applications, if the affective state of user is available. Few examples can be the self-regulation of someone's mental health for stress and for the driver assistance techniques. There exist various types of signals, which are used to record the bio-signals produced by the human's nervous system. These signals can be represented by any emotion representation model. The most common model for this is the dimensional model which provides arousal and valence values for a given range. The commonly used signals for the emotion detection task are [5].

Electroencephalography (EEG) sensor records the change in the voltage, which occurs in the neurons when current flows through them. The recorded signal is divided into five different waves based on their frequency range [3]. Delta waves (1–4 Hz) which occur from the unconscious mind. Theta waves (4–7 Hz) occurs when the mind is at subconscious state like dreaming. Alpha waves (8–13 Hz) are associated with an aware and relaxed mind. Beta (13–30 Hz) and gamma (more than 30 Hz) waves are recorded during focused mental activity and hyper brain activity respectively. These signals are recorded by placing electrodes on the scalp of the person. The location to place these electrodes is predefined by standards such as International 10–20 system, where 10–20 refers to the distance between adjacent electrodes should be 10% or 20% to the front back or right left distance of the skull.

Electrodermal Activity (EDA) also known as *Galvanic Skin Response (GSR)* measures the skin conductance caused by sweating. Apart from the external factors like temperature, human's body sweating is regulated by the autonomic nervous system. The sweat is generated whenever the nervous system becomes aroused with states such as stress and fear. EDA signals can successfully distinguish between anger and fear, which is a difficult in emotion detection system [5]. To record EDA signals, electrodes are placed on the fingers. These electrodes need to be calibrated before using to make them invariant to the external environment.

Electromyography (EMG) sensor records the electric activities of muscles, controlled by a motor neuron in human's nervous system. The activated motor neurons transmit signals which cause muscles to contract. EMG records these signals which can be used to identify the behaviour of muscle cells which varies in case of the positive or negative emotions. EMG signals can be used to identify the presence of stress in a person. These signals are recorded by using surface electrodes which record the muscle activities above the skin surface. The recording can also be performed by inserting an electrode inside the skin depending on muscle location.

Electrocardiogram (ECG) sensor records small electric changes that occur with each heartbeat. The autonomic nervous system consists of a sympathetic system which stimulate differently on the presence of a particular emotion. The stimulations include dilation of coronary blood vessels, increased force of contraction of the cardiac fibres, faster conduction of the SA node (natural pacemaker), etc. [1]. The ECG signals are recorded by placing electrodes on a person's chest. 12-Lead ECG system is a predefined standard followed to record ECG signals.

Blood Volume Pulse (BVP) captures the amount of blood flow that runs through the blood vessels across different emotions. A photoplethysmogram (PPG) device is used to measure the BVP. PPG is an optical sensor which emits a light signal which get reflects to the skin indicating the blood flow. *Skin temperature* of the body differs on the presence of different emotions. The temperature of the skin varies due to the flow of the blood in the blood vessels which contracts on the occurrence of any emotions. This measure of emotion provides a slow indicator corresponding to any emotion. The arousal of the emotion can also be noticed by the *respiration pattern* of the person which is recorded by placing a belt around a person's chest [71].

EEG signals have different attributes like alpha, beta, theta, gamma, spectral power of each electrode, etc. For respiration pattern, average respiration signal, band energy ratio, etc. can be extracted. Further details of these features can be found in [71]. The features from different sensors are fused depending on the fusion technique. Verma et al. [138] discussed the multimodal fusion framework for physiological signals for the emotion detection task. Different non-deep learning and deep learning based learning algorithm can be applied to train the model. Most of the current studies use LSTM to learn the patterns present in the data obtained from the sensors [114].

3.9 Fusion Methods Across Modalities

As discussed in the previous sections, different modalities are useful for an emotion detection system. Features are extracted from each modality independently as the type of data present in each modality differs from the other. To leverage the features learned from each modality, a fusion technique is used to combine the learned information. The resulting system can identify the emotion of a person using the different type of data.

Commonly, two types of fusion methods: feature level and decision level are used. *Feature level fusion* combines the features extracted from each modality to create a single feature. Different feature level fusion operations can be used to accomplish this task such as addition, concatenation, multiplication, selection of maximum value, etc. The classifier is then applied to single high dimensional feature vector.

Feature level fusion combines the discriminative features learned by each modality, resulting in efficient emotion detection model. However, the method has some practical limitations [140]. Training a classifier on a high dimensional feature space may lacks to perform well due to the curse of dimensionality. On the presence of high dimensional data, the classifier can perform differently as compared to the low dimensional data. Also, the combined features may require high computation resources. Hence, the performance of the feature level fusion methods lies in the efficient feature selection system from individual modality and on the classifier used.

In *decision level fusion* method, a classifier is employed on each modality independently. The decision of each classifier is then merged together. In this fusion, different classification systems can also be used for each modality based on the type of the data. This is different from the feature level fusion, where only one classifier is trained for all the types of data. Decision level fusion performs well on the presence of complex data. This is due to the fact that multiple classifiers can learn better representations for different data distributions rather than the single classifier.

Wagner et al. [140] proposed different decision level fusion methods to solve the problem of missing data. As different classifiers can have different priorities associated to them, the authors proposed various methods to combine these decisions. Operations like weighted majority voting, weighted average, selection of maximum, minimum, median supports, etc. are used for the fusion. The decision can also be combined by identifying the expert data points for each class and then using this information for the ensemble. Fan et al. [42] performed decision level fusion to combine the audio-visual information. CNN-RNN and 3-D convolutions were used corresponding to frame wise data and SVM was learned for the audio data.

Features learned from one modality can also be used to predict the emotions from a different modality. This interesting concept was proposed by Albanie et al. [4] where a cross modal distillation neural network was learned for facial data. The learned model was further used for the prediction of the audio data.

3.10 Applications of Automatic Emotion Recognition

Health Care and Well-being—Monitoring the emotional reaction of the person can help doctors to understand the symptoms and to remove the identity biased verbal reaction. The emotion of the person can also be analyzed to estimate their mental health [66]. The early symptoms of many psychological diseases can be identified by analyzing the person's emotions for some time. Disorders like Parkinson's disease, autism, borderline personality disorder, schizophrenia, etc. affect the person's ability to interpret own or others emotions. Continuous monitoring of a person's emotions can help their family members to understand the feelings of the patients.

Security and Surveillance—Automatic emotion detection can be used to find the behaviour of the crowd for any abnormal activity. The expressions of a person with their speech can be analyzed to predict any kind of violent behaviour in a group of people. The emotion of the person can also be analyzed by any self operating machine available in public areas. The machines can detect any negative emotion in the user and can contact the concerned authority.

Human Machine Interaction—The presence of emotional information to robots or similar devices can make them understand a person's state of the mind. The understanding of emotion can improve the smart personal assistant softwares like Alexa, Cortana, Siri, etc. to understand the emotion of the user from their speech. Suggestions can be provided by the personal assistant software to relax a person such as music options depending on the mood, making a call to someone, etc.

Estimating User Feedback—The emotion of a person can be used to provide genuine feedback for any product. It can change the current shopping style where one of the possibility is to estimate a person's choice by analyzing their behaviour. Emotions can also be analyzed to obtain the review of a visual content like movies, advertisements and video games etc.

Education—Emotion of a person can also depict their engagement level. This can be used in online or class room teaching to provide real time feedback to the teacher to improve the learning of the students.

Driver Distraction—The emotional state of a driver is an important factor to ensure their safety. It is useful to be able to identify any distraction which can be there due to fatigue, drowsiness and yawning. The emotion detection model can identify these categories of distraction to set a warning.

3.11 Privacy in Affective Computing

With the progress in AI, important questions are being raised about the privacy of users. It is required that the model creators follow these ethics. A technology is developed to improve the lifestyle of human's directly or indirectly. Certainly, the techniques produced to do so should follow and respect a person's sentiments and privacy. Emotion recognition system requires the data of different modalities to be able to produce efficient and generalized prediction system. A person's face, facial expressions, voice, written text and physiological information, all are recorded independently or in a combined form as a data collection process. Therefore, the data needs to be secured in both raw and processed forms. Issues are being raised and researchers have proposed possible solutions for this problem [90, 112].

Several studies are now focusing on capturing data without recording the identity specific information. Visual information can be recorded by the thermal cameras as they only record the change in the heat distribution of the scene. It is non-trivial to identify the subject in such data. However, cost associated with the data collection by thermal cameras and the inferior performance of the current emotion recognition methods on such data implies that more research is required in this direction.

Other than facial information, the health related information from various physiological sensors is also a concern from the privacy perspective. Current methods can predict the subject information like heart rate, blood pressure, etc. by just pointing a regular camera towards the face of a person. Such techniques require to record the changes in the skin color, which is caused by the blood circulation to make such prediction. To keep such information private and avoid any chance by which it can be misused, Chen et al. [22] proposed an approach to eliminate the physiological details from a facial video. The videos produced from this method doesn't have any physiological details without effecting the visual appearance of the video.

3.12 Ethics and Fairness in Automatic Emotion Recognition

Recently, automatic emotion recognition methods have been applied to different use cases such as analysis of a person during an interview or analyzing students in a classroom. This raises an important question about the validity, scope and fair usage of these models in different out of the lab environments. In a recent study, Rhue [113] show that such processes can make a negative impact on a person such as fault perception, emotional pressure on an individual, etc. The study also shows that most of the current emotion recognition systems are biased towards the person's race to interpret emotions. In some sensitive cases, the model predictions can prove to be dangerous for the well-being of a person. From a different perspective, according to ecological model of social perception, humans always judge others on the basis of physical appearance and the issue arises whenever someone overgeneralize others [11]. It is a challenge to develop affect sensing systems, which are able to learn emotions without bias towards age, ethnicity and gender.

3.13 Conclusion

The progress of deep learning based methods has changed the way automatic emotion recognition based methods work. However, it is important to have an understanding of different feature extraction ways to be able to create a suitable model for emotion detection. The advancements in face detection, face tracking, facial landmark prediction methods have made it possible to preprocess the data efficiently. Feature extractor methods in visual, speech, text and physiological based data can be easily used in real time. Both deep learning and traditional machine learning based methods have been used successfully to learn emotion specific information based on the complexity of the data available. All these techniques have improved the emotion detection process to a greater extent from the last decade. The current challenge lies to make the process more generalized such that machines can identify emotions on par with humans. Ethics related to the affect prediction need to be defined and followed to create an automatic emotion recognition system without compromising with the human's sentiments and privacy.

References

1. Agrafioti, F., Hatzinakos, D., Anderson, A.K.: ECG pattern analysis for emotion detection. IEEE Trans. Affect. Comput. **3**(1), 102–115 (2012)
2. Ahonen, T., Hadid, A., Pietikainen, M.: Face description with local binary patterns: application to face recognition. IEEE Trans. Pattern Anal. Mach. Intell. **12**, 2037–2041 (2006)
3. Alarcao, S.M., Fonseca, M.J.: Emotions recognition using EEG signals: a survey. IEEE Trans. Affect. Comput. (2017)
4. Albanie, S., Nagrani, A., Vedaldi, A., Zisserman, A.: Emotion recognition in speech using cross-model transfer in the wild. arXiv preprint arXiv:1808.05561 (2018)
5. Ali, M., Mosa, A.H., Al Machot, F., Kyamakya, K.: Emotion recognition involving physiological and speech signals: a comprehensive review. In: Recent Advances in Nonlinear Dynamics and Synchronization, pp. 287–302. Springer (2018)
6. Asghar, N., Poupart, P., Hoey, J., Jiang, X., Mou, L.: Affective neural response generation. In: European Conference on Information Retrieval, pp. 154–166. Springer (2018)
7. Asthana, A., Zafeiriou, S., Cheng, S., Pantic, M.: Incremental face alignment in the wild. In: Computer Vision and Pattern Recognition, pp. 1859–1866. IEEE (2014)
8. Bachorowski, J.A.: Vocal expression and perception of emotion. Curr. Direct. Psychol. Sci. **8**(2), 53–57 (1999)
9. Baltrusaitis, T., Zadeh, A., Lim, Y.C., Morency, L.P.: Openface 2.0: Facial behavior analysis toolkit. In: 13th International Conference on Automatic Face & Gesture Recognition (FG 2018), pp. 59–66. IEEE (2018)
10. Bänziger, T., Mortillaro, M., Scherer, K.R.: Introducing the geneva multimodal expression corpus for experimental research on emotion perception. Emotion **12**(5), 1161 (2012)

11. Barber, S.J., Lee, H., Becerra, J., Tate, C.C.: Emotional expressions affect perceptions of younger and older adults' everyday competence. Psychol. Aging **34**(7), 991 (2019)
12. Basbrain, A.M., Gan, J.Q., Sugimoto, A., Clark, A.: A neural network approach to score fusion for emotion recognition. In: 10th Computer Science and Electronic Engineering (CEEC), pp. 180–185 (2018)
13. Batliner, A., Hacker, C., Steidl, S., Nöth, E., D'Arcy, S., Russell, M.J., Wong, M.: "You Stupid Tin Box" Children Interacting with the AIBO Robot: A Cross-linguistic Emotional Speech Corpus. Lrec (2004)
14. Bosch, A., Zisserman, A., Munoz, X.: Representing shape with a spatial pyramid Kernel. In: 6th ACM international conference on Image and video retrieval, pp. 401–408. ACM (2007)
15. Bou-Ghazale, S.E., Hansen, J.H.: A comparative study of traditional and newly proposed features for recognition of speech under stress. IEEE Trans. Speech Audio Process. **8**(4), 429–442 (2000)
16. Busso, C., Bulut, M., Lee, C.C., Kazemzadeh, A., Mower, E., Kim, S., Chang, J.N., Lee, S., Narayanan, S.S.: IEMOCAP: interactive emotional dyadic motion capture database. Lang. Resour. Eval. **42**(4), 335 (2008)
17. Busso, C., Deng, Z., Yildirim, S., Bulut, M., Lee, C.M., Kazemzadeh, A., Lee, S., Neumann, U., Narayanan, S.: Analysis of emotion recognition using facial expressions, speech and multimodal information. In: 6th International Conference on Multimodal Interfaces, pp. 205–211. ACM (2004)
18. Busso, C., Parthasarathy, S., Burmania, A., AbdelWahab, M., Sadoughi, N., Provost, E.M.: MSP-IMPROV: an acted corpus of dyadic interactions to study emotion perception. IEEE Trans. Affect. Comput. **8**(1), 67–80 (2017)
19. Cairns, D.A., Hansen, J.H.: Nonlinear analysis and classification of speech under stressed conditions. J. Acoust. Soc. Am. **96**(6), 3392–3400 (1994)
20. Cambria, E.: Affective computing and sentiment analysis. Intell. Syst. **31**(2), 102–107 (2016)
21. Chen, J., Chen, Z., Chi, Z., Fu, H.: Dynamic texture and geometry features for facial expression recognition in video. In: International Conference on Image Processing (ICIP), pp. 4967–4971. IEEE (2015)
22. Chen, W., Picard, R.W.: Eliminating physiological information from facial videos. In: 12th International Conference on Automatic Face and Gesture Recognition (FG 2017), pp. 48–55. IEEE (2017)
23. Cho, K., Van Merriënboer, B., Gulcehre, C., Bahdanau, D., Bougares, F., Schwenk, H., Bengio, Y.: Learning phrase representations using RNN encoder-decoder for statistical machine translation. arXiv preprint arXiv:1406.1078 (2014)
24. Cootes, T.F., Edwards, G.J., Taylor, C.J.: Active appearance models. IEEE Trans. Pattern Anal. Mach. Intell. **6**, 681–685 (2001)
25. Correa, J.A.M., Abadi, M.K., Sebe, N., Patras, I.: AMIGOS: A dataset for affect, personality and mood research on individuals and groups. IEEE Trans. Affect. Comput. (2018)
26. Cowie, R., Douglas-Cowie, E., Tsapatsoulis, N., Votsis, G., Kollias, S., Fellenz, W., Taylor, J.G.: Emotion recognition in human-computer interaction. IEEE Signal Process. Mag. **18**(1), 32–80 (2001)
27. Dalal, N., Triggs, B.: Histograms of oriented gradients for human detection. In: International Conference on Computer Vision & Pattern Recognition (CVPR'05), vol. 1, pp. 886–893. IEEE Computer Society (2005)
28. Davison, A., Merghani, W., Yap, M.: Objective classes for micro-facial expression recognition. J. Imaging **4**(10), 119 (2018)
29. Davison, A.K., Lansley, C., Costen, N., Tan, K., Yap, M.H.: SAMM: a spontaneous micro-facial movement dataset. IEEE Trans. Affect. Comput. **9**(1), 116–129 (2018)
30. Dhall, A., Asthana, A., Goecke, R., Gedeon, T.: Emotion recognition using phog and lpq features. In: Face and Gesture 2011, pp. 878–883. IEEE (2011)
31. Dhall, A., Goecke, R., Gedeon, T.: Automatic group happiness intensity analysis. IEEE Trans. Affect. Comput. **6**(1), 13–26 (2015)

32. Dhall, A., Goecke, R., Lucey, S., Gedeon, T., et al.: Collecting large, richly annotated facial-expression databases from movies. IEEE Multimedia **19**(3), 34–41 (2012)
33. Dhall, A., Kaur, A., Goecke, R., Gedeon, T.: Emotiw 2018: audio-video, student engagement and group-level affect prediction. In: International Conference on Multimodal Interaction, pp. 653–656. ACM (2018)
34. Du, S., Tao, Y., Martinez, A.M.: Compound facial expressions of emotion. Natl. Acad. Sci. **111**(15), E1454–E1462 (2014)
35. Ekman, P., Friesen, W.V.: Unmasking the face: a guide to recognizing emotions from facial clues. Ishk (2003)
36. Ekman, P., Friesen, W.V., Hager, J.C.: Facial Action Coding System: The Manual on CD ROM, pp. 77–254. A Human Face, Salt Lake City (2002)
37. El Ayadi, M., Kamel, M.S., Karray, F.: Survey on speech emotion recognition: features, classification schemes, and databases. Pattern Recogn. **44**(3), 572–587 (2011)
38. Ertugrul, I.O., Cohn, J.F., Jeni, L.A., Zhang, Z., Yin, L., Ji, Q.: Cross-domain au detection: domains, learning approaches, and measures. In: 14th International Conference on Automatic Face & Gesture Recognition, pp. 1–8. IEEE (2019)
39. Eyben, F., Scherer, K.R., Schuller, B.W., Sundberg, J., André, E., Busso, C., Devillers, L.Y., Epps, J., Laukka, P., Narayanan, S.S., et al.: The geneva minimalistic acoustic parameter set (GeMAPS) for voice research and affective computing. IEEE Trans. Affect. Comput. **7**(2), 190–202 (2016)
40. Eyben, F., Weninger, F., Gross, F., Schuller, B.: Recent developments in Opensmile, the Munich open-source multimedia feature extractor. In: 21st ACM international conference on Multimedia, pp. 835–838. ACM (2013)
41. Fabian Benitez-Quiroz, C., Srinivasan, R., Martinez, A.M.: Emotionet: An accurate, real-time algorithm for the automatic annotation of a million facial expressions in the wild. In: Computer Vision and Pattern Recognition, pp. 5562–5570. IEEE (2016)
42. Fan, Y., Lu, X., Li, D., Liu, Y.: Video-based emotion recognition using CNN-RNN and C3D hybrid networks. In: 18th ACM International Conference on Multimodal Interaction, pp. 445–450. ACM (2016)
43. Filntisis, P.P., Efthymiou, N., Koutras, P., Potamianos, G., Maragos, P.: Fusing body posture with facial expressions for joint recognition of affect in child-robot interaction. arXiv preprint arXiv:1901.01805 (2019)
44. Friesen, E., Ekman, P.: Facial action coding system: a technique for the measurement of facial movement. Palo Alto **3**, (1978)
45. Ganchev, T., Fakotakis, N., Kokkinakis, G.: Comparative evaluation of various MFCC implementations on the speaker verification task. SPECOM **1**, 191–194 (2005)
46. Ghimire, D., Lee, J., Li, Z.N., Jeong, S., Park, S.H., Choi, H.S.: Recognition of facial expressions based on tracking and selection of discriminative geometric features. Int. J. Multimedia Ubiquitous Eng. **10**(3), 35–44 (2015)
47. Ghosh, S., Dhall, A., Sebe, N.: Automatic group affect analysis in images via visual attribute and feature networks. In: 25th IEEE International Conference on Image Processing (ICIP), pp. 1967–1971. IEEE (2018)
48. Girard, J.M., Chu, W.S., Jeni, L.A., Cohn, J.F.: Sayette group formation task (GFT) spontaneous facial expression database. In: 12th International Conference on Automatic Face & Gesture Recognition (FG 2017), pp. 581–588. IEEE (2017)
49. Goodfellow, I., Bengio, Y., Courville, A.: Deep Learning. MIT Press (2016). http://www.deeplearningbook.org
50. Goodfellow, I.J., Erhan, D., Carrier, P.L., Courville, A., Mirza, M., Hamner, B., Cukierski, W., Tang, Y., Thaler, D., Lee, D.H., et al.: Challenges in representation learning: a report on three machine learning contests. Neural Netw. **64**, 59–63 (2015)
51. Graves, A., Schmidhuber, J.: Framewise phoneme classification with bidirectional LSTM and other neural network architectures. Neural Netw. **18**(5–6), 602–610 (2005)
52. Gunes, H., Pantic, M.: Automatic, dimensional and continuous emotion recognition. Int. J. Synth. Emotions (IJSE) **1**(1), 68–99 (2010)

53. Haggard, E.A., Isaacs, K.S.: Micromomentary facial expressions as indicators of ego mechanisms in psychotherapy. In: Methods of research in psychotherapy, pp. 154–165. Springer (1966)
54. Han, J., Zhang, Z., Ren, Z., Schuller, B.: Implicit fusion by joint audiovisual training for emotion recognition in mono modality. In: International Conference on Acoustics, Speech and Signal Processing (ICASSP), pp. 5861–5865. IEEE (2019)
55. Han, J., Zhang, Z., Schmitt, M., Ren, Z., Ringeval, F., Schuller, B.: Bags in bag: generating context-aware bags for tracking emotions from speech. Interspeech **2018**, 3082–3086 (2018)
56. Happy, S., Patnaik, P., Routray, A., Guha, R.: The Indian spontaneous expression database for emotion recognition. IEEE Trans. Affect. Comput. **8**(1), 131–142 (2017)
57. Harvill, J., AbdelWahab, M., Lotfian, R., Busso, C.: Retrieving speech samples with similar emotional content using a triplet loss function. In: International Conference on Acoustics, Speech and Signal Processing (ICASSP), pp. 7400–7404. IEEE (2019)
58. He, K., Zhang, X., Ren, S., Sun, J.: Deep residual learning for image recognition. In: Computer vision and pattern recognition, pp. 770–778. IEEE (2016)
59. Hu, P., Ramanan, D.: Finding tiny faces. In: Computer vision and pattern recognition, pp. 951–959. IEEE (2017)
60. Huang, G., Liu, Z., Van Der Maaten, L., Weinberger, K.Q.: Densely connected convolutional networks. In: Computer vision and pattern recognition, pp. 4700–4708. IEEE (2017)
61. Huang, Y., Yang, J., Liu, S., Pan, J.: Combining facial expressions and electroencephalography to enhance emotion recognition. Future Internet **11**(5), 105 (2019)
62. Hussein, H., Angelini, F., Naqvi, M., Chambers, J.A.: Deep-learning based facial expression recognition system evaluated on three spontaneous databases. In: 9th International Symposium on Signal, Image, Video and Communications (ISIVC), pp. 270–275. IEEE (2018)
63. Jack, R.E., Blais, C., Scheepers, C., Schyns, P.G., Caldara, R.: Cultural confusions show that facial expressions are not universal. Curr. Biol. **19**(18), 1543–1548 (2009)
64. Jack, R.E., Sun, W., Delis, I., Garrod, O.G., Schyns, P.G.: Four not six: revealing culturally common facial expressions of emotion. J. Exp. Psychol. Gen. **145**(6), 708 (2016)
65. Jiang, B., Valstar, M.F., Pantic, M.: Action unit detection using sparse appearance descriptors in space-time video volumes. In: Face and Gesture, pp. 314–321. IEEE (2011)
66. Joshi, J., Goecke, R., Alghowinem, S., Dhall, A., Wagner, M., Epps, J., Parker, G., Breakspear, M.: Multimodal assistive technologies for depression diagnosis and monitoring. J. Multimodal User Interfaces **7**(3), 217–228 (2013)
67. Jyoti, S., Sharma, G., Dhall, A.: Expression empowered residen network for facial action unit detection. In: 14th International Conference on Automatic Face and Gesture Recognition, pp. 1–8. IEEE (2019)
68. Kaiser, J.F.: On a Simple algorithm to calculate the 'Energy' of a Signal. In: International Conference on Acoustics, Speech, and Signal Processing, pp. 381–384. IEEE (1990)
69. King, D.E.: Dlib-ML: A machine learning toolkit. J. Mach. Learn. Res. **10**, 1755–1758 (2009)
70. Knyazev, B., Shvetsov, R., Efremova, N., Kuharenko, A.: Convolutional neural networks pretrained on large face recognition datasets for emotion classification from video. arXiv preprint arXiv:1711.04598 (2017)
71. Koelstra, S., Muhl, C., Soleymani, M., Lee, J.S., Yazdani, A., Ebrahimi, T., Pun, T., Nijholt, A., Patras, I.: DEAP: a database for emotion analysis; using physiological signals. IEEE Trans. Affect. Comput. **3**(1), 18–31 (2012)
72. Kratzwald, B., Ilić, S., Kraus, M., Feuerriegel, S., Prendinger, H.: Deep learning for affective computing: text-based emotion recognition in decision support. Decis. Support Syst. **115**, 24–35 (2018)
73. Krizhevsky, A., Sutskever, I., Hinton, G.E.: Imagenet classification with deep convolutional neural networks. In: Advances in Neural Information Processing Systems, pp. 1097–1105 (2012)
74. Latif, S., Rana, R., Khalifa, S., Jurdak, R., Epps, J.: Direct modelling of speech emotion from raw speech. arXiv preprint arXiv:1904.03833 (2019)

75. Lee, C.M., Narayanan, S.S., et al.: Toward detecting emotions in spoken dialogs. IEEE Trans. Speech Audio Process. **13**(2), 293–303 (2005)
76. Lee, J., Kim, S., Kim, S., Park, J., Sohn, K.: Context-aware emotion recognition networks. In: The IEEE International Conference on Computer Vision (ICCV) (2019)
77. Li, S., Deng, W.: Deep facial expression recognition: a survey. arXiv preprint arXiv:1804.08348 (2018)
78. Li, S., Deng, W., Du, J.: Reliable crowdsourcing and deep locality-preserving learning for expression recognition in the wild. In: Computer Vision and Pattern Recognition, pp. 2852–2861. IEEE (2017)
79. Li, W., Xu, H.: Text-based emotion classification using emotion cause extraction. Expert Syst. Appl. **41**(4), 1742–1749 (2014)
80. Lian, Z., Li, Y., Tao, J.H., Huang, J., Niu, M.Y.: Expression analysis based on face regions in read-world conditions. Int. J. Autom. Comput. 1–12
81. Liao, S., Jain, A.K., Li, S.Z.: A fast and accurate unconstrained face detector. IEEE Trans. Pattern Anal. Mach. Intell. **38**(2), 211–223 (2016)
82. Lienhart, R., Maydt, J.: An extended set of haar-like features for rapid object detection. In: Proceedings of International Conference on Image Processing, vol. 1, p. I. IEEE (2002)
83. Liu, X., Zou, Y., Kong, L., Diao, Z., Yan, J., Wang, J., Li, S., Jia, P., You, J.: Data augmentation via latent space interpolation for image classification. In: 24th International Conference on Pattern Recognition (ICPR), pp. 728–733. IEEE (2018)
84. Livingstone, S.R., Russo, F.A.: The Ryerson Audio-Visual Database of Emotional Speech and Song (RAVDESS): a dynamic, multimodal set of facial and vocal expressions in North American English. PloS One **13**(5), e0196391 (2018)
85. Lotfian, R., Busso, C.: Building naturalistic emotionally balanced speech corpus by retrieving emotional speech from existing podcast rRecordings. IEEE Trans. Affect. Comput. (2017)
86. Lowe, D.G.: Distinctive image features from scale-invariant keypoints. Int. J. Comput. Vis. **60**(2), 91–110 (2004)
87. Lowe, D.G., et al.: Object recognition from local scale-invariant features. ICCV **99**, 1150–1157 (1999)
88. Lucas, B.D., Kanade, T., et al.: An iterative image registration technique with an application to stereo vision (1981)
89. Lucey, P., Cohn, J.F., Kanade, T., Saragih, J., Ambadar, Z., Matthews, I.: The extended Cohn-kanade dataset (ck+): a complete dataset for action unit and emotion-specified expression. In: Computer Vision and Pattern Recognition Workshops (CVPRW), pp. 94–101. IEEE (2010)
90. Macías, E., Suárez, A., Lacuesta, R., Lloret, J.: Privacy in affective computing based on mobile sensing systems. In: 2nd International Electronic Conference on Sensors and Applications, p. 1. MDPI AG (2015)
91. Makhmudkhujaev, F., Abdullah-Al-Wadud, M., Iqbal, M.T.B., Ryu, B., Chae, O.: Facial expression recognition with local prominent directional pattern. Signal Process. Image Commun. **74**, 1–12 (2019)
92. Mandal, M., Verma, M., Mathur, S., Vipparthi, S., Murala, S., Deveerasetty, K.: RADAP: regional adaptive affinitive patterns with logical operators for facial expression recognition. IET Image Processing (2019)
93. Martin, O., Kotsia, I., Macq, B., Pitas, I.: The eNTERFACE'05 audio-visual emotion database. In: 22nd International Conference on Data Engineering Workshops (ICDEW'06), pp. 8–8. IEEE (2006)
94. Mavadati, S.M., Mahoor, M.H., Bartlett, K., Trinh, P., Cohn, J.F.: DISFA: a spontaneous facial action intensity database. IEEE Trans. Affect. Comput. **4**(2), 151–160 (2013)
95. McDuff, D., Amr, M., El Kaliouby, R.: AM-FED+: an extended dataset of naturalistic facial expressions collected in everyday settings. IEEE Trans. Affect. Comput. **10**(1), 7–17 (2019)
96. McGilloway, S., Cowie, R., Douglas-Cowie, E., Gielen, S., Westerdijk, M., Stroeve, S.: Approaching automatic recognition of emotion from voice: a rough benchmark. In: ISCA Tutorial and Research Workshop (ITRW) on Speech and Emotion (2000)

97. McKeown, G., Valstar, M., Cowie, R., Pantic, M., Schroder, M.: The SEMAINE database: annotated multimodal records of emotionally colored conversations between a person and a limited agent. IEEE Trans. Affect. Comput. **3**(1), 5–17 (2012)
98. Mehrabian, A.: Pleasure-arousal-dominance: a general framework for describing and measuring individual differences in temperament. Curr. Psychol. **14**(4), 261–292 (1996)
99. Mehrabian, A., Ferris, S.R.: Inference of attitudes from nonverbal communication in two channels. J. Consult. Psychol. **31**(3), 248 (1967)
100. Mikolov, T., Sutskever, I., Chen, K., Corrado, G.S., Dean, J.: Distributed representations of words and phrases and their compositionality. In: Advances in Neural Information Processing Systems, pp. 3111–3119 (2013)
101. Moffat, D., Ronan, D., Reiss, J.D.: An evaluation of audio feature extraction toolboxes (2015)
102. Mollahosseini, A., Hasani, B., Mahoor, M.H.: Affectnet: A database for facial expression, valence, and arousal computing in the wild. arXiv preprint arXiv:1708.03985 (2017)
103. Munezero, M.D., Montero, C.S., Sutinen, E., Pajunen, J.: Are they different? Affect, feeling, emotion, sentiment, and opinion detection in text. IEEE Trans. Affect. Comput. **5**(2), 101–111 (2014)
104. Murray, I.R., Arnott, J.L.: Toward the simulation of emotion in synthetic speech: a review of the literature on human vocal emotion. J. Acoust. Soc. Am. **93**(2), 1097–1108 (1993)
105. Nwe, T.L., Foo, S.W., De Silva, L.C.: Speech emotion recognition using hidden Markov models. Speech Commun. **41**(4), 603–623 (2003)
106. Ojansivu, V., Heikkilä, J.: Blur insensitive texture classification using local phase quantization. In: International Conference on Image and Signal Processing, pp. 236–243. Springer (2008)
107. Ou, J., Bai, X.B., Pei, Y., Ma, L., Liu, W.: Automatic facial expression recognition using gabor filter and expression analysis. In: 2nd International Conference on Computer Modeling and Simulation, vol. 2, pp. 215–218. IEEE (2010)
108. Pan, X., Guo, W., Guo, X., Li, W., Xu, J., Wu, J.: Deep temporal-spatial aggregation for video-based facial expression recognition. Symmetry **11**(1), 52 (2019)
109. Parkhi, O.M., Vedaldi, A., Zisserman, A., et al.: Deep face recognition. BMVC **1**, 6 (2015)
110. Rabiner, L., Schafer, R.: Digital Processing of Speech Signals. Prentice Hall, Englewood Cliffs (1978)
111. Rassadin, A., Gruzdev, A., Savchenko, A.: Group-level emotion recognition using transfer learning from face identification. In: 19th ACM International Conference on Multimodal Interaction, pp. 544–548. ACM (2017)
112. Reynolds, C., Picard, R.: Affective sensors, privacy, and ethical contracts. In: CHI'04 Extended Abstracts on Human Factors in Computing Systems, pp. 1103–1106. ACM (2004)
113. Rhue, L.: Racial influence on automated perceptions of emotions. Available at SSRN 3281765, (2018)
114. Ringeval, F., Eyben, F., Kroupi, E., Yuce, A., Thiran, J.P., Ebrahimi, T., Lalanne, D., Schuller, B.: Prediction of asynchronous dimensional emotion ratings from audiovisual and physiological data. Pattern Recogn. Lett. **66**, 22–30 (2015)
115. Ringeval, F., Schuller, B., Valstar, M., Cummins, N., Cowie, R., Tavabi, L., Schmitt, M., Alisamir, S., Amiriparian, S., Messner, E.M., et al.: AVEC 2019 workshop and challenge: state-of-mind, detecting depression with AI, and cross-cultural affect recognition. In: 9th International on Audio/Visual Emotion Challenge and Workshop, pp. 3–12. ACM (2019)
116. Ringeval, F., Sonderegger, A., Sauer, J., Lalanne, D.: Introducing the RECOLA multimodal corpus of remote collaborative and affective interactions. In: 10th International Conference and Workshops on Automatic Face and Gesture Recognition (FG), pp. 1–8. IEEE (2013)
117. Russell, J.A.: A circumplex model of affect. J. Pers. Soc. Psychol. **39**(6), 1161 (1980)
118. Sabour, S., Frosst, N., Hinton, G.E.: Dynamic routing between capsules. Adv. Neural Inform. Process. Syst. 3856–3866 (2017)
119. Saragih, J.M., Lucey, S., Cohn, J.F.: Face alignment through subspace constrained mean-shifts. In: 12th International Conference on Computer Vision, pp. 1034–1041. IEEE (2009)
120. Sariyanidi, E., Gunes, H., Cavallaro, A.: Automatic analysis of facial affect: a survey of registration, representation, and recognition. IEEE Trans. Pattern Anal. Mach. Intell. **37**(6), 1113–1133 (2015)

121. Schuller, B., Steidl, S., Batliner, A., Vinciarelli, A., Scherer, K., Ringeval, F., Chetouani, M., Weninger, F., Eyben, F., Marchi, E., et al.: The INTERSPEECH 2013 computational paralinguistics challenge: social signals, conflict, emotion, Autism. In: 14th Annual Conference of the International Speech Communication Association (2013)
122. Sebe, N., Cohen, I., Gevers, T., Huang, T.S.: Emotion recognition based on joint visual and audio cues. In: 18th International Conference on Pattern Recognition, vol. 1, pp. 1136–1139. IEEE (2006)
123. Seyeditabari, A., Tabari, N., Zadrozny, W.: Emotion detection in text: a review. arXiv preprint arXiv:1806.00674 (2018)
124. Shi, J., Tomasi, C.: Good Features to Track. Tech. rep, Cornell University (1993)
125. Siddharth, S., Jung, T.P., Sejnowski, T.J.: Multi-modal approach for affective computing. arXiv preprint arXiv:1804.09452 (2018)
126. Sikka, K., Dykstra, K., Sathyanarayana, S., Littlewort, G., Bartlett, M.: Multiple Kernel learning for emotion recognition in the wild. In: 15th ACM on International Conference on Multimodal Interaction, pp. 517–524. ACM (2013)
127. Simonyan, K., Zisserman, A.: Very deep convolutional networks for large-scale image recognition. arXiv preprint arXiv:1409.1556 (2014)
128. Sneddon, I., McRorie, M., McKeown, G., Hanratty, J.: The Belfast induced natural emotion database. IEEE Trans. Affect. Comput. **3**(1), 32–41 (2012)
129. Soleymani, M., Lichtenauer, J., Pun, T., Pantic, M.: A multimodal database for affect recognition and implicit tagging. IEEE Trans. Affect. Comput. **3**(1), 42–55 (2012)
130. Strapparava, C., Mihalcea, R.: Learning to identify emotions in text. In: ACM Symposium on Applied Computing, pp. 1556–1560. ACM (2008)
131. Strapparava, C., Valitutti, A., et al.: Wordnet affect: an affective extension of wordnet. In: Lrec, vol. 4, p. 40. Citeseer (2004)
132. Teager, H.: Some observations on oral air flow during phonation. IEEE Trans. Acoust. Speech Signal Process. **28**(5), 599–601 (1980)
133. Thoits, P.A.: The sociology of emotions. Annu. Rev. Sociol. **15**(1), 317–342 (1989)
134. Tomasi, C., Detection, T.K.: Tracking of point features. Tech. rep., Tech. Rep. CMU-CS-91-132, Carnegie Mellon University (1991)
135. Torres, J.M.M., Stepanov, E.A.: Enhanced face/audio emotion recognition: video and instance level classification using ConvNets and restricted boltzmann machines. In: International Conference on Web Intelligence, pp. 939–946. ACM (2017)
136. Trigeorgis, G., Ringeval, F., Brueckner, R., Marchi, E., Nicolaou, M.A., Schuller, B., Zafeiriou, S.: Adieu features? End-to-end speech emotion recognition using a deep convolutional recurrent network. In: International Conference on Acoustics, Speech and Signal Processing (ICASSP), pp. 5200–5204. IEEE (2016)
137. Tulyakov, S., Liu, M.Y., Yang, X., Kautz, J.: MoCoGAN: decomposing motion and content for video generation. In: Computer Vision and Pattern Recognition, pp. 1526–1535. IEEE (2018)
138. Verma, G.K., Tiwary, U.S.: Multimodal fusion framework: a multiresolution approach for emotion classification and recognition from physiological signals. NeuroImage **102**, 162–172 (2014)
139. Viola, P., Jones, M., et al.: Rapid object detection using a boosted cascade of simple features. CVPR **1**(1), 511–518 (2001)
140. Wagner, J., Andre, E., Lingenfelser, F., Kim, J.: Exploring fusion methods for multimodal emotion recognition with missing data. IEEE Trans. Affect. Comput. **2**(4), 206–218 (2011)
141. Wagner, J., Vogt, T., André, E.: A systematic comparison of different HMM designs for emotion recognition from acted and spontaneous speech. In: International Conference on Affective Computing and Intelligent Interaction, pp. 114–125. Springer (2007)
142. Wang, S., Liu, Z., Lv, S., Lv, Y., Wu, G., Peng, P., Chen, F., Wang, X.: A natural visible and infrared facial expression database for expression recognition and emotion inference. IEEE Trans. Multimedia **12**(7), 682–691 (2010)

143. Warriner, A.B., Kuperman, V., Brysbaert, M.: Norms of valence, arousal, and dominance for 13,915 English lemmas. Behav. Res. Methods **45**(4), 1191–1207 (2013)
144. Wiles, O., Koepke, A., Zisserman, A.: Self-supervised learning of a facial attribute embedding from video. arXiv preprint arXiv:1808.06882 (2018)
145. Wu, S., Falk, T.H., Chan, W.Y.: Automatic speech emotion recognition using modulation spectral features. Speech Commun. **53**(5), 768–785 (2011)
146. Wu, T., Bartlett, M.S., Movellan, J.R.: Facial expression recognition using gabor motion energy filters. In: Computer Vision and Pattern Recognition-Workshops, pp. 42–47. IEEE (2010)
147. Wu, Y., Kang, X., Matsumoto, K., Yoshida, M., Kita, K.: Emoticon-based emotion analysis for Weibo articles in sentence level. In: International Conference on Multi-disciplinary Trends in Artificial Intelligence, pp. 104–112. Springer (2018)
148. Xingjian, S., Chen, Z., Wang, H., Yeung, D.Y., Wong, W.K., Woo, W.c.: Convolutional LSTM network: a machine learning approach for precipitation nowcasting. In: Advances in Neural Information Processing Systems, pp. 802–810 (2015)
149. Yan, W.J., Li, X., Wang, S.J., Zhao, G., Liu, Y.J., Chen, Y.H., Fu, X.: CASME II: an improved spontaneous micro-expression database and the baseline evaluation. PloS One **9**(1), e86041 (2014)
150. Yan, W.J., Wu, Q., Liang, J., Chen, Y.H., Fu, X.: How fast are the leaked facial expressions: the duration of micro-expressions. J. Nonverbal Behav. **37**(4), 217–230 (2013)
151. Yin, L., Wei, X., Sun, Y., Wang, J., Rosato, M.J.: A 3D facial expression database for facial behavior research. In: 7th International Conference on Automatic Face and Gesture Recognition, pp. 211–216. IEEE (2006)
152. Zafeiriou, S., Kollias, D., Nicolaou, M.A., Papaioannou, A., Zhao, G., Kotsia, I.: Aff-wild: valence and arousal'In-the-wild'challenge. In: Computer Vision and Pattern Recognition Workshops, pp. 34–41. IEEE (2017)
153. Zamil, A.A.A., Hasan, S., Baki, S.M.J., Adam, J.M., Zaman, I.: Emotion detection from speech signals using voting mechanism on classified frames. In: International Conference on Robotics, Electrical and Signal Processing Techniques (ICREST), pp. 281–285. IEEE (2019)
154. Zhalehpour, S., Onder, O., Akhtar, Z., Erdem, C.E.: BAUM-1: a spontaneous audio-visual face database of affective and mental states. IEEE Trans. Affect. Comput. **8**(3), 300–313 (2017)
155. Zhang, K., Zhang, Z., Li, Z., Qiao, Y.: Joint face detection and alignment using multitask cascaded convolutional networks. IEEE Signal Process. Lett. **23**(10), 1499–1503 (2016)
156. Zhang, Z., Girard, J.M., Wu, Y., Zhang, X., Liu, P., Ciftci, U., Canavan, S., Reale, M., Horowitz, A., Yang, H., et al.: Multimodal spontaneous emotion corpus for human behavior analysis. In: Computer Vision and Pattern Recognition, pp. 3438–3446. IEEE (2016)
157. Zhang, Z., Luo, P., Loy, C.C., Tang, X.: From facial expression recognition to interpersonal relation prediction. Int. J. Comput. Vis. **126**(5), 550–569 (2018)
158. Zhao, G., Huang, X., Taini, M., Li, S.Z., PietikäInen, M.: Facial expression recognition from near-infrared videos. Image Vis. Comput. 607–619 (2011)
159. Zhao, G., Pietikainen, M.: Dynamic texture recognition using local binary patterns with an application to facial expressions. IEEE Trans. Pattern Anal. Mach. Intell. **6**, 915–928 (2007)
160. Zhong, P., Wang, D., Miao, C.: An affect-rich neural conversational model with biased attention and weighted cross-entropy loss. arXiv preprint arXiv:1811.07078 (2018)
161. Zhou, G., Hansen, J.H., Kaiser, J.F.: Nonlinear feature based classification of speech under stress. IEEE Trans. Speech Audio Process. **9**(3), 201–216 (2001)

Chapter 4
"Speech Melody and Speech Content Didn't Fit Together"—Differences in Speech Behavior for Device Directed and Human Directed Interactions

Ingo Siegert and Julia Krüger

Abstract Nowadays, a diverse set of addressee detection methods is discussed. Typically, wake words are used. But these force an unnatural interaction and are error-prone, especially in case of false positive classification (user says the wake up word without intending to interact with the device). Therefore, technical systems should be enabled to perform a detection of device directed speech. In order to enrich research in the field of speech analysis in HCI we conducted studies with a commercial voice assistant, Amazon's ALEXA (Voice Assistant Conversation Corpus, VACC), and complemented objective speech analysis with subjective self and external reports on possible differences in speaking with the voice assistant compared to speaking with another person. The analysis revealed a set of specific features for device directed speech. It can be concluded that speech-based addressing of a technical system is a mainly conscious process including individual modifications of the speaking style.

4.1 Introduction

Voice assistant systems recently receive increased attention. The market for commercial voice assistants is rapidly growing: e.g. Microsoft Cortana had 133 million active users in 2016 (cf. [37]), the Echo Dot was the best-selling product on all of Amazon in the 2017 holiday season (cf. [11]). Furthermore, 72% of people who own a voice-activated speaker say their devices are often used as part of their daily routine (cf. [25]). Already in 2018 approximately 10% of the internet population used voice control according to [23]. Mostly the ease of use is responsible for the

I. Siegert (✉)
Mobile Dialog Systems, Otto von Guericke University Magdeburg, Universitätsplatz 2, 39106 Magdeburg, Germany
e-mail: ingo.siegert@ovgu.de

J. Krüger
Department of Psychosomatic Medicine and Psychotherapy, Otto von Guericke University Magdeburg, Leipziger Str. 44, 39120 Magdeburg, Germany
e-mail: julia.krueger@med.ovgu.de

G. Phillips-Wren et al. (eds.), *Advances in Data Science: Methodologies and Applications*, Intelligent Systems Reference Library 189,
https://doi.org/10.1007/978-3-030-51870-7_4

attractiveness of today's voice assistant systems. By simply using speech commands, users can play music, search the web, create to-do and shopping lists, shop online, get instant weather reports, and control popular smart-home products.

Besides enabling an as simple as possible operation of the technical system, voice assistants should allow a natural interaction. A natural interaction is characterized by the understanding of natural actions and the engagement of people into a dialog, while allowing them to interact naturally with each other and the environment. Furthermore, users don't need to use additional devices or learn any instruction, as the interaction respects the human perception. Correspondingly, the interaction with such systems is easy and seductive for everyone (cf. [63]). To fulfill these properties, cognitive systems, which are able to perceive their environment and are working on the basis of gathered knowledge and model-based recognition, are needed. In contrast, today's voice assistant's system functionality is still very limited and not seen as a natural interaction. Especially, when navigating the nuances of human communication, today's voice assistants still have a long way to go. They are still incapable of handling other expressions that have semantic similarity, still based on the evaluation of pre-defined keywords, and still unable to interpret prosodic variations.

Another important aspect on the way towards a natural interaction with voice assistants, is the interaction initiation. Nowadays two solutions have become established to initiate an interaction with a technical system: push-to-talk and wake words. In research also other methods have been evaluated, e.g. look-to-talk [36].

In push-to-talk systems, the user has to press a button, wait for a (mostly acoustic signal) and can then start to talk. The conversation set-up time can be reduced using buffers and contextual analyzes for the initial speech burst [65]. Push-to-talk systems are mostly used in environments where a error-free conversation initiation is needed, e.g. telecommunication systems or cars [10]. The false acceptance rate is nearly zero, only rare cases of wrong button pushes have to bet taken into account. But, this high robustness is at the expense of the naturalness of the interaction initiation. Therefore, in voice assistants the wake-word method is more common.

For the wake-word technique, the user has to say a pre-defined keyword to activate the voice assistant and afterwards the speech command can be uttered. Each voice assistant has its own unique wake work[1] which can sometimes be selected from a short list of (pre-defined) alternatives. This approach of calling your device by a name is more natural than the push-to-talk solution, but far away from a human-like interaction, as every dialog has to be initiated with the wake-word. Only for a few exceptions the wake-word can be neglected. Therefore, developers use a simple trick and extend the time-span the device is listening after a dialog [56]. But, the currently preferred wake word method is still error-prone. The voice assistant is still not able to detect when it is addressed and when it is just talked about him. This can result in users' confusion, e.g., when the wake word has been said but no interaction with the system was intended by the user. Especially for voice assistant systems that are already able to buy products automatically and in future should be enabled to

[1]The wake word to activate Amazon's ALEXA from its "inactive" state to be able to make a request is 'Alexa' by default.

autonomously make decisions it is crucial to only react when truly intended by the user.

The following examples show how wake words already led to errors. The first example went through the news in January 2017. At the end of a news story the presenter remarked: "I love the little girl, saying 'ALEXA order me a dollhouse.'" Amazon Echo owners who were watching the broadcast found that the remark triggered orders on their own devices (cf. [31]). Another wake word failure highlights the privacy issues of voice assistants. According to the KIRO7 news channel, a private conversation of a family was recorded by Amazon's ALEXA and sent to the phone of a random person, who was in the family's contact list. Amazon justified this misconduct as follows: ALEXA woke up due to a word in the background conversation sounding like 'ALEXA', the subsequent conversation was heard as a "send message" request, the customer's contact name and the confirmation to send the message (cf. [21]). A third example illustrated the malfunctions of smart home services using Apple's Siri. A neighbor of a house owner who had equipped its house with a smart lock and the apple HomeKit was able let himself in by shouting, "Hey Siri, unlock the front door." [59]. These examples illustrate that today's solution of using a wake word is in many ways insufficient. Additional techniques are needed to detect whether the voice assistant is (properly) addressed (by the owner) or not. One possibility is the development of a reliable Addressee Detection (AD) technique implemented in the system itself. Such systems, will only react when the (correct) user addresses the voice assistant with the intention to talk to the device.

Regarding AD research various aspects have already been investigated so far, cf. Sect. 4.2. But, previous research concentrated on the analyzes of observable users' speech characteristics in the recorded data and the subsequent analyzes of external ratings. The question whether users themselves recognize differences or even perhaps deliberately change their speaking style when interacting with a technical system (and potential influencing factors for this change) have not been evaluated so far. Furthermore, a comparison between self reported modifications in speech behavior and externally as well as automatic identificated modifications seems promising in case of fundamental research.

In this chapter, an overview of recent advances in AD research will be given. Furthermore, changes in speaking style will be identified by analyzing modifications of conversation factors during a multi-party human-computer interaction (HCI). The remainder of the chapter is structured as follows: In Sect. 4.2 previous work on related AD research is presented and discussed. In Sect. 4.3 the experimental setup of the utilized dataset and the participant description is presented. In Sect. 4.4 the dimensions under analyze "automatic", "self" and "external" are introduced. The results regarding these dimensions are then presented in Sect. 4.5. Finally, Sect. 4.6 concludes the chapter and presents an outlook.

4.2 Related Work

Many investigations for an improved AD make use of the full set of modalities human conversation offers and investigate both human-human interactions (HHIs) as well as HCIs. Within these studies, most authors use either eye-gaze, language related features (utterance length, keyword, trigram-model), or a combination of both. But, as this chapter deals with voice assistant systems, which are speech activated, only related work considering the acoustic channel are reported.

Another issue is that most of the AD studies for speech enabled systems utilize self-recorded databases. Thereby either interactions of one human and a technical system or groups of humans (mostly two) interacting with each other and a technical system [1, 5, 45, 46, 61, 62, 64], teams of robots and teams of humans [12], elderly people, or children [44]. These studies are mostly done using one specific scenario, just a few researchers analyze how people interact with technical systems in different scenarios [4, 30] or comparing different datasets [2].

Regarding acoustic AD systems, researchers employ different mostly not comparable tasks, as there are no generally accepted benchmark data, except the Interspeech-2017 paralinguistics challenge dealing with the AD between children and adults [44].

In [5], the authors utilize the SmartWeb database to distinguish "on-talk" (utterances directed to the device) and off-talk (every utterance not directed towards the system). This database contains 4 hours of spontaneous conversations of 96 speakers interacting with a mobile system, recorded under a Wizard-of-Oz (WOZ)-technique. As features the authors used duration features, energy features, F0 features and length of pauses features. Using an LDA-classifier and Leave-One-Speaker-Out (LOSO) validation their best averaged recognition rate to distinguish On-Talk and Off-Talk is 74.2% using only acoustic features. A recent study utilizing the same dataset and problem description achieves up to 82.2% Unweighted Average Recall (UAR) using the IS13_ComParE feature set (reduced to 1000 features using feature selection) with a Support Vector Machine (SVM) [1].

In [64] 150 multiparty interactions of 2–3 people playing a trivia question game with a computer are utilized. The dataset comprises audio, video, beamforming, system state and ASR information. The authors extracted a set of 308 expert features from 47 basic features utilizing seven different modalities and knowledge sources in the system. Using random forests models with the expert features the authors achieved an Equal Error Rate (EER) of 10.8% best. The same data is used in [61]. For acoustic analyzes, energy, energy change and temporal shape of speech contour features, in total 47 features, are used to train an adaboost classifier. The authors of [61] achieved an EER of 13.88%.

The authors of [62] used a WOZ setup to collect 32 dialoges of human-human-computer interaction. Comparing the performance of gaze, utterance length and dialog events using a naive bayes classifier, the authors stated that for their data the approach "the addressee is where the eye is" gives the best result of 0.75 area under the curve (AUC).

The position paper of [12] describes an approach to build spoken dialogue systems for multiple mixed human and robot conversational partners. The dataset is gathered during the Mars analog field tests and comprises 14,103 utterances. The authors argue that the dialog context provides valuable information to distinguish between human-directed and robot-directed utterances.

In [45], data of 38 sessions of two people interacting in a more formal way with a "Conversational Browser" are recorded. Using energy, speaking rate as well as energy contour features to train a Gaussian Mixture Model (GMM) together with linear logistic regression and boosting, the authors achieved an EER of 12.63%. The same data is used in [46]. Their best acoustic EER of 12.5% is achieved using a GMM with adaptive boosting of energy contour features, voice quality features, tilt features, and voicing onset/offset delta features. Also the authors of [30] used this data and used a language model-based score computation for AD recognition, using the assumption that device-directed speech produces less error for an automatic speech recognition (ASR)-system. Using just this information an EER of 12.2% on manual transcripts and 26.1% using an ASR-system could be achieved.

The authors of [4] used two different experimental settings (standing and sitting) of a WOZ data collection with 10 times two speakers interacting with an animated character. The experimental setup was about two decision-making sessions with formalized commandos. They employed an SVM and four supra-segmental speech features (F_0, intensity, speech rate and duration) as well as two speech features describing the difference for a speaker from all speakers's average for F_0 and intensity. The reported acoustic accuracy is 75.3% for the participants standing and 80.7% for the participants sitting.

A very recent work by researchers of AMAZON (cf. [33]) uses long short-term memory neural networks trained on acoustic features, ASR decoder, and 1-best hypotheses of automatic speech recognition output with an EER of 10.9% (acoustic alone) and 5.2% combined for the recognition of device directed utterances. As dataset 250 hours (350k utterances) of natural human interactions with a voice controlled far-field devices are used for training.

Furthermore, in [54] it could be shown that an AD system based on acoustic features only achieves an outstanding classification performance (>84%), also for inter-speaker groups across age, sex and technical affinity using data from a formal computer interaction [39] and a subsequently conducted interview [29].

One assumption that all of these investigations have, is the (simulated) limited ability of the technical system in comparison with the human conversational partner. Despite the vocabulary, also the complexity and the structure of the utterances as as well as the topics of the dialogs are limited for the technical system in comparison to a human conversational partner. Therefore, the AD problem complexity is always biased. To overcome this issue, in [53] another dataset is presented comprising content-identical human-human and human-machine conversations. In [2] data augmentation techniques are used achieving an UAR of 62.7% in comparison to the recognition rate of 60.54% gained using human raters. The so far reported research concentrated on analyzing observable users' speech characteristics in the recorded data. Regarding research on how humans identify the addressee during interactions, most studies rely on visual cues (eye-gaze) and lexical cues (markers of addressee), cf. [7, 24, 66]. Only few studies analyze acoustic cues.

In [57] the human classification rate using auditory and visual cues is analyzed. The authors analyzed conversations between a person playing as a clerk of travel agency and two people playing as customers and reported that the tone of voice was useful for human evaluators to identify the addressee in their face-to-face multiparty conversations. Analyzing the judgments of human evaluators in correctly identifying the addressee, the authors stated that the combination of audio and video presentation gave the best performance of 64.2%. Auditory and visual information alone resulted in a somewhat poorer performance of 53.0% and 55.8%, respectively. Both results are still well above chance level, which was 33.3%.

The authors of [32] investigated how people identified the addressee in human-computer multiparty conversations. To this avail, the authors recorded videos of three to four people sitting around a computer display. The computer system answered questions from the users. Afterwards human annotators were asked to identify the addressee of the human conversation partners by watching the videos. Additionally, the annotators should rate the importance of lexical, visual and audio cues for their judgment. The list of cues comprise fluency of speech, politeness terms, conversational/command style, speakers' gaze, peers' gaze, loudness, careful pronunciation, and tone of voice. An overall judgment of 63% identifying the correct human of the group of interlocutors and of 86% identifying the computer addressee was reported. This emphasizes the difficulty of the AD task. The authors furthermore reported that both audio and visual information are useful for humans to predict the addressee even when both modalities—audio and video—are present. The authors additionally stated that the subjects performed the judgment significantly faster based on the audio information than on the visual information. Regarding the importance of the different cues, the most informative cues are intonation and speakers' gaze (cf. [32]).

In summary, the studies so far examined identified acoustic cues as meaningful as visual cues for human evaluation. But, these studies analyzed only a few acoustic characteristics. Furthermore, it must be stated that previous studies are based on the judgments of evaluators, never on the statements of the interacting speakers themselves. Although [27] describes that users change their speaking style when interacting with technical systems. In the following, study we wanted to explore in detail what differences in their speech speakers themselves awarely recognize and compare these reports with external perspectives by human annotators and automatic detection on their speaking styles.

4.3 The Voice Assistant Conversation Corpus (VACC)

In order to analyze the speakers' behavior during a multi-party HCI, the Voice Assistant Conversation Corpus (VACC) was utilized, see [51]. VACC consists of recordings of interaction experiments between a participant, a confederate speaker[2] and Amazon's ALEXA. Furthermore, questionnaires presented before and after the experiment are used to get insights about the speakers' addressee behavior, see Fig. 4.1.

4.3.1 Experimental Design

The interactions with ALEXA consisted of two modules, (I) the Calendar module and (II) the Quiz module. Each of them has a pre-designed conversation type. The arrangement is done according to their complexity level. Additionally, each module was conducted in two conditions to test the influence of the confederate speaker. Thus, each participant conducted four "rounds" of interactions with ALEXA. A round was finished when the aim was reached or broken up to avoid frustration if hardly any success could be realized.

The *Calendar Module* represents a formal interaction. The task of the participant is to make appointments with the confederate speaker. As the participant's calendar was stored online and was only accessible via ALEXA, the participants had to interact with ALEXA to get their calendar. The two conditions now describe how the participant get the information about the confederate's available dates. In condition C_A ("alone") the participant only got written information about the confederate's available dates. In condition C_C ("with confederate") the confederate speaker was

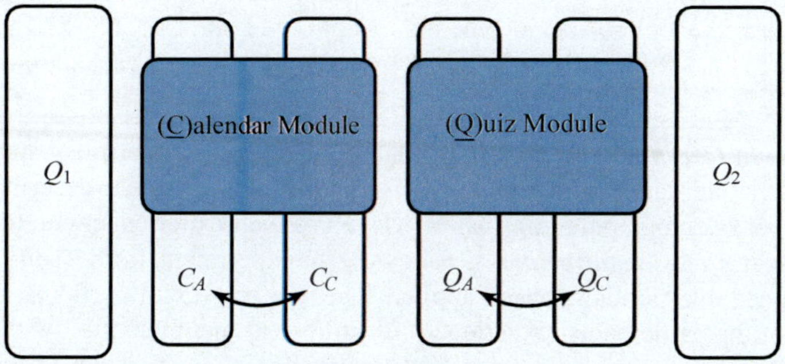

Fig. 4.1 The experimental procedure of VACC. Q_1 and Q_2 are the questionnaire rounds. The order of the scenarios (Calendar Module and Quiz Module) is fixed. A and C denote the experimental conditions alone and together with an confederate respectively

[2]The confederate speaker was introduced to the participants as "Jannik".

present and could give the information by himself. Thus, the participant had to interact with both, ALEXA and the confederate, to find available time slots. The confederate speaker was part of the research team and was instructed to interact only with the participant, not with ALEXA.

In the *Quiz Module*, the participant had to answer questions of a quiz (e.g., "How old was Martin Luther King?"). The questions were designed in such a way, that ALEXA is not able to give the full answer, see Dialog 4.1 for an example dialog. It could only solve partial steps or answer a reformulated question. In condition Q_A the participant had to find a strategy to solve the questions on its own. In condition Q_C the participant and the confederate speaker built up a team to discuss about an optimal strategy. Thus, these conversations were more informal than the previous calendar task. The confederate (here again only interacting with the participant, not with ALEXA) was instructed to make command proposals to the participant if frustration due to failures was imminent. The questions in Q_C was more sophisticated than in Q_A to force cooperation between the two speakers.

Dialog 4.1 Example of a question from the Quiz Module and a possible solution.

```
QUESTION: How old was Martin Luther King? (
    Wie alt wurde Martin Luther King?)
USER: Alexa how old was Martin Luther King
    when he died? (Alexa wie alt war Martin
    Luther King als er starb?)
ALEXA: I'm not sure (Ich bin mir nicht sicher)
USER: Alexa when was Martin Luther King born?
    (Alexa wann ist Martin Luther King geboren
    ?)
ALEXA: The date of birth of Martin Luther King
     is 15. January 1929 (Das Geburtsdatum von
    Martin Luther King ist 15. Januar 1929)
USER: Alexa when did Martin Luther King die? (
    Alexa wann ist Martin Luther King gestorben
    ?)
ALEXA: Martin Luther King died on 4. April
    1968 (Martin Luther King ist am 4. April
    1968 gestorben)
```

In *Questionnaire Round 1*, filled out before the experiment starts, a self-defined computer-aided questionnaire as used in [43] was utilized to describe the participants' socio-demographic information as well as their experience with technical systems.

In *Questionnaire Round 2* following the experiment, further self-defined computer-aided questionnaires were applied. The first one (Q2-1 participants' version) asked for the participants' experiences regarding (a) the interaction with ALEXA and the confederate speaker in general, (b) possible changes in voice and speaking style while interacting with ALEXA and the confederate speaker. The second questionnaire (Q2-2 participants' version) asked for recognized differences in the specific prosodic characteristics (choice of words, sentence length, monotony, melody, syllable/word stress, speech rate). According to the so-called principle of openness in examining subjective experiences (cf. [20]), the formulation of questions developed from higher openness and a free, non-restricted answering format in the first questionnaire (e.g., "If you compare your speaking style when interacting with ALEXA or with the confederate speaker—did you recognize differences? If yes, please describe the differences when speaking with ALEXA!") to lower openness and more structured answering formats in the second questionnaire (e.g., "Did your speed of speech differ when interacting with ALEXA or with the confederate speaker? Yes or no? If yes, please describe the differences!").

A third questionnaire focused on previous experiences with voice assistants. Lastly, AttrakDiff, see [19], was used to supplement the open questions on self-evaluation of the interaction by a quantifying measurement of the quality of the interaction with ALEXA (hedonic and pragmatic quality).

In total, this dataset contains recordings of 27 participants with a total duration of approx. 17 hours. The recordings were conducted in a living-room-like surrounding, in order to avoid that the participants may be distracted by a laboratory surrounding and to underline a natural interaction atmosphere. As voice assistant system, the Amazon ALEXA Echo Dot (2nd generation) was utilized. It was decided to use this system to create a fully free interaction with a currently available commercial system. The speech of the participant and the confederate speaker was recorded using two high-quality neckband microphones (Sennheiser HSP 2-EW-3). Additionally, a high-quality shotgun microphone (Sennheiser ME 66) captured the overall acoustics and the output of Amazon's ALEXA. The recordings were stored uncompressed in WAV-format with 44.1 kHz sample rate and 16 bit resolution.

The interaction data were further processed. The recordings were manually separated into utterances with additional information about the belonging speaker (participant, confederate speaker, ALEXA). Using a manual annotation the addressee of each utterance was identified—human directed (HD) for statements addressing the confederate, device directed (DD) for statements directed to ALEXA. Consequently, all statements not directed towards a specific speaker or soliloquies are marked as off-talk (OT) and parts were simultaneous utterances are occurring are marked as cross-talk (CT).

4.3.2 Participant Characterization

All participants were German-speaking students at the Otto von Guericke University Magdeburg. The corpus is nearly balanced regarding sex (13 male, 14 female). The mean age is 24.11 years, ranging from 20 to 32 years. Furthermore, the dataset is not biased towards technophilic students, as different study courses are covered, including computer science, engineering science, humanities and medical sciences.

The participants reported to have at least heard of Amazon's ALEXA before, only six participants specified that they had used ALEXA prior to this experiment. Only one participant specified that he uses ALEXA regularly—for playing music. Regarding the experience with other voice assistants, in total 16 out of 27 participants reported to have at least basic experience with voice assistants. Overall this dataset

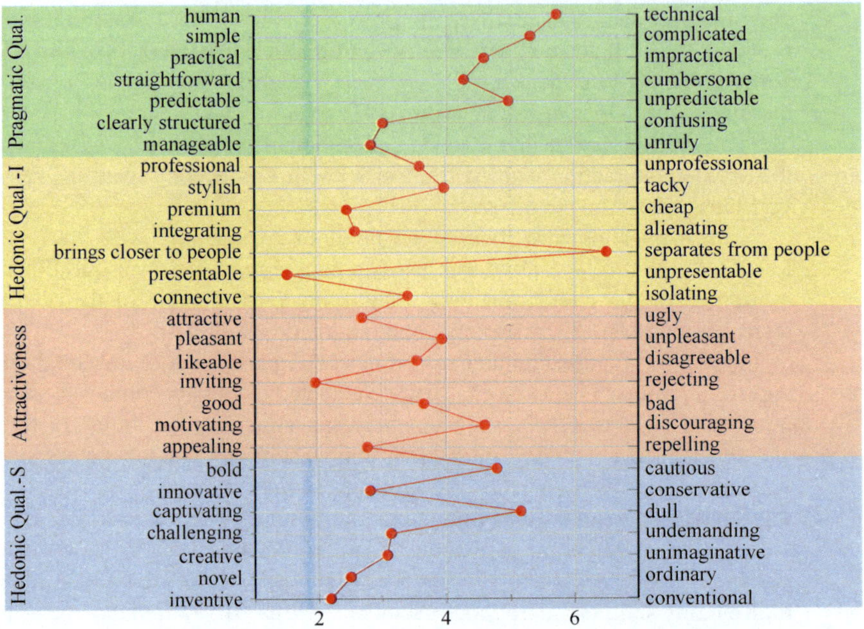

Fig. 4.2 Participants' evaluation of the AttrakDiff questionnaire for ALEXA after completing all tasks (—●—)

represents a heterogeneous set of participants, which is representative for younger users with an academic background.

AttrakDiff is employed to understand how participants evaluate the usability and design of interactive products (cf. [19]). It distinguishes four aspects (pragmatic quality (PQ), hedonic Quality (HQ)—including the sub-qualities identity (HQ-I) and stimulation (HQ-S), as well as attractiveness (ATT)). For all four aspects no significant difference between technology experienced and technology unexperienced participants could be observed, see Fig. 4.2. Moreover, PQ, HQ-I, and ATT are overall perceived as neutral with only one outlier for "separates me from people". Regarding HQ-S, a slight negative assessment can be observed, showing that the support of the own needs was inappropriate. This can be justified by difficulties of the calendar task where ALEXA has deficits. But overall, it can be assumed that ALEXA provides useful features and allows participants to identify themselves with the interaction.

4.4 Methods for Data Analyzes

Speech behavior was analyzed on the basis of three different dimensions.
The *self perspective* relies on the participants post-experiments questionnaires, which are part of VACC (see Sect. 4.3.1). It comprises open and structured self reports.

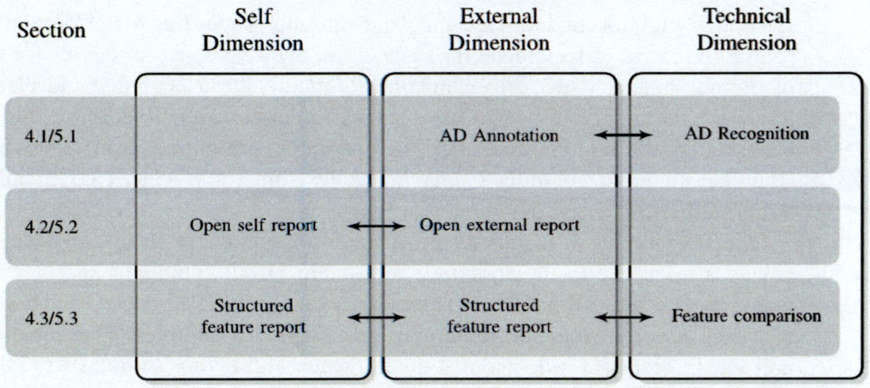

Fig. 4.3 Overview of the three utilized perspectives, and the relation of their different analyzes

The *external perspective* comprises the annotation of DD or HD samples as well as the post-annotation open and structured self report (annotators version). This annotation will be described in detail in Sects. 4.4.1 and 4.4.2.

The *technical dimension* compromises the automatic DD/HD recognition and a statistical feature comparison. Fig. 4.3 depicts the relation between the different analysis methods along the different dimensions. It will be described in Sect. 4.4.1, too. This approach enables to draw connections between external evaluations, technical evaluations and can point this back to the self-evaluations. This approach has not been used before in speaker behavior analyzes for AD research.

4.4.1 Addressee Annotation and Addressee Recognition Task

In order to substantiate the assumption that humans are speaking differently to human and technical conversation partners, an AD task was conducted, as a first external test. Hereby, both human annotators and a technical recognition system, have to evaluate selected samples of the VACC in terms of DD or HD.

For the *human annotation*, ten native German-speaking and ten non-German-speaking annotators took part in the annotation. This approach allows to evaluate the influence of the speech content on the annotation by analyzing the difference in the annotation performance and questionnaire answers of these two groups, see Sects. 4.4.1 and 4.4.2. The annotators’ age ranges from 21 to 29 (mean of 24.9 years) for the German-speaking annotators and from 22 to 30 (mean of 25.4 years) for the non-German-speaking annotators. The sex balance is 3 male and 7 female German-speaking annotators and 6 male and 4 female non-German-speaking annotators. Just the minority of both annotator groups has experience with voice assistants. The German-speaking annotators are coming from various study courses, including engineering science, humanities and business administration. The non-German-speaking annotators all had a technical background. According to the Common European Framework of Reference for Languages, the language proficiency for German of the non-German-speaking annotators is mainly on a beginner and elementary level (8 annotators), only two have an intermediate level. Regarding the cultural background, the non-German-speaking annotators are mainly from Southeast Asia (Bangladesh, Pakistan, India, China) and a few from South America (Colombia) and Arabia (Egypt).

The 378 samples were selected by a two-step approach. In the first step, all samples having a reference to the addressee or contain offtalk, apparent laughter etc., are manually omitted. Afterwards, 7 samples are randomly selected for each experiment and each module from the remaining set of cleaned samples. The samples were presented in random order, so that the influence of previous samples from the same speaker is minimized. The annotation was conducted with ikannotate2 [8, 55]. The task for the annotators was to listen to these samples and decide whether the sample contains DD speech or HD speech. To state the quality of the annotations, in terms of interrater reliability, Krippendorff's alpha is calculated, cf. [3, 48] for an overview. It determines the extent to which two or more coders obtain the same result when measuring a certain object [17]. We use the MATLAB-macro kriAlpha.m to measure the IRR as used in the publication [13]. To interpret the IRR values, the interpretation scheme by [28] is used, defining values between 0.2 and 0.4 as fair and values between 0.4 and 0.6 as moderate. For the evaluation and comparison with the automatic recognition results, the overall UAR and Unweighted Average Precision (UAP) as well as the individual results for the Calendar and Quiz module were calculated.

The *automatic recognition experiments* used the same two-class problem as for the human annotation: detecting whether an utterance is DD or HD. Utterances containing off-talk or laughter are skipped as before. The "emobase" feature set of OpenSMILE was utilized, as this set provides a good compromise between feature size and feature accuracy and has been successfully used for various acoustic recognition systems: dialog performance (cf. [40]), room-acoustic analyzes (cf. [22]), acoustic scene classification (cf. [34]), surround sound analyzes (cf. [49]), user satisfaction prediction (cf. [14]), humor recognition (cf. [6]), spontaneous speech analyzes (cf. [60]), physical pain detection (cf. [38]), compression influence analyzes (cf. [47, 52]), emotion recognition (cf. [9, 15]), and AD recognition (cf. [54]). Differences between the data samples of different speakers were eliminated using standardization [9]. As recognition system, an SVM with linear kernel and a cost factor of 1 was utilized with WEKA [18]. The same classifier has already been successfully used for an AD problem archiving very good results (>86% UAR) [50, 54]. A LOSO validation was applied and the overall UAR and UAP as well as the individual results for the Calendar and Quiz module as the average over all speakers were calculated. This strategy allows to revise the generalization ability of the actual experiment and compare it with the human annotation.

4.4.2 Open Self Report and Open External Report

The first questionnaire in Questionnaire Round 2—Q2-1 (participant's version, see Fig. 4.1)—asked for the participants' overall experiences in and evaluation of the interaction with ALEXA and the confederate speaker as well as for differences in their speaking style while interacting with these two counterparts. By adopting an open format for collecting and analyzing data, the study complements others in the field exploring speech behavior in HCI (cf. [32, 57]). The formulation of questions and the answering format allowed the participants to set individual relevance when telling about their subjective perceptions.

The German- and non-German-speaking annotators answered an adapted version of this questionnaire (Q2-1, annotator's version, e.g. "On the basis of which speech characteristics of the speaker did you notice that he/she addressed a technical system?"). Besides the focus of the initial open questions dealing with what in general was useful to differentiate between DD and HD, the annotators' version differed from the participants' ones in giving up to ask for melody as a speech feature (the analysis of the participants' version revealed that people had problems in differentiating melody and monotony and often answered similarly regarding both features). Results from analyzing and comapring the open self and open external reports contribute to basic research on speaking style in HD and DD.

The answers in the open self and open external reports were analyzed using qualitative content analysis, see [35], in order to summarize the material sticking close to the text. At the beginning, the material was broken down into so-called meaning units. These are text segments, which are understandable by themselves, represent a single idea, argument or information and vary between word groups and text paragraphs in length (cf. [58]). These meaning units were paraphrased, generalized, and reduced in accordance with the methods of summarizing qualitative content analysis. Afterwards, they were grouped according to similarities and differences across each group (participants, German-speaking annotators, Non-German-speaking annotators).

4.4.3 Structured Feature Report and Feature Comparison

The second questionnaire in Questionnaire Round 2—Q2-2 participant's version (see Fig. 4.1)—asked for differences in speaking style between the interaction with ALEXA and the confederate speaker in a more structured way. Each question aimed at one specific speech characteristic. The answering format included closed questions (e.g. "Did you recognize differences in sentence length between the interaction with ALEXA and the interaction with Jannik?"—"yes – no – I do not know") accompanied by open questions where applicable (e.g. "If yes, please describe to what extent the sentence length was different when talking to ALEXA."). See Table 4.1 for an overview of requested characteristics.

Table 4.1 Overview of requested characteristics for the structured reports. ∗non-German-speaking annotators were not asked for choice of words

Self report	External report
Choice of words	Choice of words∗
Sentence length	Sentence length
Monotony	Monotony
Intonation (word/syllable accentuation)	Intonation (word/syllable accentuation)
Speaking rate	Speaking rate
Melody	–
–	Content
–	Loudness

This more structured answering format guaranteed subjective evaluations for each speech feature the study was interested in and allowed to complement the results of the open self reports by statistical analysis. Comparing answers of the open and the more structured self reports yields the participants' level of awareness of differences in their speaking style in both interactions.

Again, in line with interests in basic research, the German and non-German-speaking annotators answered an adapted version of this questionnaire (Q2-2 annotator's version, e.g. "You have just decided for several recordings whether the speaker has spoken with a technical system or with a human being. Please evaluate whether and, if so, to what extent the sentence length of the speaker was important for your decision.—"not – slightly – moderately – quite – very", "If the sentence length was slightly, moderately, quite or very important for your decision, please describe to what extent the sentence length was different when talking to the technical system."). In addition to the features asked for in the self reports the feature list was extended by loudness of speech. It was considered as meaningful in speech behavior decisions regarding DD and HD based on the feature comparison and participants reports. In order to control possible influences of the speech content on the annotation decision (DD or HD) the feature list also included this characteristic. See Table 4.1, for an overview of requested characteristics. The answers in the self and external structured feature reports were analyzed using descriptive statistics, especially frequency analysis. Results from both reports were compared with each other as well as compared with results from automatic feature analysis.

The feature comparison is based on statistical comparisons of various acoustic characteristics. The acoustic characteristics were automatically extracted using openSMILE (cf. [16]). As the related work does not indicate specific feature sets distinctive for AD, a broad set of features extractable with openSMILE was utilized. For feature extraction, it is differentiated between Low-Level-Descriptors (LLDs) and functionals. Low-Level-Descriptors (LLDs) comprise the sub-segmental acoustic characteristics extractable for a specific short-time window (usually 25–40 ms), while functionals represent super-segmental contours of the LLDs regarding a specific cohesive course (usually an utterance or turn). In Table 4.2, the used LLDs and functionals are shortly described. For reproducibility, the same feature identifier notifiers as supplied by openSMILE are used.

Table 4.2 Overview of investigated LLDs and functionals

Name	Description
alphaRatio	Ratio between energy in low frequency region and high frequency region
F0	Fundamental frequency
F0_env	Envelope of the F0-contour
F0semitone	Logarithmic F0 on a semitone frequency scale, starting at 27.5 Hz
F*X*amplitude	Formant *X* amplitude in relation to the F0 amplitude
F*X*frequency	Centre frequency of 1st, 2nd, and 3rd formant
F*X*bandwidth	Bandwidth of 1st, 2nd, and 3rd formant
lspFreq[0-7]	Line spectral pair frequencies
mfcc_[1-12]	Mel-Frequency cepstral coefficients 1-12
pcm_intensity	Mean of the squared windowed input values
pcm_loudness	Normalized intensity
pcm_zcr	Zero-crossing rate of time signal (frame-based)
slope0-500	Linear regression slope of the logarithmic power spectrum within 0-500Hz
slope500-1500	Linear regression slope of the logarithmic power spectrum within 500-1500Hz
jitterLocal	Deviations in consecutive F0 period lengths
shimmerLocal	Difference of the peak amplitudes of consecutive F0 periods

(a) Short description of utilized LLDs

Name	Description
amean	Arithmetic mean
stddev	standard deviation
max	maximum value
maxPos	abs. position of max (frames)
min	minimum value
minPos	abs. position of min (frames)
range	max-min
quartile1	first quartile
quartile2	second quartile
quartile3	third quartile
percentile50.0	50% percentile
percentile80.0	50% percentile
iqrY-X	Inter-quartile range: quartileX-quartileY
pctlrange0-2	inter-percentile range: 20%-80%
skewness	skewness (3rd order moment)
kurtosis	Kurtosis (4th order moment)
linregc1	Slope (m) of a linear approximation
linregc2	Offset (t) of a linear approximation
linregerrA	Linear error computed as the difference of the linear approximation
linregerrQ	Quadratic error computed as the difference of the linear approximation
meanFallingSlope	Mean of the slope of falling signal parts
meanRisingSlope	Mean of the slope of rising signal parts

(b) Short description of utilized functionals

In order to identify the changed acoustic characteristics, statistical analyzes were conducted utilizing the previously automatically extracted features. To this avail, for each feature the distribution across the samples of a specific condition were compared to the distribution across all samples of another condition by applying a non-parametric U-Test. The significance level was set to $\alpha = 0.01$. This analysis was performed independently for each speaker of the dataset. Afterwards, a majority voting (qualified majority: 3/4) of the analyzed features was applied over all speakers within each dataset. Features with a p-value below the threshold α in the majority of the speakers are identified as changed between the compared conditions.

4.5 Results

4.5.1 Addressee Annotation and Addressee Recognition Task

4.5.1.1 Human AD Annotation

To first test the quality of annotations in terms of the interrater reliability, Krippendorff's alpha is calculated. The differences between the Calendar and the Quiz Module for German-speaking annotators are marginal with around 0.55. For non-German-speaking annotators the IRR is only 0.168 and only 0.255 for the Calendar module and the Quiz module respectively. According to the interpretation scheme of [28], this means a slight to fair IRR value for the non-German-speaking annotators and a moderate IRR value for the German-speaking annotators. These numbers already show that the task leaves space for interpretations by the annotators. Especially some of the non-German-speaking annotators are faced with difficulties.

Regarding the human annotated AD task, the results are presented in Fig. 4.4. It can be seen that in general, German-speaking annotators are roughly 10% better in correctly identifying the addressee than non-German-speaking annotators. This underlines to a certain degree the importance of the speech content. Furthermore, the variance between the German-speaking annotators regarding the two modules Cal-

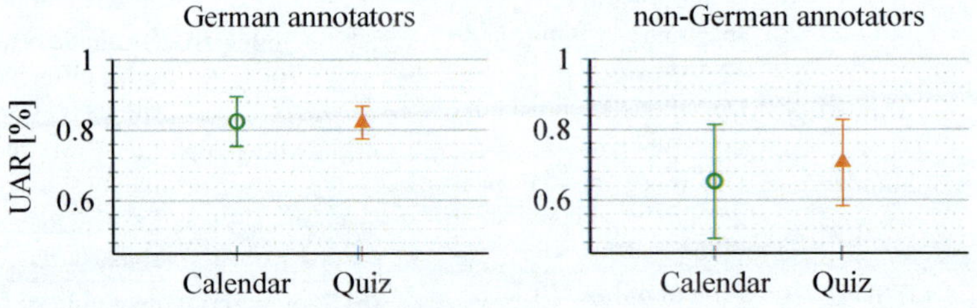

Fig. 4.4 Mean and standard deviation of the UAR values for the German and non-German-speaking annotators according to the two modules of VACC

endar and Quiz is much less than for the non-German-speaking ones with approx. 6% and 14% respectively. Regarding the two modules of VACC representing different conversational styles, it can be seen that the more formal calendar task complicates the correct annotation for the non-German-speaking annotators, the average is 65.39% and 70.61% for calendar and quiz task, respectively. The German-speaking annotators did not show these difficulties.

4.5.1.2 Automatic AD Recognition

Regarding the automatic recognition results, it can be seen that even with a very simple recognition system, a performance of 81.97% was archived in distinguishing DD and HD utterances in the Calendar module, see Fig. 4.5. Regarding the more informal conversation during the Quiz module, an even improved performance of 88.24% could be achieved. Comparing the result of the automatic classification with the human annotation baseline, it can be seen that the acoustic characteristics reveal a lot of information for the classifier. For the Quiz module, even the annotation for German-speaking annotators can be outperformed. Hereby it has to be noted that a fair comparison is only valid against the non-German-speaking annotators, as the utilized classifier does not incorporate any speech content, which is considerably improved for the human annotation.

In comparison to the classification results of more sophisticated classifiers reported in the related work chapter (Sect. 4.2) being around 87% the best classification result of approx. 88% is already outstanding. Applying a mixup data augmentation approach method for the AD problem of VACC increases the performance significantly with an even better UAR of 90.01% over both modules, see [2].

Additionally, a classifier was developed to analyze the discriminative power of the acoustic characteristics in recognizing if the participant is interacting with ALEXA alone or with the presence of the confederate speaker. For this case the classifier is just slightly above chance level for both modules, with 59.63% and 66.87% respectively. This suggests, that the influence of the second speaker for the interaction with

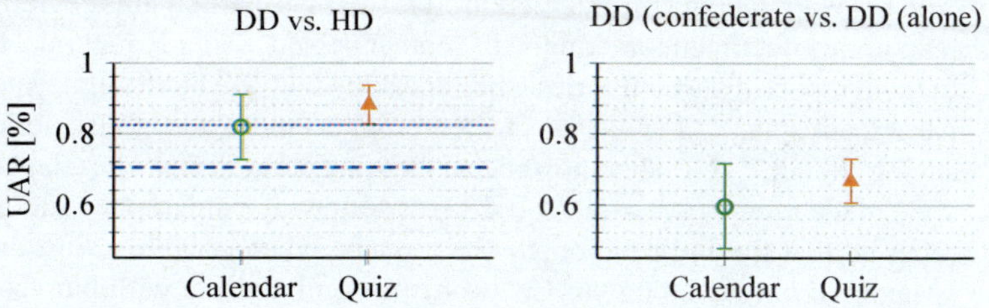

Fig. 4.5 Mean and standard deviation of the UAR values for the automatic AD classification according to the two different modules and the influence of the confederate speaker onto the DD utterances of the participants. For comparison the best annotation results are indicated (........ German-speaking ---- non-German-speaking)

ALEXA is nearly not given for the Calendar task. But, for the Quiz task an influence can be observed.

4.5.2 Open Self Report and Open External Report

The analyzes of the open self reports and external reports concentrated on the first questionnaire, Q2-1 participant's and annotator's version. The first two questions in the participant's version of the questionnaire asking for descriptions of the experienced cooperation with ALEXA and Jannik were not taken into account, because there were no comparable questions in the annotator's version. Thus, the non-restricted answers of the following questions were taken into account:

- Self-report: one question asking for differences in speaking with ALEXA compared to speaking with the confederate speaker, and questions regarding subjective thoughts and decisions about the speaking behavior in interacting with both.
- External report: one question asking for possible general decision criteria considered in annotating (DD or HD), and questions asking which speech characteristics helped the annotator in his/her decision.

The participants and annotators used headwords or sentences to answer these questions. In the self reports these texts made up a total number of 2068 words. In the external reports there was a total number of 535 words and 603 words.

4.5.2.1 Open Self Report

Subjective experiences of the interaction with ALEXA and with the confederate speaker

In general, all 27 participants recognized differences in their speaking style. The interaction with the confederate speaker is described as "free and reckless" (B[3]) and "intuitive" (X). Participants stated that they "spoke like [they] always do" (G) and "did not worry about" the interaction style (M). The participants explain this behavior by pointing out that interacting with another person is simply natural. However, some of them reported particularities when speaking with the confederate speaker, e.g. one participant stated: "I spoke much clearer with Jannik, too. I also addressed him by saying 'Jannik'" (C). This showed that there are participants who adapt their speaking style in the interaction with ALEXA (see following paragraph). Another participant reported that the information can be reduced when speaking with the confederate speaker: "I only need one or two words to communicate with him and speak about the next step" (H). Altogether, interacting with the confederate speaker is described as "more personal" (E) and "friendly" (E) than interacting with ALEXA.

[3]Participants were anonymized by using letters in alphabetic order.

Speaking with ALEXA is described as more extensively. Only a few participants experienced it as "intuitive" (AB) and spoke without worrying about their speaking style: "I did not worry about the intonation, because ALEXA understood me very well" (Y). Another one did think about how to speak with ALEXA only when ALEXA did not understand him (B). Besides these few exceptions, all of the other participants report about differences in their voice and speaking style when interacting with ALEXA. The interaction is described as "more difficult" (P), "not that free" (B), "different to interacting with someone in the real world" (M); there is "no real conversation" (I), "no dialog" (J) and "speaking with Jannik was much more lively" (AB).

Subjective experiences of changes in the speaking style characteristics

Differences are reported in relation to choice of words, speaking rate, sentence length (complexity), loudness, intonation (word/syllable accentuation), and rhythm (monotony, melody). Regarding reported differences in *choice of words*, the participants described that they repeated words, avoided using slang or abbreviations, and used synonyms and key words, e.g. "Usually one does not address persons with their first name at the beginning of each new sentence. This is certainly a transition with ALEXA." (M). Furthermore, participants reported that they had to think about how to formulate sentences properly and which words to use in order to "formulate as precise and unambiguous as possible" (F) taking into account what they thought ALEXA might be able to understand. Some of them reported to "always use commands and no requests" (W) when addressing ALEXA. Regarding the *speaking rate* many participants reported to speak slower with ALEXA than with the confederate speaker or even "slower [. . .] than usual" (C). Furthermore, participants described that they avoided *complex sentences*: "You have to think about how to formulate a question as short and simple as possible" (O), "in order to enable ALEXA to understand [what I wanted to ask]" (P). Some of the participants stated that they preformulated the sentences in the mind or sometimes only used keywords instead of full sentences, so that "you [. . .] do not speak fluently" (X). Many participants emphasized that they had to use different formulations until they got the answers or the information they wanted. Once participants noticed how sentences have to be formulated they used the same sentence structures in the following: "You have to use the routine which is implemented in the system" (O). Thus, participants learned from their mistakes at the beginning of the interaction (I) and adapted to ALEXA. In the case of *loudness*, participants reported to "strive much more to speak louder" (J) with ALEXA, e.g. because "I wanted that it replied directly on my first interaction" (M). In combination with reflections upon *intonation* one participant said: "I tried to speak particularly clearly and a little bit more louder, too. Like I wanted to explain something to a child or asked it for something." (W). Furthermore, many participants stated that they stressed single words, e.g. "important keywords" (V), and speak "as clearly and accurately as possible" (G), e.g. "to avoid misunderstandings" (F). However, a few participants explained that they did not worry about intonation (Q, Y) or only worried about it, if ALEXA did not understand them (B, O). Regarding *melody* and *monotony*, participants emphasized to speak in a staccato-like style because of the slowness and aspired clearness of speaking, the repetition of words, and the worrying about how to further formulate the sentences.

4.5.2.2 Open External Report

The German and the non-German-speaking annotators slightly vary in their open reports on what helped them to decide if a speaker interacted with another person or with a technical system. Besides mentioning special speech features they describe their decision bases metaphorically: For example, DD was described as "more sober" (B*[4]) and "people speak more flat with machine" (E**[5]), whereas "sentences sound more natural" (D**), "speech had more emotions" (I**), was "more lively" (E*) and "not that inflexible" (G*) in HD.

In their open reports both groups furthermore refer on nearly each of the characteristics listed later on in the structured questions (see Sect. 4.4.3). First, this indicates that the annotators are aware of these characteristics to be relevant for their decision process. However, the differing number of annotators referring to each of the features showed that there are differences regarding relevance setting in the annotator groups. This mirrors the means presented in the structured report (see Sect. 4.4.3): The non-German-speaking annotators did not mention length of sentences in their free answers regarding their general decision making. Furthermore, when specially asked for aspects helping them to decide whether the speaker interacted with a technical system, they did not mention speech content. In addition, when deciding for DD, the loudness was not mentioned by the German-speaking annotators. Interestingly, both annotator groups bring up emotionality of speech as a characteristic that helped them in their decision, without explaining in detail what they meant by this term.

In the following, each feature referred to by the annotators will be examined in more detail based on the open reports regarding helpful aspects for deciding if the speaker interacted with another person or with a technical system (first three questions from Q2-1) and the open, but more specialized questions regarding differences in preformulated speech characteristics (the remaining questions from Q2-1). Nearly all of German-speaking annotators deal with the *choice of words*. They describe, that compared to the interaction with another person, when speaking with a technical system, the speaker adopt no or only less colloquial speech or dialectal speech, politeness forms, personal forms like personal pronoun "you" or filler words like "ehm". One participant describes a "more direct choice of words without beating about the bush" (D*). There are only a few non-German-speaking annotators referring to choice of words by themselves. These describe an "informal way of talking" (I**)

[4]German-speaking annotators were anonymized by using letters in alphabetic order including *.

[5]Non-German-speaking annotators were anonymized by using letters in alphabetic order including **.

and the use of particles as hints for HD, whereas the speaker avoids casual words when speaking to a technical system. Regarding the *speaking rate* both annotator groups describe a slow to moderate speaking rate ("calmer", F*) in the interaction with a technical system, whereas the speaking rate is faster in the interaction with another person, however, hesitation or pauses appear ("[speaker] is stopping and thinking", C**). If the speaker speaks *loudly* and/or on a constant volume level, this points to DD ("as loudness will allow better speech recognition", K**). On the contrary, a low volume and variations in the volume level indicate an interaction with another person. Interestingly, loudness was brought up more frequently by the group of non-German-speaking annotators. On the contrary, *monotonous speech* was important for both groups. Participants' speech in DD was experienced as much more monotonous than in HD ("the more lively the speech the more it counts for [speaking with another] person", C*, "[HHI is] more exciting", E**), whereby the German-speaking annotators recognized a variation of tone at the end of questions in DD. As possible reasons for this observation the annotators state "monotonous speech [...] so that the electronic device is able to identify much better [...] like the [speech] of the devices themselves" (F*), and "because there is no emotional relationship with the counterpart" (D*). *Words and syllables are accentuated* more or even "overaccentuated" (J*), syllables and end of words are not "slurred" (C*, H*) and speech "sounds clearer" (D**) in DD ("pronounce each word of a sentence separately, like when you try to speak with someone who doesn't understand you", H**). This impression can be found in both of the annotator groups. However, German-speaking annotators reflect much more about the accentuation in their free answering than non-German ones. *Speech content* is mentioned solely by German-speaking annotators. They recognized precise questions, "without any information which is unnecessary" (I*), focused on specific topics in DD, whereas in HD utterances entailed "answering" (K*)/referencing to topics mentioned before during the conversation, "positive feedback regarding the understanding" (H*), or even "mistakes and uncertainties [...if] the speaking person flounders and nevertheless goes on speaking" (E*). One of the German-speaking annotators recognized "melody and content of speech didn't fit together" (E*). Many of the non-German-speaking annotators explained that they didn't take speech content into account because there were not able to understand the German language. In answering what helped deciding between DD and HD the *length of sentences* was only infrequently mentioned by some of the German-speaking annotators. None of the non-German-speaking ones referred to this characteristic. Only when directly asked for aspects relevant regarding this feature. The annotators in both groups showed contrary assessments regarding sentences in DD indicating them as being longer ("like an artificial prolongation", I*) or shorter ("as short as possible, only saying the important words" (G**) than those in HD. Finally, both annotator groups indicate emotionality of speech as being relevant in their decision-making process. They experienced speaking with another person as (more) emotional ("emotional – for me this means human being", J*). As an example for emotionality both annotator groups bring up giggling or "voice appearing more empathic", H*).

4.5.3 Structured Feature Report and Feature Comparison

Besides gaining individual non-restricted reports about participants' and annotators' impressions regarding the speech characteristics in the interaction with ALEXA and with the confederate speaker, a complementary structured questioning with prescribed speech features of interest should allow statistical analysis and comparisons.

4.5.3.1 Structured Feature Self Report

In the more closed answering format of the second questionnaire (Q2-2), the participants should assess variations of different speaking style characteristics between the interaction with the confederate speaker and with ALEXA. Thereby it was explicitly asked for separate assessments of the Calendar and Quiz module. Table 4.3 shows the response frequencies.

It could be seen that all participants indicate to deliberately have changed speaking style characteristics. Only in the Quiz module two participants denied changes in all speaking style characteristics or indicated that they do not know if they changed the characteristic asked for (K, AB). In the Calendar module all participants answered at least one time with "yes" when asking for changes in speaking style characteristics. Furthermore, in the Quiz module more differences were individually recognized by the participants than in the Calendar module.

4.5.3.2 Structured Feature External Report

The following table shows the mean ratings of the German-speaking and non-German-speaking annotators regarding prescribed features (Table 4.4).

Table 4.3 Response frequencies for the self-assessment of different speaking style characteristics for the Calendar module (first number) and the Quiz module (second number). Given answers are: **R**eported difference, **N**o difference, I don't **K**now, **I**nvalid answer

Characteristic	R	N	K	I
Choice of words	24/24	3/0	0/3	0/0
Sentence length	18/19	5/3	3/4	1/1
Monotony	19/19	6/6	2/2	0/0
Intonation (word/syllable accentuation)	16/17	7/5	4/4	0/1
Speaking rate	17/20	8/4	1/2	1/1
Melody	10/11	8/7	7/7	2/2

Table 4.4 Ratings base on a 5-point-Likert-Scale ("1 – not important" to "5 – very important"). The two most important characteristics for each group are highlighted. *non-German-speaking annotators were not asked for choice of words

Characteristic	German-speaking annotators	non-German-speaking annotators
Choice of words	4.6 (0.52)	–*
Sentence length	3.3 (1.34)	4.0 (0.82)
Monotony	4.0 (1.55)	3.7 (1.25)
Intonation (word/syllable accentuation)	3.7 (1.49)	3.9 (1.10)
Speaking rate	4.3 (0.67)	3.8 (1.23)
Content	4.0 (1.15)	2.6 (1.58)
Loudness	3.1 (1.29)	3.2 (1.62)

Choice of words was most important for the German-speaking annotators to decide if a speaker interacted with a technical system or with another person, the sentence length was most important for the non-German ones. The German ones rated speech content as quite important, whereas the non-German-speaking annotators expectedly were not able to use this feature for their decision. Interestingly, the relevance set regarding loudness and monotony and speech rate did not differ highly between both annotator groups indicating that these features are important no matter if the listener is familiar with the speaker's language or not. Although, the German-speaking annotators did not indicate loudness as an important characteristic in their open report. In general the characteristics does not differ significantly between both groups, except for the content ($F = 5.1324$, $p = 0.0361$, one-way Anova). But, as the language proficiency for the non-German-speaking annotators is on a beginner level, this result was excepted and to a certain degree provoked.

4.5.3.3 Statistical DD/HD-Feature Differences

In the statistical analysis of the features between speakers' DD and HD utterances, there are only significant differences for a few feature descriptors in the Calendar module, cf. Table 4.5. Primarily, characteristics from the group of energy related descriptors (*pcm*_intensity, *pcm*_loudness) were significantly larger when the speakers are talking to ALEXA. Regarding the functionals, this applies to the absolute value (mean) as well as the range-related functionals (*stddev*, *range*, *iqr's*, and *max*). This shows that the participants were in general speaking significantly louder towards ALEXA than to the confederate speaker. The analysis of the data revealed that the participants start uttering their commands very loud but the loudness drops to the end of the command. As further distinctive descriptors only spectral characteristic *lspFreq[1]* and *lspFreq[2]* were identified, having a significantly smaller first quartile.

Table 4.5 Overview of identified distinctive LLDs ($p<0.05$) for both modules of VACC independently

	Calendar	Quiz
	DD versus HD	DD versus HD
Identified distinctive LLDs	*pcm*_intensity, *pcm*_loudness	*lspFreq[0-6]*, *mfcc[2,4]*, *pcm*_intensity, *pcm*_loudness, *pcm*_zcr, *alphaRatio*, *F0semitone*, *F2amplitude*, *F3amplitude*

In contrast to the Calendar module, several features in the Quiz module showed a significant difference between DD and HD utterances of the participants. This comprises energy related descriptors (*pcm*_intensity, *pcm*_loudness, *alphaRatio*) partly identified in the Calendar module as well as spectral characteristics (*lspFreq[0-6]*, *mfcc[2,4]*, *F0semitone*, *F2amplitude*, *F3amplitude*) and the *pcm*_zcr as a measure for the "percussiveness". The energy-based features behave in the same way as in the Calendar module: the participants generally speak louder. For the group of spectral descriptors the distribution over almost all examined functionals is changed, i.e. here the articulation is strongly different in the addressing of ALEXA (DD) and the confederate (HD).

4.6 Conclusion and Outlook

The presented study analyzes subjective and objective changes in speaking style characteristics when addressing humans or technical systems. Therefore, the VACC is utilized providing real-life multi-party HCI of one participant interacting with ALEXA alone and with another confederate speaker in two different task settings. This dataset comprises a more natural interaction than most of the previous investigations, as a real system is used and the interaction took place in a usual, unconventional living room environment. Furthermore, due to the two different tasks with distinguishing conversational styles a broad variety of interactions is covered. Additionally, besides audio recordings of the interaction this dataset additionally provides self-assessments of the participants and external assessments of annotators allowing to reveal insights of experiences in the interaction with ALEXA and the confederate speaker.

The open reports of participants as well as annotators revealed that speech interaction and addressee detection are highly intuitive processes mirrored in metaphorical descriptions made. However, they operate on a high level of awareness. Participants and annotators resemble one another in material of the open reports pointing to four main characteristics indicating differences for HD and DD speech:

- Naturalness: Participants as well as annotators indicated that speaking with another person "sound[s] more natural" (D**). They describe the interaction as "intuitive" (X), e.g. by speaking "free and easy" (B) in the way they usually speak or interacted with each other. This includes that it is unnecessary to think about how or what to say or to hesitate or pause during an utterance. Regarding accentuation and volume one participant resumed: "[they] were rather different to how I would have done it with someone in the real world" (M).
- Emotionality: Compared to interpersonal communication, speaking to a technical system is described as less emotional or even without any emotion ("I often assessed utterances addressing a machine, which were rather unemotional", E*). On the contrary, speaking with another person is experienced as "more affectionate [...with] more emotion inside" (E**), e.g. indicated by laughing, and with a more empathic voice.
- Relatedness: Whereas speaking with a technical system is "no real conversation" (I) without "unnecessary information" (I*) and with "more content-related speech" (E*), interpersonal communication is characterized by being "friendly" (S), using politeness forms, and "more personal" (E), e.g. by the use of the personal pronoun "you" (I*) indicating to relate to the conversation partner. Moreover, relatedness is represented by "referring to a topic mentioned before" (K*) and "positive feedback regarding the understanding" (H*). HD in contrast is described as being "more personal" (E) and "voice appears more empathic" (H*). The speaker refers to "a topic mentioned before" (K*) and "answers" (K*) to his/her conversation partner. Furthermore, "positive feedback regarding the understanding" (H*). However, DD is described as being less dialogical, "no real conversation" (I).
- Liveliness: Speaking with another person is experienced as "much more lively" (AB), whereas "people speak more flat with the machine" (E**). This is reflected in the variations in nearly each of the reported speech characteristics (choice of words, speech rate, volume, monotony, speech content, word-/syllable accentuation): People speak "more prudently" (B*), speech sounds "very clocked" (I*) and formulations are "direct... straight and narrow" (D*) when speaking with the technical system. In interpersonal communication the speaker e.g. "rises and downs the tone" (C**), varies the volume, or is "stopping and thinking" (C**) and "hesitates" (J*) during his utterances.

By the more structured questions regarding certain speech features participants and annotators are forced to analyze this intuition. Thereby it becomes obvious, that decisions about how to speak (participants) and decisions about differentiation between HD and DD are highly complex in the combination of nearly all speech features taken into account in this study. Considering the answers of the participants and the annotators, it is surprising that differences are described in very detail already in the open answering format. All speech and voice characteristics explicitly asked for in Q2-2 (choice of word, sentence length, speaking rate, loudness, intonation, monotony, and melody) were brought up by them. However, it has to be emphasized that in the open answering format none of the participants described differences in all of these characteristics. When asked for differences more precisely during the second questionnaire differences regarding a variety of speech and voice characteristics come to mind and could be described.

In comparison between the subjective and objective analyzes, it can be stated that humans are aware of their different addressing behavior, which can be assessed by human evaluators. Furthermore, these differences are distinctive enough to archive adequate recognition results of over 81% already for even simple classifier systems. To compare the self-assessments and external assessments regarding the different speaking styles with the automatic extractable acoustic characteristics, the description of [40] is used: Intonation and stress are related to the basic functionals (mean, minimum, maximum, range, standard deviation) of the fundamental frequency and energy related descriptors. Melody and monotony as categories of the speech rhythm are related to changes in functionals describing the mean distance, the mean deviation and, the range and quartile-ranges of fundamental frequency's semitones, formant frequencies, formant bandwidths descriptors. Changes in the range of spectral descriptors describe the tendency of a monotonic voice. According to this comparison of acoustic descriptors, the subjective self-assessments and external assessments are supported by the objective statistical feature-distribution comparison in general. Amongst the prosodic evaluations the majority of participants and annotators indicated a change in intonation and rhythm. But, the objective analyzes revealed that these perceived changes are not reflected equally for every type of interaction. Within the formal Calendar module differences are nearly only identifiable within energy related descriptors (intonation, loudness) and much less within rhythm related descriptors. Whereas, within the Quiz module several prosodic characteristics changed between speaking with ALEXA and speaking with the confederate speaker (intonation, loudness, rhythm). Additionally, it has to be noted that neither in the Calendar module nor in the Quiz module distinctive changes of the fundamental frequency could be observed. Reported individually recognized changes seem to be due to interaction approximating to interpersonal interaction (and experienced as "more lively" (AB)) because of the less structured interaction context.

Before discussing future work, some remarks have to be made about limitations of the present study. A main limitation of this work can be seen in the relatively small number of participants and annotators, preventing sub-group analyzes e.g. regarding regular usage of voice assistants and a bigger variety in the open answers. Furthermore, due to the limited number of cases it is difficult to structure some terms used by the laymen in the open question part. Also, the terms used for the characteristics can lead to misinterpretations due to its simplicity. Furthermore, the interaction initiation with ALEXA using a wake word impairs the naturalness of the interaction, which may be an additional factor for the differences in the addressing behavior.

Future work will deal with the identification of a general set of characteristics that distinguish human addressed from system addressed utterances. Hereby the influence of different factors of the technical system (voice, wake word, artificial presence) and of the participants (technical affinity, age, prior experience) will be analyzed. Furthermore, the influence of the confederate speaker will be analyzed. Especially the participants' individuality onto their accomodation behavior, first insights have already been reported in [41, 42], have to be examined in detail. Also in-depth analyzes of reported individual changes in comparison to their objectively measurable characteristics have to be conducted to further get insights on user specific addressee behavior. Thereby, a special focus will be laid on the subjectively reported motives for changing speaking style. During the analyzes of the reports, we observed that the annotators started to imagine the origin of the situations they evaluated, e.g. "[...] sometimes I wondered whether the questions were part of a e.g. game. In any case, this should contribute to the transmission of information" (H). To improve the utilized type of open questions, the influence of these imaginations on the reports should be investigated. Another important issue is the mind-set of the participants about the abilities of the technical system, see [27]. This as well will be evaluated in future.

From the main characteristics differencing between HD and DD in the annotators' and speakers' reports implications for further research and development of automatic processing and addressee detection can be derived: Detecting the presence of emotional speech seem to be promising for AD as for DD less emotional speech is reported ("emotionality", "liveliness"). This affects features already considered in our analyzes, e.g. monotony or volume. Furthermore, pauses within an utterance seem promising, too (naturalness). When incorporating ASR-techniques, choice of words seems to be a promising feature for further investigation, too (relatedness), e.g. regarding the use of politeness forms, personal pronouns or content-relational speech. The development of a proper AD system is one component in the further development from limited assistance systems towards cognitive assistants. A robust AD allows voice assistant systems to offer a real conversation mode, which is not only based on the simple continuation of listening after certain dialog steps (asking for the weather, setting up shopping lists, etc.) and reacting to a stop word as it is implemented actually in Google Now [26]. Furthermore, a proper AD for device directed utterances also allows voice assistants to take part and support trustworthy multi-user cooperative tasks with future cognitive systems.

References

1. Akhtiamov, O., Sidorov, M., Karpov, A., Minker, W.: Speech and text analysis for multimodal addressee detection in human-human-computer interaction. In: Proceedings of the INTERSPEECH-2017, pp. 2521–2525 (2017)
2. Akhtiamov, O., Siegert, I., Minker, W., Karpov, A.: Cross-corpus data augmentation for acoustic addressee detection. In: 20th Annual SIGdial Meeting on Discourse and Dialogue (2019)
3. Artstein, R., Poesio, M.: Inter-coder agreement for computational linguistics. Comput. Linguist. **34**, 555–596 (2008)
4. Baba, N., Huang, H.H., Nakano, Y.I.: Addressee identification for human-human-agent multiparty conversations in different proxemics. In: Proceedings of the 4th Workshop on Eye Gaze in Intelligent Human Machine Interaction, pp. 6:1–6:6 (2012)
5. Batliner, A., Hacker, C., Nöth, E.: To talk or not to talk with a computer. J. Multimodal User Interfaces **2**, 171–186 (2008)

6. Bertero, D., Fung, P.: Deep learning of audio and language features for humor prediction. In: Proceedings of the 10th LREC, Portorož, Slovenia (2016)
7. Beyan, C., Carissimi, N., Capozzi, F., Vascon, S., Bustreo, M., Pierro, A., Becchio, C., Murino, V.: Detecting emergent leader in a meeting environment using nonverbal visual features only. In: Proceedings of the 18th ACM ICMI, pp. 317–324. ICMI 2016 (2016)
8. Böck, R., Siegert, I., Haase, M., Lange, J., Wendemuth, A.: ikannotate—a tool for labelling, transcription, and annotation of emotionally coloured speech. In: Affective Computing and Intelligent Interaction, LNCS, vol. 6974, pp. 25–34. Springer (2011)
9. Böck, R., Egorow, O., Siegert, I., Wendemuth, A.: Comparative study on normalisation in emotion recognition from speech. In: Horain, P., Achard, C., Mallem, M. (eds.) Proceedings of the 9th IHCI 2017, pp. 189–201. Springer International Publishing, Cham (2017)
10. DaSilva, L.A., Morgan, G.E., Bostian, C.W., Sweeney, D.G., Midkiff, S.F., Reed, J.H., Thompson, C., Newhall, W.G., Woerner, B.: The resurgence of push-to-talk technologies. IEEE Commun. Mag. **44**(1), 48–55 (2006)
11. Dickey, M.R.: The echo dot was the best-selling product on all of amazon this holiday season. TechCrunch (December 2017). Accessed 26 Dec 2017
12. Dowding, J., Clancey, W.J., Graham, J.: Are you talking to me? dialogue systems supporting mixed teams of humans and robots. In: AIAA Fall Symposium Annually Informed Performance: Integrating Machine Listing and Auditory Presentation in Robotic System, Washington, DC, USA (2006)
13. Eggink, J., Bland, D.: A large scale experiment for mood-based classification of TV programmes. In: Proceedings of ICME, pp. 140–145 (2012)
14. Egorow, O., Siegert, I., Wendemuth, A.: Prediction of user satisfaction in naturalistic human-computer interaction. Kognitive Systeme **1** (2017)
15. Eyben, F., Scherer, K.R., Schuller, B.W., Sundberg, J., André, E., Busso, C., Devillers, L.Y., Epps, J., Laukka, P., Narayanan, S.S., Truong, K.P.: The geneva minimalistic acoustic parameter set (gemaps) for voice research and affective computing. IEEE Trans. Affect. Comput. **7**(2), 190–202 (2016)
16. Eyben, F., Wöllmer, M., Schuller, B.: openSMILE—the Munich versatile and fast open-source audio feature extractor. In: Proceedings of the ACM MM-2010 (2010)
17. Gwet, K.L.: Intrarater reliability, pp. 473–485. Wiley, Hoboken, USA (2008)
18. Hall, M., Frank, E., Holmes, G., Pfahringer, B., Reutemann, P., Witten, I.: The WEKA data mining software: an update. SIGKDD Explor. Newsl. **11**(1), 10–18 (2009)
19. Hassenzahl, M., Burmester, M., Koller, F.: AttrakDiff: Ein Fragebogen zur Messung wahrgenommener hedonischer und pragmatischer Qualität. In: Szwillus, G., Ziegler, J. (eds.) Mensch & Computer 2003, Berichte des German Chapter of the ACM, vol. 57, pp. 187–196. Vieweg+Teubner, Wiesbaden, Germany (2003)
20. Hoffmann-Riem, C.: Die Sozialforschung einer interpretativen Soziologie - Der Datengewinn. Kölner Zeitschrift für Soziologie und Sozialpsychologie **32**, 339–372 (1980)
21. Horcher, G.: Woman says her amazon device recorded private conversation, sent it out to random contact. KIRO7 (2018). Accessed 25 May 2018
22. Höbel-Müller, J., Siegert, I., Heinemann, R., Requardt, A.F., Tornow, M., Wendemuth, A.: Analysis of the influence of different room acoustics on acoustic emotion features. In: Elektronische Sprachsignalverarbeitung 2019. Tagungsband der 30. Konferenz, pp. 156–163, Dresden, Germany (2019)
23. Jeffs, M.: Ok google, siri, alexa, cortana; can you tell me some stats on voice search? The Editr Blog (2017). Accessed 8 Jan 2018
24. Jovanovic, N., op den Akker, R., Nijholt, A.: Human perception of intended addressee during computer-assisted meetings. In: Proceedings of the 11th EACL, pp. 169–176 (2006)
25. Kleinberg, S.: 5 ways voice assistance is shaping consumer behavior. Think with Google (2018). Accessed Jan 2018
26. Konzelmann, J.: Chatting up your google assistant just got easier. The Keyword, blog.google (2018). Accessed 21 June 2018

27. Krüger, J.: Subjektives Nutzererleben in derMensch-Computer-Interaktion: Beziehungsrelevante Zuschreibungen gegenüber Companion-Systemen am Beispiel eines Individualisierungsdialogs. Qualitative Fall- und Prozessanalysen. Biographie – Interaktion – soziale Welten, Verlag Barbara Budrich (2018). https://books.google.de/books?id=v6x1DwAAQBAJ
28. Landis, J.R., Koch, G.G.: The measurement of observer agreement for categorical data. Biometrics **33**, 159–174 (1977)
29. Lange, J., Frommer, J.: Subjektives Erleben und intentionale Einstellung in Interviews zur Nutzer-Companion-Interaktion. Proceedings der 41. GI-Jahrestagung, Lecture Notes in Computer Science, vol. 192, pp. 240–254. Bonner Köllen Verlag, Berlin, Germany (2011)
30. Lee, H., Stolcke, A., Shriberg, E.: Using out-of-domain data for lexical addressee detection in human-human-computer dialog. In: Proceedings of NAACL, Atlanta, USA, pp. 221–229 (2013)
31. Liptak, A.: Amazon's alexa started ordering people dollhouses after hearing its name on tv. The Verge (2017). Accessed 7 Jan 2017
32. Lunsford, R., Oviatt, S.: Human perception of intended addressee during computer-assisted meetings. In: Proceedings of the 8th ACM ICMO, Banff, Alberta, Canada, pp. 20–27 (2006)
33. Mallidi, S.H., Maas, R., Goehner, K., Rastrow, A., Matsoukas, S., Hoffmeister, B.: Device-directed utterance detection. In: Proceedings of the INTERSPEECH'18, pp. 1225–1228 (2018)
34. Marchi, E., Tonelli, D., Xu, X., Ringeval, F., Deng, J., Squartini, S., Schuller, B.: Pairwise decomposition with deep neural networks and multiscale kernel subspace learning for acoustic scene classification. In: Proceedings of the Detection and Classification of Acoustic Scenes and Events 2016 Workshop (DCASE2016), pp. 543–547 (2016)
35. Mayring, P.: Qualitative Content Analysis: Theoretical Foundation, Basic Procedures and Software Solution. SSOAR, Klagenfurt (2014)
36. Oh, A., Fox, H., Kleek, M.V., Adler, A., Gajos, K., Morency, L.P., Darrell, T.: Evaluating look-to-talk. In: Proceedings of the Extended Abstracts on Human Factors in Computing Systems (CHI EA '02), pp. 650–651 (2002)
37. Osborne, J.: Why 100 million monthly cortana users on windows 10 is a big deal. TechRadar (2016). Accessed 20 July 2016
38. Oshrat, Y., Bloch, A., Lerner, A., Cohen, A., Avigal, M., Zeilig, G.: Speech prosody as a biosignal for physical pain detection. In: Proceedings of Speech Prosody, pp. 420–424 (2016)
39. Prylipko, D., Rösner, D., Siegert, I., Günther, S., Friesen, R., Haase, M., Vlasenko, B., Wendemuth, A.: Analysis of significant dialog events in realistic human-computer interaction. J. Multimodal User Interfaces **8**, 75–86 (2014)
40. Ramanarayanan, V., Lange, P., Evanini, K., Molloy, H., Tsuprun, E., Qian, Y., Suendermann-Oeft, D.: Using vision and speech features for automated prediction of performance metrics in multimodal dialogs. ETS Res. Rep. Ser. **1**, (2017)
41. Raveh, E., Siegert, I., Steiner, I., Gessinger, I., Möbius, B.: Three's a crowd? Effects of a second human on vocal accommodation with a voice assistant. In: Proceedings of Interspeech 2019, pp. 4005–4009 (2019). https://doi.org/10.21437/Interspeech.2019-1825
42. Raveh, E., Steiner, I., Siegert, I., Gessinger, I., Móbius, B.: Comparing phonetic changes in computer-directed and human-directed speech. In: Elektronische Sprachsignalverarbeitung 2019. Tagungsband der 30, Konferenz, Dresden, Germany, pp. 42–49 (2019)
43. Rösner, D., Frommer, J., Friesen, R., Haase, M., Lange, J., Otto, M.: LAST MINUTE: a multimodal corpus of speech-based user-companion interactions. In: Proceedings of the 8th LREC, Istanbul, Turkey, pp. 96–103 (2012)
44. Schuller, B., Steid, S., Batliner, A., Bergelson, E., Krajewski, J., Janott, C., Amatuni, A., Casillas, M., Seidl, A., Soderstrom, M., Warlaumont, A.S., Hidalgo, G., Schnieder, S., Heiser, C., Hohenhorst, W., Herzog, M., Schmitt, M., Qian, K., Zhang, Y., Trigeorgis, G., Tzirakis, P., Zafeiriou, S.: The interspeech 2017 computational paralinguistics challenge: Addressee, cold & snoring. In: Proceedings of the INTERSPEECH-2017, Stockholm, Sweden, pp. 3442–3446 (2017)
45. Shriberg, E., Stolcke, A., Hakkani-Tür, D., Heck, L.: Learning when to listen: detecting system-addressed speech in human-human-computer dialog. In: Proceedings of the INTERSPEECH'12, Portland, USA, pp. 334–337 (2012)

46. Shriberg, E., Stolcke, A., Ravuri, S.: Addressee detection for dialog systems using temporal and spectral dimensions of speaking style. In: Proceedings of the INTERSPEECH'13, Lyon, France, pp. 2559–2563 (2013)
47. Siegert, I., Lotz, A., Egorow, O., Wendemuth, A.: Improving speech-based emotion recognition by using psychoacoustic modeling and analysis-by-synthesis. In: Proceedings of SPECOM 2017, 19th International Conference Speech and Computer, pp. 445–455. Springer International Publishing, Cham (2017)
48. Siegert, I., Böck, R., Wendemuth, A.: Inter-rater reliability for emotion annotation in human-computer interaction—comparison and methodological improvements. J. Multimodal User Interfaces **8**, 17–28 (2014)
49. Siegert, I., Jokisch, O., Lotz, A.F., Trojahn, F., Meszaros, M., Maruschke, M.: Acoustic cues for the perceptual assessment of surround sound. In: Karpov, A., Potapova, R., Mporas, I. (eds.) Proceedings of SPECOM 2017, 19th International Conference Speech and Computer, pp. 65–75. Springer International Publishing, Cham (2017)
50. Siegert, I., Krüger, J.: How do we speak with alexa—subjective and objective assessments of changes in speaking style between hc and hh conversations. Kognitive Systeme **1** (2019)
51. Siegert, I., Krüger, J., Egorow, O., Nietzold, J., Heinemann, R., Lotz, A.: Voice assistant conversation corpus (VACC): a multi-scenario dataset for addressee detection in human-computer-interaction using Amazon's ALEXA. In: Proceedings of the 11th LREC, Paris, France (2018)
52. Siegert, I., Lotz, A.F., Egorow, O., Wolff, S.: Utilizing psychoacoustic modeling to improve speech-based emotion recognition. In: Proceedings of SPECOM 2018, 20th International Conference Speech and Computer, pp. 625–635. Springer International Publishing, Cham (2018)
53. Siegert, I., Nietzold, J., Heinemann, R., Wendemuth, A.: The restaurant booking corpus—content-identical comparative human-human and human-computer simulated telephone conversations. In: Berton, A., Haiber, U., Wolfgang, M. (eds.) Elektronische Sprachsignalverarbeitung 2019. Tagungsband der 30. Konferenz. Studientexte zur Sprachkommunikation, vol. 90, pp. 126–133. TUDpress, Dresden, Germany (2019)
54. Siegert, I., Shuran, T., Lotz, A.F.: Acoustic addressee-detection – analysing the impact of age, gender and technical knowledge. In: Berton, A., Haiber, U., Wolfgang, M. (eds.) Elektronische Sprachsignalverarbeitung 2018. Tagungsband der 29. Konferenz. Studientexte zur Sprachkommunikation, vol. 90, pp. 113–120. TUDpress, Ulm, Germany (2018)
55. Siegert, I., Wendemuth, A.: ikannotate2—a tool supporting annotation of emotions in audio-visual data. In: Trouvain, J., Steiner, I., Möbius, B. (eds.) Elektronische Sprachsignalverarbeitung 2017. Tagungsband der 28. Konferenz. Studientexte zur Sprachkommunikation, vol. 86, pp. 17–24. TUDpress, Saarbrücken, Germany (2017)
56. Statt, N.: Amazon adds follow-up mode for alexa to let you make back-to-back requests. The Verge (2018). Accessed 8 Mar 2018
57. Terken, J., Joris, I., De Valk, L.: Multimodalcues for addressee-hood in triadic communication with a human information retrieval agent. In: Proceedings of the 9th ACM ICMI, Nagoya, Aichi, Japan, pp. 94–101 (2007)
58. Tesch, R.: Qualitative Research Analysis Types and Software Tools. Palmer Press, New York (1990)
59. Tilley, A.: Neighbor unlocks front door without permission with the help of apple's siri. Forbes (2017). Accessed 17 Sept 2017
60. Toyama, S., Saito, D., Minematsu, N.: Use of global and acoustic features associated with contextual factors to adapt language models for spontaneous speech recognition. In: Proceedings of the INTERSPEECH'17, pp. 543–547 (2017)
61. Tsai, T., Stolcke, A., Slaney, M.: Multimodal addressee detection in multiparty dialogue systems. In: Proceedings of the 40th ICASSP, Brisbane, Australia, pp. 2314–2318 (2015)
62. van Turnhout, K., Terken, J., Bakx, I., Eggen, B.: Identifying the intended addressee in mixed human-human and human-computer interaction from non-verbal features. In: Proceedings of the 7th ACM ICMI, Torento, Italy, pp. 175–182 (2005)
63. Valli, A.: Notes on natural interaction. Technical Report, University of Florence, Italy (09 2007)

64. Vinyals, O., Bohus, D., Caruana, R.: Learning speaker, addressee and overlap detection models from multimodal streams. In: Proceedings of the 14th ACM ICMI, Santa Monica, USA, pp. 417–424 (2012)
65. Weinberg, G.: Contextual push-to-talk: a new technique for reducing voice dialog duration. In: MobileHCI (2009)
66. Zhang, R., Lee, H., Polymenakos, L., Radev, D.R.: Addressee and response selection in multi-party conversations with speaker interaction RNNs. In: Proceedings of the 2016 Conference on Empirical Methods in Natural Language Processing, pp. 2133–2143 (2016)

Chapter 5
Methods for Optimizing Fuzzy Inference Systems

Iosif Papadakis Ktistakis, Garrett Goodman, and Cogan Shimizu

5.1 Introduction

The world is inundated with data. For any definition of data, too, the amount generated per second is incredible. With the explosion of the Internet through the World Wide Web in the 1990s and early 2000s, as well as the more recent *exponential* explosion of the Internet of Things, it is without a doubt that making *sense* of this data is a primary research question of this century.

In answer to that, Data Science, as a field, emerged as a new sub-discipline of Computer Science. This new profession is expected to make sense of those vast stores of data. However, due to the field's nascent nature, exactly what it means to "make sense" of data, the techniques to do so, and the body of curricula that comprises the field are ill-defined. That is not to say it is not a rich field of study with a common basis among definitions. During its initial conceptualization, perhaps it was most accurate to say that Data Science is a coupling of statistics and computer science. In fact, in 2001, William S. Cleveland publishes "Data Science: An Action Plan for Expanding the Technical Areas of the Field of Statistics", where he describes several areas of statistics that could be enhanced by applying data processing methods from computer science.

I. P. Ktistakis (✉)
ASML LLC US, Wilton, CT 06851, USA
e-mail: sktistakisp@gmail.com

G. Goodman
Center of Assistive Research Technologies, Wright State University, Dayton, OH 45435, USA
e-mail: garrett.goodman@wright.edu

C. Shimizu
Data Semantics Lab, Wright State University, Dayton, OH 45435, USA
e-mail: shimizu.5@wright.edu

G. Phillips-Wren et al. (eds.), *Advances in Data Science: Methodologies and Applications*, Intelligent Systems Reference Library 189,
https://doi.org/10.1007/978-3-030-51870-7_5

Nowadays, Data Science means more than just statistics, instead referring to anything that has something to do with data. While statistical rigor is there, the field has grown to encompass much more, from collecting data to analyzing it to produce a model, to drawing from other fields for imparting context to the data (e.g. business intelligence and analytics).

Data scientists combine entrepreneurship with patience, an immense amount of exploration, the willingness to build data products, and an ability to iterate over a solution. This has grown the field such that there are now a multitude of interdisciplinary application areas that can reasonably fall under the purview of Data Science.

Unfortunately, the field was outgrowing its ability to provide guidance for developing applications [1]. This leads to teams carrying out ad hoc fashion data analysis and a time-consuming process of trial and error for identifying the right tool for the job [2]. Thus, in order to preserve its internal coherence, data scientists have placed an increased emphasis on developing methodologies, best practices, and how they interact in order to provide solutions.

In response, a methodology for Data Science is described by John Rollins, a data scientist at IBM Analytics [3]. Rollin outlines an iterative process of 10 stages, starting with solution conception and concludes with solution deployment, feedback solicitation, and refinement. Figure 5.1 provides a graphical overview of this methodology.

We may consider this to be a *foundational* methodology, as it provides an overarching strategy for solving problems in Data Science, in general. Furthermore, such methodologies are independent of specific technologies or tools and should provide a framework for adapting to specific problems by incorporating methods and processes best suited to the domain.

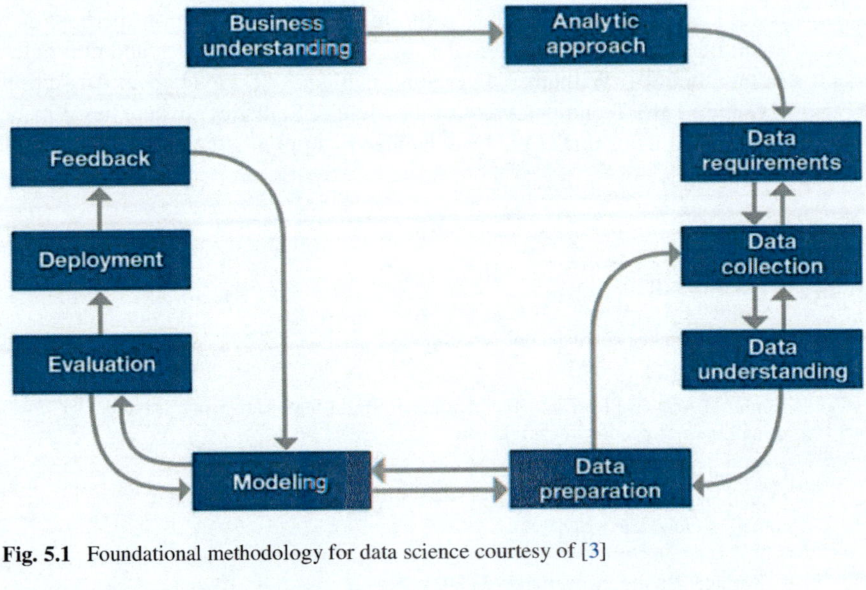

Fig. 5.1 Foundational methodology for data science courtesy of [3]

In this methodology, there are parts that remain unchanged and those that must be adapted. For example, the first stage, "Business understanding," is a critical aspect to the development of any application, proprietary or otherwise. On the other hand, the seventh stage, "Modelling," is very much domain and application specific stage. Perhaps the application is to build a repository of knowledge—should this be a graph structure? Tabular? Or, perhaps the application is to develop a predictive model the model is mathematical in nature, but should it be developed empirically? These are important questions as a methodology is adapted for a problem space.

In broad terms, this chapter considers how data science has come to play a crucial role in Artificial Intelligence. As an intersection of domain expertise, data engineering and visualization, and advanced computational methods equipped with statistical rigor, Data Science is uniquely suited to complement the field of Artificial Intelligence and its subfields, machine learning and soft computing but also the area of robotics as mentioned in [4]. Indeed, these fields provide models and the appropriate machinery for several useful tasks, such as predicting or classifying outcomes based on complex input or discovering or learning underlying patterns in the data. Further, many of these models can incorporate their insights to improve their future outcomes.

However, this chapter will focus on one methodology for predicting outcomes, the Fuzzy Inference System (FIS) [5]. Conventional methods, such as artificial neural networks (and their more exotic brethren) or unsupervised statistical methods (e.g. clustering), are not very resistant to noise in the data. That we live in a world filled with noisy, fuzzy data and the labels that must be applied or the categories into which data must be sorted are imprecise and vague (but must be evaluated anyway) has created a need for an alternative methodology [6].

Perhaps the most well-known example for introducing fuzzy inference systems is "How to hard boil an egg." The example was developed for the Computational Intelligence Society's 2011 Fuzzy Logic Video Competition and can be found online.[1] The video's premise is to use a small robot cooker to intelligently cook an egg. To do so, it uses some rules:

If the size of the egg is small, then boil for less than 5 min.
If the size of the egg is large, then boil for more than 5 min.

In essence, we represent domain knowledge, e.g. a recipe for a hardboiled egg taken from a chef, attempt to convert them to if-then rules, and apply fuzzification. Fuzzy logic, fuzzy sets, and fuzzy inference systems are covered in more detail, along with relevant literature in the following section.

However, given a particular data set, domain knowledge may be yet insufficient. It provides a basis for forming the rules, but perhaps the exact nature of the fuzziness or ambiguity of terms remains obfuscated. In order to improve the performance of such a hindered system, it is possible to attempt to optimize the system via application of a genetic algorithm (GA).

A GA refers to a family of computational models that have been inspired by evolution as mentioned by Darrell Whitley [7]. A GA is a metaheuristic inspired by the

[1] https://www.youtube.com/watch?v=J_Q5X0nTmrA.

process of natural selection that is part of larger category called Evolutionary Algorithms (EA) [8]. They rely on bio-inspired operators such as mutation, crossover and selection as they have been introduced by John Holland in 1960 based on Darwin's theory of evolution. GAs are covered in more detail with their relevant literature in the following section.

Additionally, there is an entire other body of work that seeks to represent domain knowledge. Formal Knowledge Representation (FKR) is a broad field that contains a variety of methods for representing knowledge in a machine-readable way. Of particular note is the term, Knowledge Graph (KG). A KG is an ambiguous term that simply means that some knowledge is represented in a graph-centric way. In the Semantic Web, for example, these KGs are called ontologies and support crisp knowledge and methods for inferring new knowledge from it.

Recently, there have been investigations into incorporating fuzzy knowledge into the logic-based formalisms that underlay the Semantic Web. These concepts are further discussed in the following section.

In this chapter, we will present an example FIS, discuss its strengths and shortcomings, and demonstrate how an FIS's performance may be improved with the use of Genetic Algorithms (GA). Additionally, we will explore potential avenues for further enhancing the FIS by incorporating additional methodologies from Artificial Intelligence, in particular Formal Knowledge Representation and the Semantic Web.

The rest of the paper is organized as follows. Section 5.2 will provide background information on each of the methods as well as representative corpora of relevant literature for topics addressed in this chapter. Section 5.3 presents the optimization of a fuzzy inference system with the use of a genetic algorithm. Section 5.4 showcases two different approaches that Formal Knowledge Representation can be integrated with an optimized fuzzy inference system with genetic algorithms. Finally, Sect. 5.5 concludes this book chapter.

5.2 Background

In this section a brief background will be given about FIS, GAs and FKR. A brief literature review is also presented to show how different methodologies have been used to evaluate and improve the FIS. As was mentioned earlier FIS is a methodology widely used to solve mainly problems with vague data producing consistent results.

5.2.1 Fuzzy Inference System

5.2.1.1 Technical Background

As discussed previously, the Mamdani type FIS is a Machine Learning technique used for a multitude of decision-making problems [9]. The algorithm operates from the premise of fuzzy logic, created by Lotfi Zadeh [10] to provide the capability of handling the uncertainty, or noise, of data. The FIS is comprised of 7 primary steps:

1. Creating a set of feature space membership functions.
2. Creating a set of fuzzy rules.
3. Fuzzification of input features.
4. Connecting the now fuzzified input features to the fuzzy rules for determining individual rule strengths.
5. Calculating the consequent of each rule from the strength of the rule against the output space membership functions.
6. Aggregating the set of consequents to form the final output distribution.
7. Defuzzification of the output distribution to obtain a crisp value.

Of these 7 steps, the final step is only necessary if the intention of the FIS is for classification in which you need to compare a crisp output to a data point label.

The first two steps are the design portion of the FIS, similar to designing the architecture to a neural network for a specific problem, these two steps are also problem specific. Beginning with step 1, the membership functions are curves on the input feature space that represent linguistic aspects of the feature. For example, an input feature of temperature can possibly have three membership functions of *cold*, *mild*, and *hot*. These membership functions that correspond to the input features and FIS output are divided into three parts which are the label, distribution form, and distribution parameters. The label depicts what the membership function represents. The distribution form is the physical distribution of the curve. While many distributions are acceptable, common ones are Gaussian, Triangular, and Trapezoidal [9]. Then, the distribution parameters are simply the values that create the curve with respect to the input feature space. So, Gaussian would require both a μ (mean) and σ^2 (variance) parameter to describe the location and width of the curve on the input feature space, respectively.

Following in step 2, we create each fuzzy rule in a linguistic manner with an **IF-THEN** structure [9]. The use of the logical **AND/OR/NOT** for combining multiple input features into a single linguistic rule allows for explainability of the overall system as for why it produced a particular output. We call the **IF** portion the antecedent and the **THEN** portion the consequent. Using the temperature example, we can now have a rule using the corresponding membership functions *cold*, *snowing*, and *winter* as follows:

IF (*temperature* is *cold*) **AND** (*weather* is *snowing*) **THEN** (*season* is *winter*)

With the linguistic fuzzy rules and membership functions created, we perform the fuzzification with the membership functions (step 3) by mapping the input features to the membership functions. Using the temperature example, an input of 40° Fahrenheit could be 65% *cold* and 35% *mild*. We follow this process by connecting the fuzzified inputs to the fuzzy rules for determining individual rule strengths (step 4). As the rules are comprised of **AND/OR/NOT** conjunctions, each have different calculations [9]. Fuzzy **AND** has two popular formulas, Zadeh's method and the Product. Where Zadeh's calculation is the minimum of *n* number of functions as $\min(m_A(x),$

$m_B(x)$), where m_A and m_B are membership function A and membership function B, respectively. The Product version is simply $m_A(x) * m_B(x)$ for *n* number of functions. For the fuzzy **OR**, Zadeh's method is the maximum of *n* number of functions as $\max(m_A(x), m_B(x))$. Once again, the Product is the second popular method for *n* number of functions where we have $(m_A(x) + m_B(x) - m_A(x) * m_B(x))$. Lastly, the fuzzy **NOT** is generally the 1 – m(x). The results of these calculations gives us the rule strengths for each rule.

Calculating the consequents of the fuzzy rule set (step 5) is done by removing the excess membership with respect to the strength of each fuzzy rule. Examining Fig. 5.1, we can see fuzzy rules interacting with the crisp inputs and removing the excess membership. The first two columns are the fuzzified input features, or antecedents, and the third column is the rule consequents. Now, we aggregate the consequents of each fuzzy rule to form an output distribution (step 6), as shown at the bottom of the third column in Fig. 5.1. With this output distribution, the final optional step 7 is to defuzzify the output distribution to obtain a crisp value. There are many ways of defuzzifying the output, though the two most popular version are to either take the centroid or bisector of the output distribution [9].

5.2.1.2 Related Literature

Rao and Zakaria utilized the Mamdani type FIS to improve the behavior control of a semi-autonomous powered wheelchair [11]. Specifically, the goal was to improve the smoothness during the switching from Follow-the-Leader and Emergency stop behaviors. Features used as input to the FIS are Distance and Angle as obtained from the laser sensor attached to the wheelchair. From these 2 input features, 9 rules are constructed, and 5 Triangular membership functions represented the output. Their results showed significant improvements to velocity and response time in switching between the Follow-the-Leader and Emergency Stop behaviors.

A similar work that utilized Mamdani FIS to improve the control of an autonomous powered wheelchair by creating 6 input features and constructing 27 rules. Followed by 3 different membership functions that represent the output that later played a crucial role in a human machine interaction scheme able to improve the movement and control of robotic arms for the daily activities of users [12–15].

An article by Mohamed et al. showed how the FIS can be used as a decision assistance tool in ranking student academic posters [16]. As research has shown that poster presentations are efficient at showing learning outcomes from students, the authors created an FIS to overcome the subjective nature of poster evaluations. The authors created 256 rules and 9 trapezoidal membership functions from 4 input variables. The results show that the FIS tool for evaluating student posters offered different results over the traditional subjective method but offered more consistent reliability in the rankings.

Pourjavad and Mayorga constructed a 2 phase Mamdani type FIS to measure the efficiency of manufacturing systems [17]. In order to assist both investors and companies in the same field, a quantitative method to measure efficiency is welcome.

To achieve this, the authors derived 11 input features, 9 of which are used in phase 1 to produce outputs used in phase 2, totaling 3 individual FIS in phase 1. The second phase consists of 1 FIS which takes in as input the outputs from the 3 phase 1 FIS and the remaining 2 input features. A total of 1216 rules were created for this 2 phase Mamdani FIS where 1024 are associated with the single phase 2 FIS. The numerical results showed a reliable way to consistently measure the efficiency of manufacturing plants, in this case, 5 manufacturing plants in Iran owned by the same company.

An article by Jain and Raheja used a FIS as a tool to assist in the diagnosis of Diabetes [18]. The authors created the FIS with the assistance of medical experts which derived 6 rules utilizing the fuzzy 'or' disjunction and a total of 23 triangular membership functions. The output of the Mamdani FIS was categorized via a threshold and converted into a natural language sentence for comparison against the medical expert's binary decision of diabetic or non-diabetic. The results show the highest overall accuracy (87.2%) as compared to 6 similar strategies of diagnosing Diabetes.

Danisman et al. applied the FIS technique to classifying genders from images of faces [19]. In this binary classification problem, the FIS provides an explainable approach to the classification process as opposed to other Machine Learning techniques such as the Artificial Neural Network. The authors extracted mustache hair length, head hair length, and vision-sensor from the image dataset. From these three features 6 rules and 8 membership functions were created. The explainability portion is derived from the rule-base. For example, **IF** the mustache hair is *long* **AND** the head hair is *short* **THEN** the gender is *male*. The results from this study show improvements over similar methodologies performed on the same facial image dataset.

Thakur, Raw, and Sharma used an FIS to assist in Thalassemia disease diagnosis [20]. An FIS is beneficial to medical experts in that the explainability provided with the result produced from the FIS assists the doctors in their final diagnosis. The authors utilized Hemoglobin, mean Corpuscular volume, and mean Corpuscular hemoglobin as the input variables to the FIS and the output being minor, intermediate, or major Thalassemia. This FIS is comprised of 15 linguistic rules and 12 membership functions. Their results showed an accuracy rating of approximately 80% in which 12 tests matched directly to the doctor's diagnosis.

5.2.2 *Genetic Algorithms*

5.2.2.1 Technical Background

A genetic algorithm is a search heuristic and reflects the process of natural selection of the fittest individuals. These individuals are selected for reproduction in order to produce offspring of the next generation. As was mentioned in the previous section genetic algorithms belong to a larger class called evolutionary algorithms. They are commonly used to generate solutions by optimizing and using operators such as mutation and crossover and selection.

In the process of finding the fittest individuals we need to start by looking into the population and start by selecting them. The fittest will produce offspring that will inherit specific characteristics of the parent that will be added to the next generation. If the parents have better fitness, their offspring will be better than the parents and will have a better chance of surviving. This is an iterative process that keeps happening and at the end a generation of the fittest individuals will be found. A search problem considers a set of solutions and selects the set of best ones out of them.

There are five phases in a genetic algorithm:

- Defining a Fitness function
- Creating an Initial Population
- Selection of Parents
- Performing Crossover
- Performing Mutation.

Starting the process, a set of individuals which is called a population is initialized and everyone is a possible solution to the problem we want to solve. An individual is characterized by a set of parameters or variables or genes. For example, each human has genes that are joined together into a string to form a chromosome which is the solution. The population has a fixed size and as new generations are formed, the individuals with least fitness die, thus providing space for the new offspring.

The set of variables of an individual is represented using a string in terms of an alphabet for example where binary values are used and that encodes the genes in a chromosome. Now, to determine how fin an individual is we need a fitness function. It gives a fitness score to each individual and the probability that an individual will be selected for reproduction is based on that fitness score.

Next phase is the selection phase and its main goal is to select the fittest individuals from the population and let them pass their characteristics into the next generation. For the chromosome example, two pairs of individuals, the parents, are selected based on the fitness score and those with the highest fitness have higher chances to be selected.

The most significant phase of a genetic algorithm is called crossover. For each pair of parents that will mate, a crossover point is chosen at a random place from within the genes. The offspring are created by exchanging the genes of the parents among themselves until that crossover point is reached and then accordingly the new offspring will be added to the population.

In some of those offspring there a low probability that their genes will be mutated. This means that some of the bits in the bit string could be flipped (binary). Mutation is a phase that occurs simply to maintain diversity within the population and also to avoid convergence prematurely.

Finally, the algorithm terminates if the population does not produce offspring that are significantly different from the previous generation anymore. This means that the population has converged and has provided a set of solutions to our problem.

Common terminating conditions are [21]:

- A solution is found
- Fixed number of generations reached
- Allocated budget met
- Highest ranking fitness solution found
- Manual inspection
- Combinations of the above.

Genetic algorithms as they were just described have several applications such as robotics, path planning, scheduling, image processing and more. Some of the most representative works are presented here for the purposes of covering all perspectives.

5.2.2.2 Related Literature

Yang and Gong used genetic algorithms for stereo image processing [22]. They used an intensity-based approach to generate a disparity map that should be smooth and detailed. They removed mismatches from occlusion by increasing the accuracy of the disparity map. They formalized stereo matching as an optimization problem and by using a genetic algorithm they optimized the compatibility between corresponding points and continuity of the disparity map. Initially, they populated the 3D disparity with dissimilarity values and defined a fitness function based on the Markov Random Field. The genetic algorithm extracts fittest population from the disparity map followed by color image segmentation and graft crossover. Their experiments proved to be more effective than existing methods.

The authors of this next paper used genetic algorithms to search large space problems in gaming. A large structure of possible traits for Quake II monsters (what is the probability of monsters running away when they are low on health, what is the probability of monsters running away when they are few in numbers, etc.) can be created and accordingly use a genetic algorithm to find the optimum combination to beat the player. The player would have to go through a level of the game and at the end, the program would pick the monsters that fared the best against the player and use those in the next generation. It is a slow procedure but after a lot of time playing reasonable traits would be evolved and moved onto the next generation [23].

Takai and Yasuda used genetic algorithms for robot path planning in unknown environment in real-time [24]. They used an algorithm to detect and find obstacles in the course of the robot and then modulated a generation of short and safe paths to avoid them. By using a genetic algorithm, they created a path as a set of orientation vectors with equal distance. By doing this they managed to composite the final path as polygonal lines and to minimize the length they changed the orientation to 5 values from $-45°$ to $45°$. They used distance parameters between goals for fitness function and the combination of roulette and elite selection. An attempt was made for the system to perform in real time.

The authors of this paper used a genetic algorithm to assign task priorities and offsets to guarantee real time constraints which is a very difficult problem in real time systems [25]. They used genetic algorithm because of its ability to generate an outcome that satisfies a subset of timing constraints in the fittest way. The mechanism of natural selection, gradually improved individual timing constraints assignment in the population.

The authors of this paper used a genetic algorithm to integrate fuzzy logic for non-linear hysteretic control devices and to find the optimal design strategy. They integrated a fuzzy controller in order to find the interactive relationships between damper forces and input voltages for MR dampers. A set of optimal solutions is created and as a result the decreasing number of dampers for the dynamic response contributed [26].

5.2.3 Formal Knowledge Representation

A fuzzy inference system is a sort of expert system. As such, there are clear parallels in the Semantic Web. Since its inception in 2001, the Semantic Web has generated many well developed and understood techniques and methodologies for modelling and leveraging complex knowledge. However, the Semantic Web, in general, deals with consistent and crisp knowledge. Unfortunately, in many cases, real world data is imperfect, noisy, and inconsistent, even while the magnitude and accessibility of available data has exploded exponentially. As such, methods for dealing with such imperfect, noisy data, are necessary. Of course, fuzzification is one such strategy. Over the years, there have been attempts at marrying concepts from both fields. Briefly, we introduce two widely used tools for knowledge representation in the Semantic Web and their fuzzy extensions or analogs.

As previously mentioned, the Semantic Web is an extension to the world wide web via standards published by the W3C. Perhaps the two most visible of these standards (and most pertinent to this chapter) are the Resource Description Framework (RDF) and the Web Ontology Language (OWL). RDF, at its simplest, is simply a method for specifying and describing information. The Web Ontology Language (OWL) is the W3C Standard for authoring ontologies and is built on top of RDF. In this chapter we specifically constrain ourselves to OWL2 DL. OWL2 DL is a maximally expressive sublanguage of OWL2 that both complete and decidable and draws its name from its close correspondence to description logic. OWL2 is the 2009 specification of OWL. Additionally, OWL may be extended with RuleML to obtain the Semantic Web Rule Language (SWRL) which is a W3C Submission. We mention these, as they both have extensions to them that incorporate fuzziness of data. For more information on the non-fuzzy versions please see [27, 28].

Since its inception, the Semantic Web has frequently struggled with expressing certain concepts: namely temporality (modality), uncertainty, provenance, and fuzziness. (Here we distinguish between uncertainty, where there is some doubt whether something may or may not be true; and fuzziness, where something may be partly true.) There have been many attempts to address these shortcomings across several domains.

In some cases, there is an attempt to simply incorporate the concepts as parts of the data model. Provenance, for example, through either use of the ontology PROV-O [29] or specific design patterns [30], attempt to model provenance as part of the data model. Fuzziness and uncertainty could be modelled in a similar fashion. However, these approaches are reifications of these aspects. They incorporate fuzziness, for example, in a crisp way. Fuzziness becomes inherent to the data model, rather than inherent to whatever it is attached to. Especially given the open world assumption, this fuzziness information may not even be specified! Thus, in parallel, some other approaches have attempted to incorporate fuzziness at a more fundamental level. For a more in-depth view of fuzzy description logics and similar, please see [31, 32] for treatment in exceptional detail.

In [33], Straccia presents fuzzy RDF where RDF triples are annotated with a real number in [0, 1] which represents a "degree of truth." In other words, we may view this as fuzzy membership. It is a very lightweight approach to indicating fuzziness. As they are annotations, the data may be treated as crisp or fuzzy, depending on the use-case. In the example below, fuzzy RDF is used to annotate the fuzziness of a person's age.

```
<externs:age>
  <Dist type="disjunctive">
    <Val Deg=0.8>28</Val>
    <Val Deg=1.0>29</Val>
    <Val Deg=0.9>30</Val>
  </Dist>
</externs:age>
```

In [34], a fuzzy extension to OWL DL is presented. While it is a preliminary work, proof of its viability was completed. In recent years, there have been implementations of fuzzy DL reasoners [27] and applications that leverage these new technologies. Indeed, [28] presents a large ontology fully specified in fuzzy OWL. The work aims to support semi-automated decision support for helping developers build large-scale software projects based on solicited system specifications.

On a different track, [35] presents a fuzzy extension to SWRL, called f-SWRL. Essentially, allows the specification of fuzzy Horn rules. Indeed, this is a very natural way of attempting to incorporate fuzzy methodologies into the Semantic Web. An example of an f-SWRL rule is as follows.

$$\mathrm{Tall}(?p) * 0.7 \wedge \mathrm{Light}(?p) * 0.8 \rightarrow \mathrm{Thin}(?p)$$

Weight of an atom is delineated by the trailing fraction, this is analogous to fuzzy membership. Examination of the semantics behind the syntax is available in [35]. As can be seen from the examples, syntactically, they are intuitive. The novelty comes from modifying the overall framework to handle these seemingly innocuous annotations.

In the same way as [33], others have attempted other strategies that incorporate annotations to specify this additional data. In fact, one such strategy, outlined in [36] builds on [33] to provide a generalized framework for specifying temporality, uncertainty, provenance, and fuzziness in annotations on RDF triples. The paper goes on to specify an extension to SPARQL (called AnQL) for the framework (aRDF) which allows for advanced querying incorporating these dimensions.

Furthermore, in [37] builds further upon [zimmerman 1] and constructs a so-called contextualized knowledge graph. The term knowledge graph was recently made popular by Google. Consensus on an exact definition for a knowledge graph is fuzzy, however, in general, it suffices to say that a knowledge graph is any data organized in a graph-centric manner [38]. Nguyen extends this concept heavily, starting with the heavy weight semantics inherent to OWL, it extends the subsequent graph to incorporate the annotations for temporality and fuzziness as described in [36]. Pan et al. [37] also provides a theoretical basis for the completeness and decidability of such a contextualized knowledge graph.

5.3 Numerical Experiment

With the knowledge to create a FIS laid out in Sect. 5.2.1.1, we prepare a numerical experiment to demonstrate the capabilities of such a system built purely from a heuristic knowledge base, then improved via GA.

5.3.1 Data Set Description and Preprocessing

The data set used in this experiment is the Craft Beer dataset [39] retrievable on Kaggle. We take the Beer Style column (Blonde Ale, Pale ale, etc.) as our target label. There is a total of 97 classes in the dataset, many with insufficient data to be differentiable. So, we limited the dataset to just 3 classes which are American Blonde Ale (ABA), American Pale Ale (APA), and American India Pale Ale (IPA). For the features, we utilized the Alcohol by Volume (ABV) and International Bitterness Units (IBU) features for the classification. The data was then randomly shuffled and split into 80/20 training and test sets, respectively. This gave us 411 training and 104 test data points.

5.3.2 FIS Construction

For the FIS construction, we will utilize MATLAB's Fuzzy Logic Designer toolbox. Here we decided on membership functions for both the ABV, IBU, and beer type FIS output as well as the fuzzy rule set. We chose to separate the ABV into 3 forms of Low,

Table 5.1 The ABV, IBU, and Beer membership functions labels, forms, and corresponding parameters

Feature	Label	Form	Parameters
ABV	Low	Gaussian	[0.0144 0.041]
	Moderate	Gaussian	[0.0144 0.090]
	High	Gaussian	[0.0144 0.065]
IBU	Low	Gaussian	[15 22.000]
	Moderate	Gaussian	[15 55.639]
	High	Gaussian	[15 81.859]
	Very-high	Gaussian	[15 108.216]
	Extreme	Gaussian	[15 129.824]
Beer	ABA	Gaussian	[0.1 0.15]
	APA	Gaussian	[0.1 0.5]
	IPA	Gaussian	[0.1 0.85]

Moderate, and High. Following, IBU was separated into 5 forms of Low, Moderate, High, Very-High, and Extreme. The output of Beer has one membership per target class, totaling to 3. For simplicity of this example, we chose the Gaussian form for all membership functions. The exact parameters that represented each membership function can be seen in Table 5.1.

With the membership functions created, we then heuristically created a fuzzy rule set utilizing both the input features and membership functions. There were 7 rules created to represent the differences between the 3 target classes, listed as follows:

1. **IF** (*ABV* is *High*) **AND** (*IBU* is *Very-High*) **THEN** (*Beer* is *IPA*)
2. **IF** (*ABV* is *Moderate*) **AND** (*IBU* is *Very-High*) **THEN** (*Beer* is *IPA*)
3. **IF** (*ABV* is *Moderate*) **AND** (*IBU* is *Extreme*) **THEN** (*Beer* is *IPA*)
4. **IF** (*ABV* is *High*) **AND** (*IBU* is *Extreme*) **THEN** (*Beer* is *IPA*)
5. **IF** (*ABV* is *Moderate*) **AND** (*IBU* is *Low*) **THEN** (*Beer* is *ABA*)
6. **IF** (*ABV* is *Low*) **AND** (*IBU* is *Moderate*) **THEN** (*Beer* is *ABA*)
7. **IF** (*ABV* is *Moderate*) **AND** (*IBU* is *Moderate*) **THEN** (*Beer* is *APA*).

Now that steps 1 and 2 as listed from Sect. 5.2.1.1 are completed, we can focus on the remaining 5 steps. We continued by setting the FIS to use Zadeh's method of calculating the logical **AND**. Also, our FIS used the defuzzification bisector method for producing a crisp value for comparison against the data. Fortunately, the MATLAB Fuzzy Logic Designer toolbox handles the calculations of steps 3 through 7.

5.3.3 GA Construction

A FIS for predicting three different beers has now been heuristically constructed. Though, due to the heuristic construction of the fuzzy rules and membership functions, optimization can be performed to improve the results. Manually performing the changes though is not practical as there are 11 membership functions with 2 parameters each. This is where the GA can be of assistance. In this example, we will utilize the GA to update the parameters of the membership functions only. Though, it is possible to also apply the GA to the **AND/OR** conjunctions as well as the **NOT** of the fuzzy rule set as well. Once again, we utilize a MATLAB toolbox, in this case the Optimization toolbox with the solver being set to GA.

There is a total of 22 membership function parameters to be updated, so a chromosome of size 22 is created. For positions 1-6, the lower and upper bounds are between 0.025 and 0.105 (ABV parameters). Following, positions 7–16 have lower and upper bounds of 1 and 150 (IBU parameters). Then, positions 17–22 have lower and upper bounds of 0.0001 and 1 (Beer parameters). We set the initial population size to 50 and have the following options for how the GA performs selection, mutation, etc.:

- Creation Function = *Uniform*: The initial population is created by randomly sampling from a uniform distribution.
- Scaling Function = *Rank*: Each chromosome in the population is scaled with respect to list sorted by fitness. Removing the clustering of raw scores and relying on an integer list instead.
- Selection Function = *Stochastic Uniform*: Selects a subset of chromosomes from the population by stepping through the rank scaled population and randomly selecting based on a uniform probability.
- Mutation Function = *Adaptive Feasible*: Mutation is applied to positions of each surviving chromosome which are feasible with respect to the constraints placed on the chromosome.
- Crossover Function = *Scattered*: Randomly creates a vector of 1's and 0's the same size as the chromosome. The 1's take the position from the first parent and the 0's take the position from the second parent.
- Function to Optimize = Sum of Squared Error (SSE).

5.3.4 Results

We present the results of both the heuristically created FIS and the GA optimized version separately. We will compare the improvements by examining the precision and recall of each target class, the prediction surface before and after GA optimization, and the SSE as this is the function we are optimizing. Beginning with the heuristically created FIS, we can see from Heuristic FIS column of Table 5.2 that the precision and recall for the ABA (0.91 and 0.77) and IPA (0.78 and 0.83) predictions is performing quite well. Though, the precision and recall for APA (0.59 and 0.55) could use improvement. We also note the SSE of the heuristic FIS is 30. From Fig. 5.2, we can see an uneven surface accompanied with unnecessary valleys. We, the authors, are not craft beer experts and thus do not know the exact ranges of ABV or IBU in which say an IPA constitutes. Though, based on the precision and recall in Table 5.1, it was a good attempt. These discrepancies in the membership functions are what cause the abnormalities found in the surface of the heuristic FIS.

Table 5.2 Precision, recall, and SSE of the heuristic FIS compared against the GA optimized FIS

	Heuristic FIS			GA FIS		
	ABA	APA	IPA	ABA	APA	IPA
Precision	0.91	0.59	0.78	0.91	0.58	0.86
Recall	0.77	0.55	0.83	0.77	0.71	0.78
SSE	30			25		

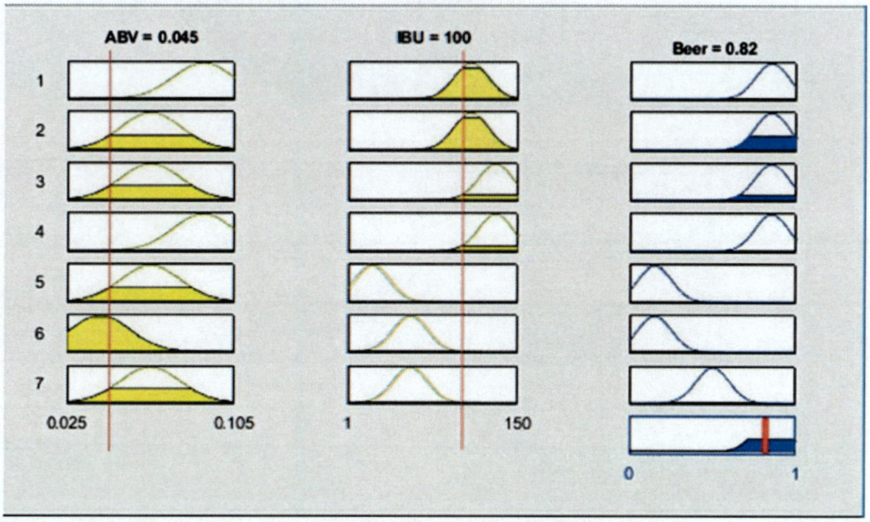

Fig. 5.2 Example of two input FIS with seven rules

Now, examining the GA optimized FIS results, the precision and recall from the GA FIS column in Table 5.1 has improved performances. Specifically, the APA recall has improved from by +0.16. Furthermore, the SSE decreased from 30 to 25. We now look at Fig. 5.3, the GA optimized FIS prediction surface. We can see an objectively smoother surface area where the valleys from Fig. 5.2 have disappeared. This is a small improvement, though the time and effort saved by using GA instead of manual iterations of membership function update and testing is priceless. The GA training time was done in 86 iterations with a literal time of approximately 2 min. Though, we note that this example problem is indeed a "toy problem" where we expect fast training times. Given a significantly larger dataset, GA training time will generally increase drastically (Fig. 5.4).

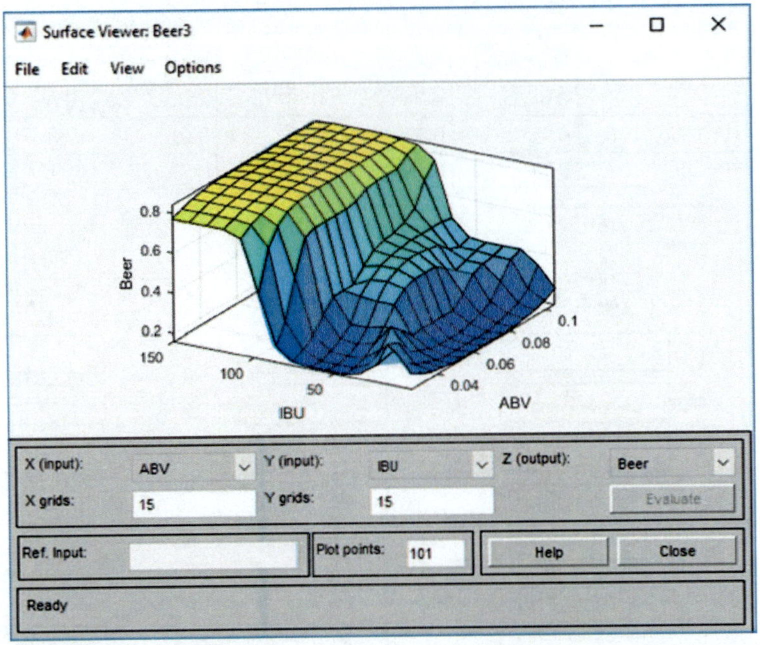

Fig. 5.3 Prediction surface of the heuristic FIS

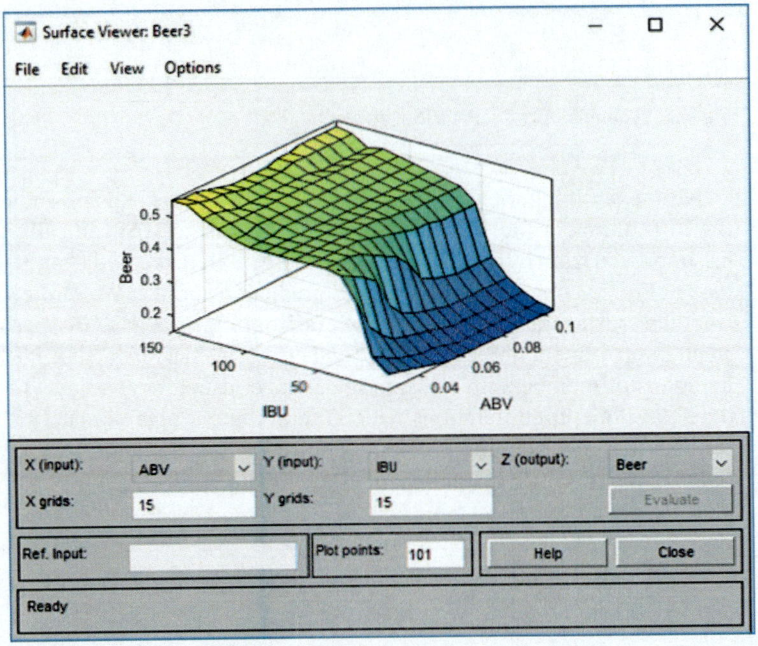

Fig. 5.4 Prediction surface of the GA optimized FIS

5.4 Advancing the Art

Given the three technologies described in this chapter, we will use this section to describe how they may be intersected to advance the state of the art. We present four hypothetical scenarios and indicate which steps are open research questions.

First, we propose an algorithm for constructing an "optimized knowledge graph."

1. Construct a rule base system with Fuzzy Logic.
2. Optimize the FIS via GA.
3. Convert the FIS rules to f-SWRL rules.
4. Reify f-SWRL rule to create a KG in OWL.

In this case, Steps 3 and 4 are the open research question. Second, we may start from a KG to construct an optimized FIS.

1. Find or construct a KG.
2. Mine rules from the KG.
3. Convert the mined rules into f-SWRL rules.
4. Convert the rule base into an FIS.
5. Optimize the FIS via GA.

The third scenario is similar—we instead start with a Fuzzy Ontology and initially attempt to mine f-SWRL rules. The pipeline would continue from Step 3, as above. For all three of the scenarios so far, it is also an open research question that all information contained in the KG can be represented via rules, as f-SWRL is an extension of SWRL and SWRL is a subset of OWL. Given that many ontologies contain axioms with existential quantifiers in the consequent, SWRL may not be wholly sufficient, but this will need further investigation.

Finally, as an FIS excels in assisting a user make an informed decision in the face uncertainty or fuzziness, we imagine a clear intersection with FIS, the above scenarios, and the nascent field of Stream Reasoning. Stream Reasoning is the study of applying inference techniques to highly dynamic data. That is, data that might change on a second to second (or faster) basis. In particular, this data may be triples about information collected from sensors. This sensor data will have uncertainty and fuzziness. A pertinent and open avenue of research would investigate how the use of an FIS (handcrafted or optimized) might complement the technologies available to the stream reasoning community.

5.5 Conclusions

In this book chapter we analyzed different methodologies on how we can optimize a broadly recognized and used fuzzy inference system. First, crucial components of how data science has affected the scientific community were presented. The application of data science and how different methodologies have played crucial role over the years for the development of the area were demonstrated. Then a background was given on fuzzy inference systems, genetic algorithms and knowledge graphs, in both technical and literature perspectives.

Accordingly, a dataset was used and rules were created for a FIS. The output of the was optimized with the use of a genetic algorithm and the results of this procedure were presented. The results showed an improvement on the recall and precision as well as a smoother convergence surface. Even though we are not experts on beer crafting the results after the use of the GA are optimized. In other words, our attempt on improving a FIS with the use of a GA worked.

Then, several different routes were proposed on how with the use of knowledge graphs we can further improve the outputs of our optimized system. The methodologies that were proposed are future work that targets the integration of three different systems into one with main goal the optimization of an FIS.

References

1. Saltz, J.S.: The need for new processes, methodologies and tools to support big data teams and improve big data project effectiveness. In: 2015 IEEE International Conference on Big Data (Big Data), pp. 2066–2071. IEEE (2015)
2. Bhardwaj, A., Bhattacherjee, S., Chavan, A., Deshpande, A., Elmore, A.J., Madden, S., Parameswaran, A.G.: Datahub: Collaborative Data Science & Dataset Version Management at Scale (2014). arXiv preprint arXiv:1409.0798
3. Rollins, J.: Why we need a methodology for data science (2015). https://www.ibmbigdatahub.com/blog/why-we-need-methodology-data-science. Accessed 06 Mar 2019
4. Papadakis Ktistakis, I.: An autonomous intelligent robotic wheelchair to assist people in need: standing-up, turning-around and sitting-down. Doctoral dissertation, Wright State University (2018)
5. Lee, C.C.: Fuzzy logic in control systems: fuzzy logic controller. II. IEEE Trans. Syst. Man Cybern. **20**(2), 419–435 (1990)
6. Abraham, A.: Adaptation of fuzzy inference system using neural learning. In: Fuzzy Systems Engineering, pp. 53–83. Springer, Berlin, Heidelberg (2005)
7. Davis, L.: Handbook of Genetic Algorithms (1991)
8. Whitley, D.: A genetic algorithm tutorial. Stat. Comput. **4**(2), 65–85 (1994)
9. Ross, T.J.: Fuzzy Logic with Engineering Applications. Wiley (2005)
10. Zadeh, L.A.: Outline of a new approach to the analysis of complex systems and decision processes. IEEE Trans. Syst. Man Cybern. **1**, 28–44 (1973)
11. Rao, J.B., Zakaria, A.: Improvement of the switching of behaviours using a fuzzy inference system for powered wheelchair controllers. In: Engineering Applications for New Materials and Technologies, pp. 205–217. Springer, Cham (2018)
12. Bourbakis, N., Ktistakis, I.P., Tsoukalas, L., Alamaniotis, M.: An autonomous intelligent wheelchair for assisting people at need in smart homes: a case study. In: 2015 6th International Conference on Information, Intelligence, Systems and Applications (IISA), pp. 1–7. IEEE (2015)
13. Ktistakis, I.P., Bourbakis, N.G.: Assistive intelligent robotic wheelchairs. IEEE Potentials **36**(1), 10–13 (2017)
14. Ktistakis, I.P., Bourbakis, N.: An SPN modeling of the H-IRW getting-up task. In: 2016 IEEE 28th International Conference on Tools with Artificial Intelligence (ICTAI), pp. 766–771. IEEE (2016)

15. Ktistakis, I.P., Bourbakis, N.: A multimodal human-machine interaction scheme for an intelligent robotic nurse. In: 2018 IEEE 30th International Conference on Tools with Artificial Intelligence (ICTAI), pp. 749–756. IEEE (2018)
16. Mohamed, S. R., Shohaimay, F., Ramli, N., Ismail, N., Samsudin, S.S.: Academic poster evaluation by Mamdani-type fuzzy inference system. In: Regional Conference on Science, Technology and Social Sciences (RCSTSS 2016), pp. 871–879. Springer, Singapore (2018)
17. Pourjavad, E., Mayorga, R.V.: A comparative study and measuring performance of manufacturing systems with Mamdani fuzzy inference system. J. Intell. Manuf. 1–13 (2017)
18. Jain, V., Raheja, S.: Improving the prediction rate of diabetes using fuzzy expert system. IJ Inf. Technol. Comput. Sci. **10**, 84–91 (2015)
19. Danisman, T., Bilasco, I.M., Martinet, J.: Boosting gender recognition performance with a fuzzy inference system. Expert Syst. Appl. **42**(5), 2772–2784 (2015)
20. Thakur, S., Raw, S.N., Sharma, R.: Design of a fuzzy model for thalassemia disease diagnosis: using Mamdani type fuzzy inference system (FIS). Int. J. Pharm. Pharm. Sci. **8**(4), 356–61 (2016)
21. Genetic Algorithm. https://en.wikipedia.org/wiki/Genetic_algorithm. Accessed 24 Mar 2019
22. Gong, M., Yang, Y.H.: Multi-resolution stereo matching using genetic algorithm. In: Proceedings IEEE Workshop on Stereo and Multi-Baseline Vision (SMBV 2001), pp. 21–29. IEEE (2001)
23. Brown, C., Barnum, P., Costello, D., Ferguson, G., Hu, B., Van Wie, M.: Quake ii as a robotic and multi-agent platform. Robot. Vis. Tech. Rep. [Digital Repository] (2004). Available at HTTP. http://hdl.handle.net/1802/1042.
24. Yasuda, G.I., Takai, H.: Sensor-based path planning and intelligent steering control of nonholonomic mobile robots. In: IECON'01 27th Annual Conference of the IEEE Industrial Electronics Society, vol. 1, pp. 317–322 (Cat. No. 37243). IEEE (2001)
25. Sandstrom, K., Norstrom, C.: Managing complex temporal requirements in real-time control systems. In: Proceedings Ninth Annual IEEE International Conference and Workshop on the Engineering of Computer-Based Systems, pp. 103–109. IEEE (2002)
26. Uz, M.E., Hadi, M.N.: Optimal design of semi active control for adjacent buildings connected by MR damper based on integrated fuzzy logic and multi-objective genetic algorithm. Eng. Struct. **69**, 135–148 (2014)
27. Bobillo, F., Straccia, U.: The fuzzy ontology reasoner fuzzyDL. Knowl.-Based Syst. **95**, 12–34 (2016)
28. Di Noia, T., Mongiello, M., Nocera, F., Straccia, U.: A fuzzy ontology-based approach for tool-supported decision making in architectural design. Knowl. Inf. Syst. 1–30 (2018)
29. Groth, W3C.: PROV-O: The PROV Ontology. https://www.w3.org/TR/prov-o/. Accessed 6 Apr 2019
30. Shimizu, C., Hitzler, P., Paul, C.: Ontology design patterns for Winston's taxonomy of part-whole-relationships. Proceedings WOP (2018).
31. Straccia, U.: Fuzzy semantic web languages and beyond. In: International Conference on Industrial, Engineering and Other Applications of Applied Intelligent Systems, pp. 3–8. Springer, Cham (2017)
32. Straccia, U.: An Introduction to Fuzzy & Annotated Semantic Web Languages (2018). arXiv preprint arXiv:1811.05724
33. Straccia, U.: A minimal deductive system for general fuzzy RDF. In: International Conference on Web Reasoning and Rule Systems, pp. 166–181. Springer, Berlin, Heidelberg (2009)
34. Straccia, U.: Towards a fuzzy description logic for the semantic web (preliminary report). In: European Semantic Web Conference, pp. 167–181. Springer, Berlin, Heidelberg (2005)
35. Pan, J.Z., Stamou, G., Tzouvaras, V., Horrocks, I.: f-SWRL: a fuzzy extension of SWRL. In: International Conference on Artificial Neural Networks, pp. 829–834. Springer, Berlin, Heidelberg (2005)
36. Lopes, N., Polleres, A., Straccia, U., Zimmermann, A.: AnQL: SPARQLing up annotated RDFS. In: International Semantic Web Conference, pp. 518–533. Springer, Berlin, Heidelberg (2010)

37. Nguyen, V.T.K.: Semantic Web Foundations for Representing, Reasoning, and Traversing Contextualized Knowledge Graphs (2017)
38. Bonatti, P.A., Decker, S., Polleres, A., Presutti, V.: Knowledge Graphs: New Directions for Knowledge Representation on the Semantic Web (Dagstuhl Seminar 18371). Schloss Dagstuhl-Leibniz-Zentrum fuer Informatik (2019)
39. Hould, J.N.: Craft Beers Dataset, Version 1. https://www.kaggle.com/nickhould/craft-cans. Accessed 10 Mar 2019 (2017)

Iosif Papadakis Ktistakis earned his Ph.D. degree in Computer Science and Engineering from Wright State University, USA. He is a member of the Center of Assistive Research Technologies and an IEEE Member. He also holds an integrated B.S. and M.S. in Mechanical Engineering from the Technical University of Crete, Greece. He is currently a Senior Mechatronics Design Engineer at ASML in Connecticut. His research interests lie in the intersection of Robotics, Assistive Technologies and Intelligent Systems.

Garrett Goodman is currently pursuing his Ph.D. at Wright State University. He is a member of the Center of Assistive Research Technologies. He earned his B.S. and M.S. degrees in Computer Science at Wright State University. His work is focused on incorporating machine learning in health care to improve the lives and the wellbeing of people in need.

Cogan Shimizu is currently pursuing his Ph.D. at Wright State University. He is a member of the Data Semantics Lab and is a DAGSI Fellow (Dayton Area Graduate Studies Institute). He earned his B.S. and M.S. in Computer Science at Wright State University. His work is focused on improving the tools and methodologies for automatic knowledge graph construction.

Chapter 6
The Dark Side of Rationality. Does Universal Moral Grammar Exist?

Nelson Mauro Maldonato, Benedetta Muzii, Grazia Isabella Continisio, and Anna Esposito

Abstract Over a century ago, psychoanalysis created an unprecedented challenge: to show that the effects of the unconscious are more powerful than those of consciousness. In an inverted scheme at present time, neurosciences challenge psychoanalysis with experimental and clinical models that are clarifying crucial aspects of the human mind. Freud himself loved to say that psychological facts do not fluctuate in the air and that perhaps one day, biologists and psychoanalysts would give a common explanation for psychic processes. Today, the rapid development of neuroimaging methods has ushered in a new season of research. Crucial questions are becoming more apparent. For instance, how can the brain generate conscious states? Does consciousness only involve limited area of the brain? These are insistent questions in a time where the tendency of neuroscience to naturalize our relationship life is ever more urgent. Consequently, these questions are also pressing: Does morality originate in the brain? Can we still say "being free" or freedom? Why does morality even exist? Lastly, is there a biologically founded universal morality? This paper will try to demonstrate how neurophysiology itself shows the implausibility of a universal morality.

N. M. Maldonato (✉)
Department of Psychology, Università degli Studi della Campania "Luigi Vanvitelli", and IIASS, Caserta, Italy
e-mail: nelsonmauro.maldonato@unina.it

B. Muzii
Department of Neuroscience and Reproductive and Odontostomatological Sciences, University of Naples Federico II, Naples, Italy
e-mail: benedetta.muzii@gmail.com

G. I. Continisio
Continuing Medical Education Unit, School of Medicine, AOU University of Naples Federico II, Naples, Italy
e-mail: continis@unina.it

A. Esposito
Department of Psychology, University of Campania "Luigi Vanvitelli", Naples, Italy
e-mail: iiass.annaesp@tin.it

G. Phillips-Wren et al. (eds.), *Advances in Data Science: Methodologies and Applications*, Intelligent Systems Reference Library 189,
https://doi.org/10.1007/978-3-030-51870-7_6

6.1 Moral Decisions and Universal Grammars

In the scientific and philosophical field, it is the prevailing opinion that morality fulfils functions necessary to our social life, allowing us to negotiate and modify individual values (and value systems) for the construction of norms and prescriptions. Human value systems are strongly influenced by emotions and feelings such as pleasure, pain, anger, disgust, fear or compassion, which strongly affect human interactions and have enormous social and legal consequences. In a juridical system, the same norms—which are then the putting into form of emotions—can serve to consider certain conducts illegitimate assuming them in the judgment of imputability in the face of criminal acts [1].

That said, what is their function in moral decisions? For "right" behaviour, it could be answered. After all, every society is full of people to whom moral prescriptions require "rightly" to act—and perhaps this even represents an argument in favour of morality as a universal institution [2]. This, however, is a moral description of the function of morality, not a response to its nature. Put in other terms, the question becomes: why are moral practices and institutions universally present? So far, theoretical research has largely privileged the idea that the basis of moral judgments is rationality [3]. in recent decades, however, experimental sciences have shown that at the origin of moral judgment there are not only significant emotional and affective components, but also rational constructions a posteriori [4, 5]. The revaluation of the social and cultural components in the formation of morality has led to redefining the very role of reasoning [6]. Alternative models to the rationalist description of morality have thus emerged. Some scholars, in particular, have argued that it has significant affinities with language. In the sense that we would be equipped with an innate sense of what is right and wrong, just as we are equipped with an innate language structure [7]. On the other hand, it is said, is it not true that a child learns to speak before knowing grammar? In short, at the base of our judgments there would be a sort of universal moral grammar, analogous to the universal Chomsky grammar. In support of the hypothesis according to which people elaborate moral judgments, even before the awareness of the relative emotional reactions, lesional clinical evidences, evolutionary data, researches of developmental psychology, neuropsychological tests [8] have been reported. It has been argued that moral judgments would originate from unconscious analysis of the effects of an action and that embarrassment, shame and guilt would come later. In short, we would have received as dowry an instinctive grammar of the action on what is right and wrong. This hypothesis would seem, however, to be confirmed by the clinical pictures of patients with lesions of the prefrontal cortex, which while maintaining intact the knowledge of moral rules manifest abnormal behaviours due to the inability to experience congruous emotions [9]. Such research delineates an extremely sophisticated system: that of a morality strongly intertwined with the neurobiological processes at the centre of which there would be emotions, although in a non-dominant position.

The impact of these researches on general issues is impressive. Studies on intraspecific and interspecific animal social behaviour have revealed the existence of behaviours aimed at the exclusive individual interest, to which others would correspond, of altruistic sign, which extend the benefits to the entire social group, even when these involve high costs for the cooperating individual [10]. Under the magnifying glass the behaviour of the fish has been placed, among others: intraspecific cleaning, cooperative collection of food, exchange of eggs between hermaphrodites, aid to the nest by non-fertile individuals, the alarm through pheromones, aggressive behaviour; the behaviour of the birds: the cooperative hunting, the call in the presence of food, the sharing of food, the alarm patterns, the aggressiveness; mammalian behaviour: mutual cleansing, warning signs, coalitions for mutual defence, alloparental care [11]; finally the behaviour and the complex organization of eusocial insects.

These are evidences in favour of the existence of an innate tendency towards altruism and mutual benefit. Of course, for a Kantian scholar—according to which a behaviour is right only if it is in accordance with the moral law—are evidences without normative force [12]. Nonetheless, if innate moral skills were proven, one could look at the themes of solidarity and violence, reciprocity and intolerance with different eyes, without being overly conditioned by religious, philosophical or cultural world views. The ontogenetic and phylogenetic evolution shows how our life would be qualitatively poor and our survival at risk without our emotional repertoires and our decision-making devices [3]. Emotions and heuristics are tools of a natural logic that help us judge our behaviour in certain circumstances, revealing to us much more quickly than an argument what we can desire, fear and more. Today we know much better than before how we creatively use the memorized experience to face new situations, using both the experiences accumulated by the species and those accumulated by the individual [13–15]. It is precisely the sensory memory—in which personal, interpersonal and nature experience are inextricably connected—that constitute the material basis of our moral identity. However, one wonders: can emotions and heuristics constitute the exclusive bases of a universal morality? There are reasons to harbor some scepticism. As we will see later, we have other tools (thought, language, culture) that allow us to juggle between the constraints of necessity and the possibilities of freedom.

6.2 Aggressiveness and Moral Dilemmas

Years ago, the reactions of subjects exposed to strong emotional suggestions and problems evoking rational responses were studied by fMRI [16]. Facing with the dilemma of killing someone with his own hands, a real conflict exploded in the brains of the interviewees between evolutionarily newer areas (medial frontal cortices) and other evolutionarily older ones (the anterior cingulate cortex). In contrast, when people were called to reflect on a situation, which did not involve actions on another individual, in their brain were activated areas usually involved in the calculation (the dorsolateral surface of the frontal lobes). It is suggestive to believe that these different answers have adaptive reasons [17].

There were probably no impersonal dilemmas at the dawn of humanity. The absolute lack of intersubjective binding norms and, above all, attributable to impersonal values, pushed our ancestors towards behaviours not mediated (or at least tempered) by rational judgment. Living in small groups, violence and hostilities of the first men inevitably manifested themselves in personal form, through the use of rudimentary weapons that struck at close range [18]. These acts of violence, and the emotions associated with them, could explain why even imagining that it would hurt someone physically causes illness. It is no coincidence, moreover, that in the wars of every time that preceded the use of ranged weapons, to neutralize the natural resistances, strategists used—in addition to ethical, juridical, economic and religious reasons—to dehumanization of the enemy, to his downgrading to an inferior race, transforming the 'natural' interspecific aggression into expressions of pure destructiveness without pietas [19].

Discussions on the relationship between morals and decisions should pass for comparison with the work of Philippa Foot, an English philosopher active in the second half of the last century in England and the United States. Her renowned experiment called Dilemma of the railway cart has influenced (and, in many ways, still influences) a lot of moral philosophies of our time [20]. This is an experiment that has sparked lively discussions and raised valuable questions. How to draw conclusion from experiments with so many variables? How do we get out of personal/impersonal opposition? Do the means always justify the ends? Is it not true that every experiment considers every action as a story in itself and must always be analysed for its actual intentions? In reality, it is one thing to save as many people as possible, another is not to harm an unsuspecting and innocent person. [21]. Furthermore, if sacrificing a person to save five has its own rationality, pushing a man off of an overpass is a repulsive action, and it is natural to refuse to do so.

Be that as it may, an innate morality should contemplate some fundamental rules: not to kill, not to steal, not to deceive, to be honest, loyal, selfless. Perhaps even trusting the ability of men to learn moral rules. Several years ago, Marc Hauser [22] recommended animal behaviour for territoriality, hierarchies, reciprocity, group dynamics, food research and more. This could help us understand human social and cultural structures, but above all to draw indications for a shared system of moral rules. For example, social reciprocity—obviously much less complex in animals than in humans—is a formidable resource [23]. Indeed, it promotes virtuous behaviour and sanctions; those that are not virtuous encourages deferral of actions over time and so forth. These dynamics would lead us to believe in a sort of universal moral norm, that is, that this social reciprocity is part of a moral in some way innate.

6.3 Is This the Inevitable Violence?

Why do people judge and sanction personal moral violations in a short time and impersonal ones for a long time? With a series of experiments, [24] showed that moral and non-moral judgments activate different area of the brain. The former involves the medial fronto-orbital cortex and the superior temporal sulcus of the left hemisphere; the latter the amygdala, the lingual gyrus and the lateral orbital gyrus [25]. Faced with such experimental evidence, can we believe in the existence of neurobiology of morality? Although the extensive production of brain imaging studies would seem to affirm this, some questions remain open. For instance, are the area involved in moral judgments the primary seat of those judgments or only the corresponding territory of a process that takes place subsequently? Can emotions intensify (and possibly to what degree?) The value of individual moral judgments? [26]. Amidst it all, the mere existence of social emotions shows that we do not act on the basis of a utilitarian moral algebra to maximize benefits and minimize pain [27]. In the course of evolution, social emotions have enabled our ancestors to understand their own kind and to build cooperative societies, thus creating productive ground for the emergence of values (and value systems) and, consequently, of social and political institutions and shared cultural activities. Even if the meaning of the violations of social norms remains unclear, the compatibility between different values, the function of violence correlates that the pain, the sense of justice, authority, purity, being part of a community have deep evolutionary roots [28]. Not only in man. In the light of theoretical reflection and empirical evidence, schematizing it could be deduced that:

1. the instinct to avoid the pain of others—which generates horror at the idea of pushing a man from a bridge (as we have seen with regard to moral dilemmas)—is widely present also in some primates. For example, they refuse to operate a lever that would bring food and an electric shock to a fellow [29];
2. the sense of justice has relations with reciprocal altruism, on condition that the act is sustainable for those who perform it and those who receive it are willing to reciprocate [30];
3. respect for authority has to do with the hierarchies of domination and submission;
4. the sense of community that drives individuals to share and sacrifice themselves for an impersonal purpose, could derive from empathy and solidarity towards kinsmen and non-blood relations [31].

Now, if the moral roots are innate, if the distinction between right and wrong is inscribed in our brains, how can we prove that events like the Holocaust and racial genocides are disgusting and abominable for all?

If we are equipped only with a rudimentary morality, it will inevitably be the experience that guides us in accordance with the values of "goodness" or "wickedness". There are, however, some elements that more than others condition the moral behaviour of an individual with social inclinations and a spirit of self-preservation. First, altruistic behaviours have better social consequences than selfish ones. Secondly, the choice not to give priority to one's own interests if one intends to be taken seriously by others.

This interchangeability of perspectives is a moral value in itself superior to the particulare which instead guides the actions of a large number of human beings and has profound consequences on the various forms of social coexistence [32]. In fact, it pushes us to consider the arguments and actions of our adversaries, even the most disconcerting ones, as something coming from people with a moral like ours and not from individuals without morals. For example, in a political competition to consider our competitor as an adversary and not as an enemy driven by dishonest motivations or criminal designs, it could be a first step towards identifying a shared ethical terrain [33].

6.4 Future Directions

In light of these considerations, what is the space of a universal morality? Of course, emotions provide us with important information to learn and act [34]. But can it really be the foundation of a universal morality? And, if so, does universal norm make morality "right"? If this were so, totalitarianisms would be earthly paradises of universal morality. Human morality, on the other hand, is tremendously vulnerable to interpretations. It often leads to confusing moral rigor with purity. To be intransigent about ideas. To place us almost always on the side of reason. To often define virtuous behaviours that are not at all. All this has profound consequences also on our rationality [35]. This comes out, in fact, weakened, more likely similar to a dimming candle light compared to the blinding power of instincts, drives and emotions.

References

1. Peter-Hagene, L.C., Salerno, J.M., Phalen, H.: Jury decision making. Psychol. Sci. Law **338** (2019)
2. Singer, N., Kreuzpointner, L., Sommer, M., Wüst, S., Kudielka, B.M.: Decision-making in everyday moral conflict situations: development and validation of a new measure. PLoS ONE **14**(4), e0214747 (2019)
3. Maldonato, M., Dell'Orco, S.: Making decisions under uncertainty emotions, risk and biases. In: Advances in Neural Networks: Computational and Theoretical Issues, pp. 293–302. Springer, Cham (2015)
4. Kahneman, D., Rosenfield, A.M., Gandhi, L., Blaser, T.: Noise: how to overcome the high, hidden cost of inconsistent decision making. Harv. Bus. Rev. **94**(10), 38–46 (2016)
5. Maldonato, M., Dell'Orco, S.: Toward an evolutionary theory of rationality. World Futures **66**(2), 103–123 (2010)
6. Maldonato, M., Dell'Orco, S.: The natural logic of action. World Futures **69**(3), 174–183 (2013)
7. Chomsky, N.: The Logical Structure of Linguistic Theory. Plenum Press, New York and London (1975)
8. Hauser, M.D., Young, L.: Modules, minds and morality. In: Hormones and Social Behaviour, pp. 1–11. Springer, Berlin, Heidelberg (2008)
9. Damasio, A.R.: The Feeling of What Happens: Body and Emotion in the Making of Consciousness. Houghton Mifflin Harcourt (1999)
10. Dugatkin, L.: Animal cooperation among unrelated individuals. Naturwissenschaften **89**(12), 533–541 (2002)
11. Seyfarth, R.M., Cheney, D.L., Bergman, T., Fischer, J., Zuberbühler, K., Hammerschmidt, K.: The central importance of information in studies of animal communication. Anim. Behav. **80**(1), 3–8 (2010)
12. Denton, K.K., Krebs, D.L.: Rational and emotional sources of moral decision-making: an evolutionary-developmental account. Evol. Psychol. Sci. **3**(1), 72–85 (2017)

13. Maldonato, M., Dell'Orco, S., Sperandeo, R.: When intuitive decisions making, based on expertise, may deliver better results than a rational, deliberate approach. In: Multidisciplinary Approaches to Neural Computing, pp. 369–377. Springer, Cham (2018)
14. Maldonato, M., Dell'Orco, S., Esposito, A.: The emergence of creativity. World Futures **72**(7–8), 319–326 (2016)
15. Oliverio, A., Maldonato, M.: The creative brain. In: 2014 5th IEEE Conference on Cognitive Infocommunications (CogInfoCom), pp. 527–532. IEEE (2014)
16. Greene, J.D., Morelli, S.A., Lowenberg, K., Nystrom, L.E., Cohen, J.D.: Cognitive load selectively interferes with utilitarian moral judgment. Cognition **107**(3), 1144–1154 (2008)
17. Wrangham, R.W.: Two types of aggression in human evolution. Proc. Natl. Acad. Sci. **115**(2), 245–253 (2018)
18. Maldonato, M.: The wonder of reason at the psychological roots of violence. In: Advances in Culturally-Aware Intelligent Systems and in Cross-Cultural Psychological Studies, pp. 449–459. Springer, Cham (2018)
19. Eibl-Eibesfeldt, I., Longo, G.: Etologia della guerra. Boringhieri (1983)
20. Foot, P.: Virtues and Vices and Other Essays in Moral Philosophy. Oxford University Press on Demand (2002)
21. Tinghög, G., Andersson, D., Bonn, C., Johannesson, M., Kirchler, M., Koppel, L., Västfjäll, D.: Intuition and moral decision-making—the effect of time pressure and cognitive load on moral judgment and altruistic behavior. PLoS ONE **11**(10), e0164012 (2016)
22. Hauser, M., Cushman, F., Young, L., Kang-Xing Jin, R., Mikhail, J.: A dissociation between moral judgments and justifications. Mind Lang. **22**(1), 1–21 (2007)
23. Hauser, M., Shermer, M.: Can science determine moral values? A challenge from and dialogue with Marc Hauser about The Moral Arc. Skeptic (Altadena, CA) **20**(4), 18–25 (2015)
24. Moll, J., Eslinger, P.J., Oliveira-Souza, R.: Frontopolar and anterior temporal cortex activation in a moral judgment task: preliminary functional MRI results in normal subjects. Arq. Neuropsiquiatr. **59**, 657–664 (2001)
25. Glannon, W.: The evolution of neuroethics. In: Debates About Neuroethics, pp. 19–44. Springer, Cham (2017)
26. Helion, C., Ochsner, K.N.: The role of emotion regulation in moral judgment. Neuroethics **11**(3), 297–308 (2018)
27. Parker, A.M., De Bruin, W.B., Fischhoff, B.: Maximizers versus satisficers: decision-making styles, competence, and outcomes. Judgm. Decis. Mak. **2**(6), 342 (2007)
28. Maldonato, M., Dell'Orco, S.: Adaptive and evolutive algorithms: a natural logic for artificial mind. In: Toward Robotic Socially Believable Behaving Systems-Volume II, pp. 13–21. Springer, Cham (2016)
29. Juavinett, A.L., Erlich, J.C., Churchland, A.K.: Decision-making behaviors: weighing ethology, complexity, and sensorimotor compatibility. Curr. Opin. Neurobiol. **49**, 42–50 (2018)
30. Feigin, S., Owens, G., Goodyear-Smith, F.: Theories of human altruism: a systematic review. J. Psychiatry Brain Funct. **1**(1), 5 (2018)
31. Pohling, R., Bzdok, D., Eigenstetter, M., Stumpf, S., Strobel, A.: What is ethical competence? The role of empathy, personal values, and the five-factor model of personality in ethical decision-making. J. Bus. Ethics **137**(3), 449–474 (2016)
32. Garfinkel, H., Rawls, A., Lemert, C.C.: Seeing Sociologically: The Routine Grounds of Social Action. Routledge (2015)
33. Portinaro, P.P.: Il realismo politico. Laterza, Roma (1999)
34. Dell'Orco, S., Esposito, A., Sperandeo, R., Maldonato, N.M.: Decisions under temporal and emotional pressure: the hidden relationships between the unconscious, personality, and cognitive styles. World Futures 1–14 (2019)
35. Maldonato, M., Dell'Orco, S.: How to make decisions in an uncertain world: heuristics, biases, and risk perception. World Futures **67**(8), 569–577 (2011)

Chapter 7
A New Unsupervised Neural Approach to Stationary and Non-stationary Data

Vincenzo Randazzo, Giansalvo Cirrincione, and Eros Pasero

Abstract Dealing with time-varying high dimensional data is a big problem for real time pattern recognition. Non-stationary topological representation can be addressed in two ways, according to the application: life-long modeling or by forgetting the past. The G-EXIN neural network addresses this problem by using life-long learning. It uses an anisotropic convex polytope, which, models the shape of the neuron neighborhood, and employs a novel kind of edge, called bridge, which carries information on the extent of the distribution time change. In order to take into account the high dimensionality of data, a novel neural network, named GCCA, which embeds G-EXIN as the basic quantization tool, allows a real-time non-linear dimensionality reduction based on the Curvilinear Component Analysis. If, instead, a hierarchical tree is requested for the interpretation of data clustering, the new network GH-EXIN can be used. It uses G-EXIN for the clustering of each tree node dataset. This chapter illustrates the basic ideas of this family of neural networks and shows their performance by means of synthetic and real experiments.

Keywords Bridge · Convex polytope · Curvilinear component analysis · Dimensionality reduction · Fault diagnosis · Hierarchical clustering · Non-stationary data · Projection · Real-time pattern recognition · Seed · Unsupervised neural network · Vector quantization

V. Randazzo (✉) · E. Pasero
DET, Politecnico di Torino, Turin, Italy
e-mail: vincenzo.randazzo@polito.it

E. Pasero
e-mail: eros.pasero@polito.it

G. Cirrincione
Université de Picardie Jules Verne, Amiens, France
e-mail: exin@u-picardie.fr

University of South Pacific, Suva, Fiji

G. Phillips-Wren et al. (eds.), *Advances in Data Science: Methodologies and Applications*, Intelligent Systems Reference Library 189,
https://doi.org/10.1007/978-3-030-51870-7_7

7.1 Open Problems in Cluster Analysis and Vector Quantization

The topological representation of data is an important challenge for unsupervised neural networks. They build a covering of the data manifold in form of a directed acyclic graph (DAG), in order to fill the input space. However, above all for high dimensional data, the covering is prone to the problem of the curse of dimensionality, which requires, in general, a large number of neural units. The nodes of the graph are given by the weight vectors of the neurons and the edges, if present, by their connections. The weight estimation, in several cases, implies the minimization of an error function based on some error (e.g. vector quantization, VQ). In other cases, only the iterative technique is given. In general, VQ is performed by using competitive learning (neural units compete for representing the input data): it can be either hard (HCL, e.g. LBG [1] and k-means [2]) or soft (SCL, e.g. neural gas [3] and Self Organizing Maps, SOM, [4]). In HCL only the winning neuron (the closest to the input in terms of weight distance) changes its weight vector. For this reason, it is also known winner-take-all. Instead, in SCL, a.k.a. winner-take-most, both the winner and its neighbors adapt their weights. This approach needs a definition of neighborhood, which requires a network topology, as a graph, whose edges are in general found by means of the Competitive Hebbian Rule (CHR [5]), as in the Topology Representing Network [6], or by back-projecting a fixed grid as in SOM.

Incremental or growing neural networks do not require a prior choice of the architecture, which is, instead, determined by the data (data-driven). All these techniques need a *novelty test* in order to decide when a new neuron has to be created. All tests demand, in general, a model representing the portion of input space explained by each unit. This model is, in general, a hypersphere, because it is as simplest as possible: only a scalar hyperparameter, its radius, is needed. All existing algorithms determine, in a way or another, this threshold. It can be set a user-dependent global parameter (IGNG [7]), or it can be automatically and locally estimated. The single-layer Enhanced Self-Organizing Incremental Neural Network (ESOINN [8]) uses a threshold for each neuron, which is defined as the largest distance from its neighbors. Furthermore, in AING [9], it is given by the sum of distances from the neuron to its data-points, plus the sum of weighted distances from its neighboring neurons, averaged on the total number of the considered distances. In both cases, the influence region of the neuron depends on the extension of its neighborhood, but not on its shape. An exhaustive description can be found in [10]. However, this simple model is isotropic, in the sense that it does not take into account the orientation of the vector connecting the new data to the winner, but only its norm. Hence, it does not consider the topology of the manifold of data of the winner Voronoi set. The use of an anisotropic criterion should be justified by the need of representing in more detail the data manifold.

Data manifolds can be stationary or time-changing (i.e., non-stationary). It should be important to have a neural network able to automatically detect the data evolution. Tracking non-stationary data distributions is an important goal. This is requested by applications like real time pattern recognition: fault diagnosis, novelty detection, intrusion detection alarm systems, speech, face and text recognition, computer vision and scene analysis and so on. The existing neural solutions tackle this problem by means of different approaches, depending both on their architecture and on the application at hand. These techniques can be mainly classified into two categories: forgetting and life-long learning networks. The first class comprises the networks with a fixed number of neurons (not incremental). Indeed, they cannot track a varying input distribution without losing the past representation (given by the old weight vectors). Furthermore, if the distribution changes abruptly (jump), they cannot track it anymore. They are used if only the most recent representation is of interest. The fastest techniques of this class are linear, like the principal component analysis (PCA) networks. However, they are not suited for non-linear problems. In this case, the best non-linear network is a variant of SOM, called DSOM [11], which is based on some changes of the SOM learning law in order to avoid a quantization proportional to the data distribution density. However, what is more interesting is the use of constant parameters (learning rate, elasticity) instead of time-decreasing ones. As a consequence, DSOM is able to promptly react to changing inputs, at the expense of forgetting the past information. Indeed, it only tracks the last changes. Forgetting networks are not suited in case the past inputs carry useful information.

Life-long learning addresses the fundamental issue of how a learning system can adapt to new information without corrupting or forgetting previously learned information, the so-called Stability-Plasticity Dilemma [12]. It should have the ability of repeatedly training a network using new data without destroying the old nodes. For this reason, they must have the capability to increase the number of neurons in order to track the distribution (the previous neurons become dead units but represent past knowledge). This kind of networks, like SOINN and its variants [8], record the whole life of the process to be modelled. The precursor is the Growing Neural Gas (GNG [13]), but it is not well suited for these problems because the instant of new node insertion is predefined by a user-dependent parameter. However, its variant GNG-U [14] is a forgetting network, which uses local utility variables to estimate the probability density of data in order to delete nodes in regions of low density.

The same observation can be repeated for the data stream clustering methods [15]. There exist techniques which can be categorized according to the nature of their underlying clustering approach, as: GNG based methods, which are incremental versions (e.g., G-Stream [16]) of the Growing Neural Gas neural network, hierarchical stream methods, like BIRCH [17] and ClusTree [18], partitioning stream methods, like CluStream [19], and density-based stream methods, like DenStream [20] and SOStream [21], which is inspired by SOM.

The first neuron layers of online Curvilinear Component Analysis (onCCA) [22] and Growing Curvilinear Component Analysis (GCCA) [23–26], use the same threshold as ESOINN, but introduce the new idea of bridge, i.e. a directed interneuron connection, which signals the presence of a possible change in the data distribution. Bridges carry information about the extent of the time change by means of its length and density and allow the outlier detection.

7.2 G-EXIN

G-EXIN [27] is an online, self-organizing, incremental neural network whose number of neurons is determined by the quantization of the input space. It uses seeds to colonize a new region of the input space, and two distinct types of links (edges and bridges), to track data non-stationarity. Each neuron is equipped with a weight vector to quantize the input space and with a threshold to represent the average shape of its region of influence. In addition, it employs a new anisotropic threshold idea, based on the shape (convex hull) of neuron neighborhood to better match data topology. G-EXIN is incremental, i.e. it can increase or decrease (pruning by age) the number of neurons. It is also online: data taken directly from the input stream are fed only once to the network. The training is never stopped, and the networks keeps adapting itself to each new datum, that is, it is stochastic in nature.

7.2.1 The G-EXIN Algorithm

The starting structure of G-EXIN is a seed (couple of neurons connected through a link) based on the first two data.

Then, each time a new datum, say x_i , is extracted from the input stream, it is fed to the network and the training algorithm, described in Fig. 7.1, is performed. All neurons are sorted according to the Euclidean distances d_i between x_i and their weights. The neuron with the shortest distance (d_I) is the *first winner*, say w_I; the one with the second shortest distance (d_2) is the *second winner*, say w_2, the third one w_3, and so on. Then, the novelty test between the new data x_i and w_I is performed. If x_i passes it, a new neuron is created; otherwise, it follows the weight adaptation, linking and doubling phase.

Novelty test. An input data x_i is considered *novel* w.r.t. the neuron γ if it satisfies two conditions: their distance d is greater than the neuron local threshold T_γ and x_i is outside the neighborhood of γ, say NG_γ.

T_γ provides the minimal resolution of the test. Indeed, if a lower threshold is not given, there is the potential risk of a too large amount of neurons. The choice of this minimum implies neighbor neurons are not too close, which results in a apriori granularity (resolution). T_γ represents the radius of a hypersphere centered on the neuron. It is given by the average of the distances between γ and its topological neighbors according to:

$$T_\gamma = \frac{1}{|NG_\gamma|} \sum_{w_i \in NG_\gamma} \gamma - w_i \tag{7.1}$$

The neighborhood NG_γ can be represented in different ways. However, if we want to respect its geometry and, in the same time, to avoid complicating too much the model, a good compromise is the convex hull (bounded convex polytope) of the

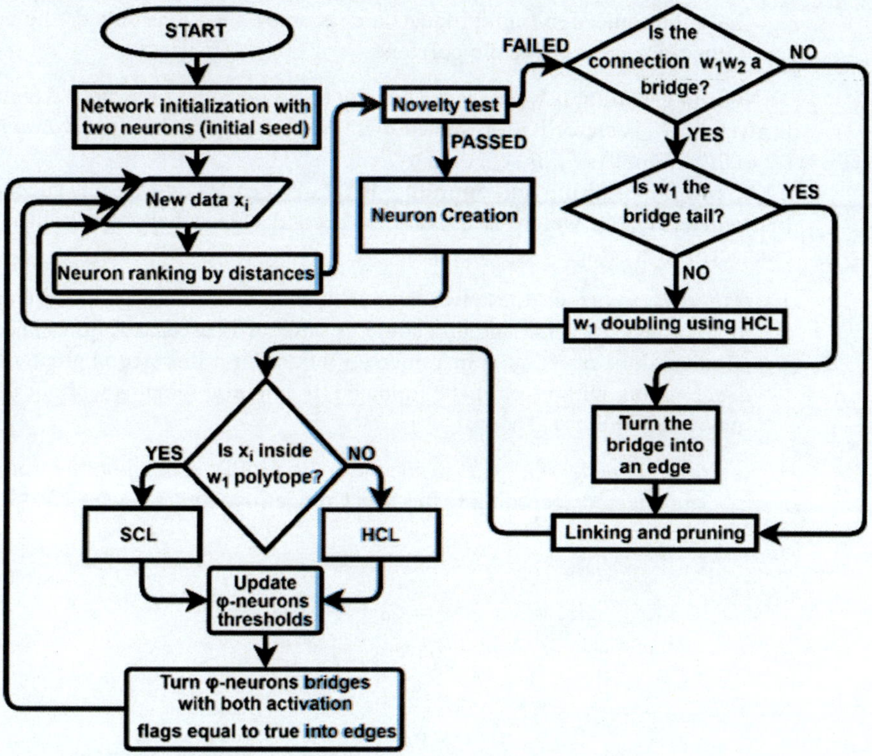

Fig. 7.1 G-EXIN flowchart

weight vector of neuron γ and the weights of its topological neighbors. Indeed, it is simple linear approach that considers not only the neighbors, but also the directionality of the corresponding edges, which implies to take into account the anisotropy of the region of influence. In this context, neurons connected through bridges are excluded, only those connected through edges are taken into account.

Depending on the network configuration, two scenarios can occur:

(1) γ has less than two topological neighbors, then, it is impossible to build the convex hull. In this case, for the novelty detection, only the isotropic hypersphere centered on γ and with radius T_γ is used. If the input data x_i is outside the sphere, then the novelty test is passed, otherwise, it is failed.

(2) γ has at least two topological neighbors then, for the novelty detection, a more sophisticated strategy is adopted. First, the convex hull of γ and its topological neurons is built. Then, if d is sufficiently big (i.e. greater than T_γ) the isotropic hypersphere with radius T_γ is replaced by the following simple and time-efficient anisotropic test to determine if x_i belongs or not to the NG_γ region. The difference vectors $\boldsymbol{\delta}_i$ between x_i and NG_γ weight vectors and their sum vector $\boldsymbol{\psi} = \Sigma\, \boldsymbol{\delta}_i$ are computed. If all the scalar products between $\boldsymbol{\delta}_i$ and $\boldsymbol{\psi}$

have the same sign (null products are ignored), then x_i is outside the polytope. Otherwise, x_i is inside the polytope.

Neuron creation. If x_i passes the novelty test, a new neuron, whose weight vector is given by x_i, is created. w_I is linked to x_i by a bridge and their *activation flags* are set to *false*. Finally, T_{xi} is set equal to d_I.

Adaptation, linking and doubling. If x_i fails the novelty test, it is checked if the first winner, whose weight is w_I, and the second winner, whose weight is w_2, are connected by a bridge:

1. *If there is no bridge*, these two neurons are linked by an edge (whose age is set to zero) and the same age procedure as onCCA is used as follows. The age of all other links of NG_{wI} is incremented by one; if a link age is greater than the age_{max} scalar parameter, it is eliminated. If a neuron remains without links, it is removed (*pruning*). Then:

   (a) *if* x_i *is inside* NG_{wI} (i.e. inside the convex hull), x_i neighbor neuron weights are adapted according to the Soft Competitive Learning (SCL):

$$\Delta w_i = \alpha_n (w_i - x_\text{i})\,\text{i} = 1 \tag{7.2a}$$

$$\Delta w_i = \alpha_n (w_i - x_\text{i})\,\text{otherwise} \tag{7.2b}$$

where $\alpha_1 = \frac{\alpha}{N_i}$ as in k-means [18] and $\alpha_n = \alpha * \exp\left(-\frac{(w_i - x_\text{i})^2}{2\sigma^2}\right)$. Here, N_i is the total number of times w_i has been the first winner and, α and σ are two user-dependent parameters.

   (b) *if* x_i *is outside* NG_{wI}, only (2a) is used (Hard Competitive Learning, HCL).

Next, for all the neurons that have been moved, i.e. whose weight vector has changed, say *φ-neurons*, their thresholds are recomputed, and their *activation flags* are set to *true*.

Finally, all the *φ-neurons* bridges, both ingoing and outgoing, are checked and all those which have both neurons at their ends with *activation flags* equal to true become edges.

2. *If there is a bridge*, it is checked if w_I is the bridge tail; in this case, step 1 is performed and the bridge becomes an edge. Otherwise, a seed is created by means of the neuron doubling:

   (a) a virtual adaptation of the w_I weight is estimated by HCL (only (2a) is used) and considered as the weight of a new neuron (doubling).

   (b) w_I and the new neuron are linked with an edge (age set to zero) and their thresholds are computed (they correspond to their Euclidean distance).

7.3 Growing Curvilinear Component Analysis (GCCA)

Dealing with time-varying high dimensional data is a big problem for real time pattern recognition. Only linear projections, like principal component analysis, are used in real time while nonlinear techniques need the whole database (offline). On the contrary, when working in real time requires a data stream, that is, a continuous input, the algorithm needs to be defined as online. This is the case, for example, of fault and pre-fault diagnosis and system modeling.

The techniques and the concepts presented above can be applied to different scenarios and applications. For instance, they can be used to perform an online quantization and dimensionality reduction (DR) of the input data, such as in the Growing Curvilinear Component Analysis (GCCA) neural network.

GCCA, whose flowchart is shown in Fig. 7.2, has a self-organized incremental (pruning by age) architecture, which adapts to the nonstationary data distribution. It performs simultaneously the data quantization and projection. The former is based on G-EXIN in the sense that it exploits the same techniques, such as seeds and bridges, to perform an online clustering of the input space. Seeds are pairs of neurons which colonize the input domain, bridges are a different kind of edge in the manifold graph,

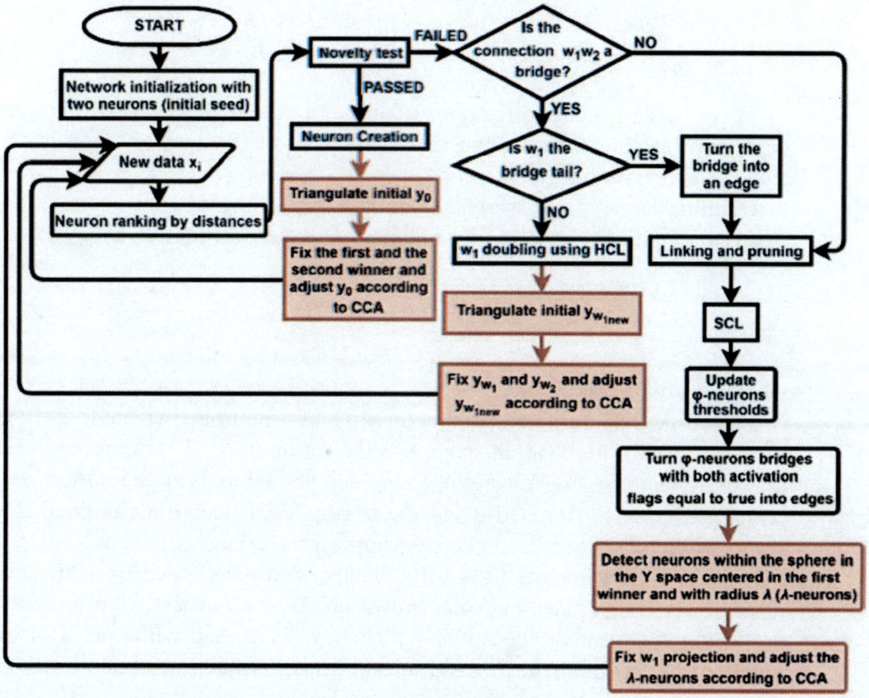

Fig. 7.2 GCCA flowchart: black blocks deal with G-EXIN quantization while red ones, specifically, with GCCA projection

signaling the data non-stationarity. The input projection is done using the Curvilinear Component Analysis (CCA), a distance-preserving reduction technique, here called *offline CCA*.

Data projection is a tool used frequently as a preprocessing stage; therefore, in a scenario such as that one characterized by an input fast-changing data stream (e.g. fault and pre-fault diagnosis), it needs to be as fast as possible. For this reason, the use of convex polytopes has been avoided and the novelty test is based only on the isotropic hypersphere whose radius is locally computed as the average of the distances from a neuron and its neighbors. The remaining has been designed as in G-EXIN with the difference that each neuron is equipped with two weight-vector, one in the input space X and one in the projected space Y. Moreover, an additional hyperparameter, λ, is needed, as in CCA, to tune the projection mechanism.

The projection works as follows. For each pair of different weight vectors in the X space (input space), a between-point distance D_{ij} is calculated as $D_{ij} = x_i - x_j$. At the same time, the distance L_{ij} of the associated Y-weights in the latent space, is computed as $L_{ij} = y_i - y_j$. CCA aims to project data such that $L_{ij} = D_{ij}$. Obviously, this is possible only if all input data lay on a linear manifold. In order to face this problem, CCA defines a distance function, which, in its simplest form, is the following:

$$F_\lambda\left(L_{ij}\right)\begin{cases} 0 \text{ if } & \lambda < L_{ij} \\ 1 \; if & \lambda \geq L_{ij} \end{cases} \tag{7.3}$$

That is a step function for constraining only the under threshold between-point distances L_{ij}. In this way, the CCA favors short distances, which implies local distance preservation.

Defining as $\mathbf{y}(j)$ the weight of the j-th projecting neuron in the Y space, the stochastic gradient algorithm for minimizing the error function follows:

$$\mathbf{y}(j) \leftarrow \mathbf{y}(j) + \alpha\left(D_{ij} - L_{ij}\right)F_\lambda\left(L_{ij}\right)\frac{\mathbf{y}(j) - \mathbf{y}(i)}{L_{ij}} \tag{7.4}$$

where α is the learning rate.

Each time a datum fails the *novelty test* a new neuron is created. As in G-EXIN, its weight vector in the input space X is the datum itself. To determine the weight in the latent space, i.e. the Y-weight, a two-step procedure is applied. First, the starting projection (y_0) is estimated using the *triangulation* technique defined in [23]. To compute y_0 the winner and second winner projections are used as the centers of the two circles, whose radii are the distances in data space from the input data, respectively. The circles intersect in two points, the farthest from the third winner projection is chosen as the initial y_0. Then, y_0 is refined with one or several CCA iterations (4), in which the first and second winner projections are considered as fixed (*extrapolation*).

The same projecting algorithm is applied in case of neuron doubling. In this case, the new neuron to be considered as input is $w_{1new,}$ that is the unit just born from the first winner w_1 doubling.

On the other side, if the datum fails the novelty test, the CHL and the SCL techniques are applied. Due to the weight updates of SCL, the first winner and its neighbors' distances D_{ij} change. Hence, the projections of the neuron whose distances from w_1 have to be updated. The CCA rule (4) is used but in an opposite way (*interpolation*). The first winner projection is fixed and the other neuron projection are moved accordingly to (4).

7.4 GH-EXIN

Hierarchical clustering is an important technique to retrieve multi-resolution information from data. It creates a tree of clusters, which corresponds to different resolution of data analysis. Generally, e.g. in data mining, the outcome is a richer information if compared with plain clustering.

The growing hierarchical GH-EXIN [28, 29] neural network builds a hierarchical tree based on a stationary variant (i.e. without bridges) of G-EXIN, called sG-EXIN. As before, the network is both incremental (data-driven) and self-organized. It is a top-down, divisive technique, in which all data start in a single cluster and, then, splits are done recursively until all clusters satisfy certain conditions.

The algorithm starts from a single root node, which has associated fictitiously the whole dataset; then, using vertical and horizontal growths, it builds a hierarchical tree (see Fig. 7.3). Vertical growth refers to the addition of further layers to leaf nodes until a higher resolution is needed; it always implies the creation of a seed, i.e. a pair of neurons, which represents the starting structure of a new sG-EXIN neural network. On the other side, horizontal growth is the process of adding further neurons to the seed. This characteristic is important in order to be able to create complex hierarchical structures; indeed, without it, it would be possible to build only binary trees. This process is performed by the neuron creation mechanism during the sG-EXIN training. As G-EXIN, GH-EXIN uses convex hull to define neuron neighborhood, which implies the anisotropic region of influence for the horizontal growth. In addition, upon time, it performs outlier detection and, when needed, it reallocates their data by using a novel simultaneous approach on all the leaves.

The GH-EXIN training algorithm starts, as already mentioned, from a single root node whose Voronoi set is the whole input dataset. It is considered as the initial father node. A father neuron Ψ is the basis for a further growth of the tree; indeed, new leaves are created (vertical growth), whose father is Ψ and whose Voronoi sets are a partition (i.e. a grouping of a set's elements into non-empty subsets, whose intersection is the empty set) of the Ψ one. More specifically, for each father neuron Ψ, which does not satisfy the vertical growth stop criterion, a new seed is created as in G-EXIN and, then, an sG-EXIN neural network is trained using the father Voronoi set as training set. The neurons yielded by the training, which defines a so-called *neural unit*, became the sons of Ψ in the tree determining a partition of its Voronoi set. If the resulting network does not satisfy the horizontal growth stop criterion, the training is repeated for further epochs (i.e. presentation of the whole Ψ dataset) until the criterion is fulfilled.

At the end of each training epoch, if a neuron remains unconnected (no neighbors) or is still lonely, it is pruned, but the associated data are analyzed and possibly reassigned as explained later in this section.

At the end of each horizontal growth, the *topology abstraction check* is performed to search for connected components within the graph of the resulting neural unit. If more than one connected component is detected, the algorithm tries to extract an abstract representation of data; at this purpose, each connected component, representing a cluster of data, is associated with a novel abstract neuron, which becomes the father node of the connected component neurons, determining a double simultaneous vertical growth. As weight vectors of the abstract neurons are used the centroids of the clusters they represent.

Then, each leaf, in the same level of the hierarchy of Ψ, that does not satisfy the vertical growth stop criterion, is considered as a father node and the growth algorithm is repeated, until no more leaves are available in that specific level.

Finally, the overall above procedure is repeated on all the leaves of the novel, deeper level yielded from the previous vertical growth; therefore, the tree can keep growing until the needed resolution is reached, that is, until the vertical growth stop criterion is satisfied for all the leaves of the tree.

It is worth to be noticed that such mechanism allows a simultaneous vertical and horizontal growth; indeed, due to node creation (seed) below a father an additional level is added to the tree (i.e. vertical growth) and, at the same time, thanks to sG-EXIN training, several nodes are added to the same level (i.e. horizontal growth).

The novelty test (Semi-Isotropic Region of Influence), the weights update (SCL) and the pruning mechanism (pruning by age) are the same as in G-EXIN. The difference is that GH-EXIN is based on sG-EXIN which, as stated above, does not have bridges; as a consequence, each time a new neuron is created along the GH-EXIN training process, it is created as a *lonely neuron*, that is a neuron with no edges. Then, in the next iterations connections may be created according to the Competitive Hebbian Rule; if, at the end of the epoch, the neuron is still lonely, it will be removed according to the pruning rule.

When a neuron is removed, its Voronoi set data remain orphans and are labelled as *potential outliers* to be checked at the end of each epoch; for each potential outlier x, i.e. each datum, GH-EXIN seeks a possible new candidate among all leaf nodes. If the closest neuron w among the remaining, i.e. the new winner, belongs to the same neural unit of x but the datum is outside its region of influence (the hypersphere and the convex-hull), x is not reassigned; otherwise, if x is within a winner region of influence within the same neural unit or in case the winner belongs to another neural unit, it is reassigned to the winner Voronoi.

The growth stop criteria are used to drive, in an adaptive way, the quantization process; for this reason, they are both based on the H index, which, depends on the application at hand, and it is used to measure clusters heterogeneity and purity, i.e.

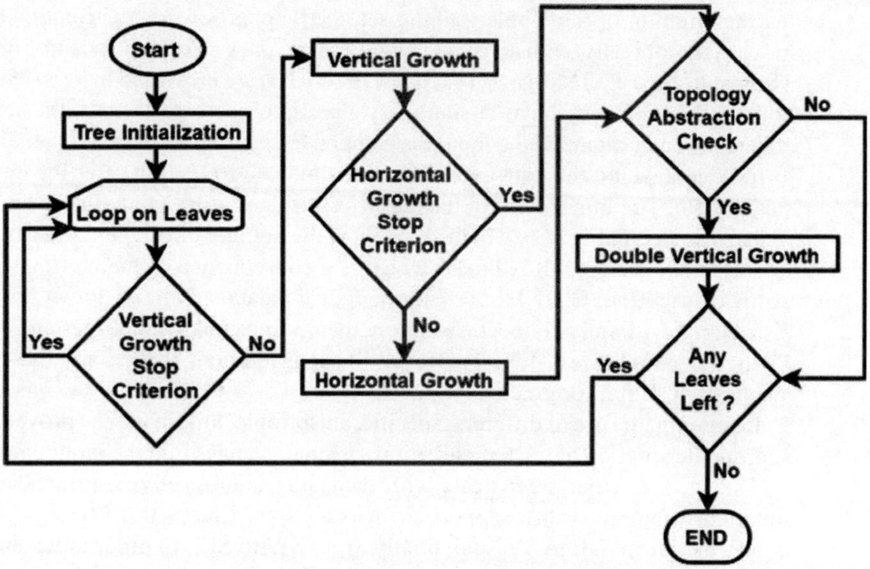

Fig. 7.3 GH-EXIN flowchart

their quality. For the horizontal growth, the idea is to check if the H average estimated value of the neurons of the neural unit being built falls below a percentage of the value of the father node. On the other side, in the vertical growth stop criterion, a global user-dependent threshold is used for H; at the same time, to avoid too small, meaningless clusters, a min_{card} parameter is used to establish the minimum cardinality of Voronoi sets, i.e. the maximum meaningful resolution.

7.5 Experiments

The performance of the above-mentioned neural networks has been tested on both synthetic and real experiments. The aim has been to check their clustering capabilities and to assess their specific abilities (e.g. projection).

7.5.1 G-EXIN

The first experiment deals with data drawn uniformly from a 5000-points square distribution, which, after an initial steady state (stationary phase), starts to move vertically (non-stationary phase). Indeed, in the beginning, the network is trained with data randomly extracted (without repetition) from the 5000-points square. Then, after the presentation of the whole training set, the (support of the) distribution starts to move monotonically, with constant velocity, along the y-axis in the positive direction. The results of G-EXIN ($age_{max}=2, \alpha=1, \sigma=0.03$) are presented in Figs. 7.4 and 7.5 both for the stationary and non-stationary phases, respectively. Firstly, the network is able to properly quantize the input distribution even along its borders; then, it is able to fully understand the data evolution over time and to track it after the end of the steady state. The importance of the density of bridges as a signal of non-stationarity is also revealed in Fig. 7.6, which shows how the number of bridges changes in time. In particular, the growth is linear, which is a consequence of the constant velocity of the distribution. G-EXIN correctly judges the data stream as drawn by a single distribution with fully connected support, thanks to its links (i.e., edges and bridges). Figure 7.5 also shows G-EXIN performs life-long learning, in the sense that previous quantization is not forgotten.

Resuming, the use of different, specific, anisotropic, links has been proved to be an appropriate solution to track non-stationary input changes into the input distribution.

The second experiment deals with data drawn uniformly from a 5000-points square distribution whose support changes abruptly (jump) three times (from NW to NE, then from NE to SW and, finally, from SW to SE), in order to test on abrupt changes. Figure 7.7 shows the results of G-EXIN ($age_{max}=9, \alpha=1, \sigma=0.06$) on such dataset, where neuron weights are represented as small dots and links as green (edges) and red segments (bridges); the same color is used for all neurons because the network does not perform any classification task.

Not only G-EXIN learns the data topology and preserves all the information without forgetting the previous history, as in the previous experiment, but it is able

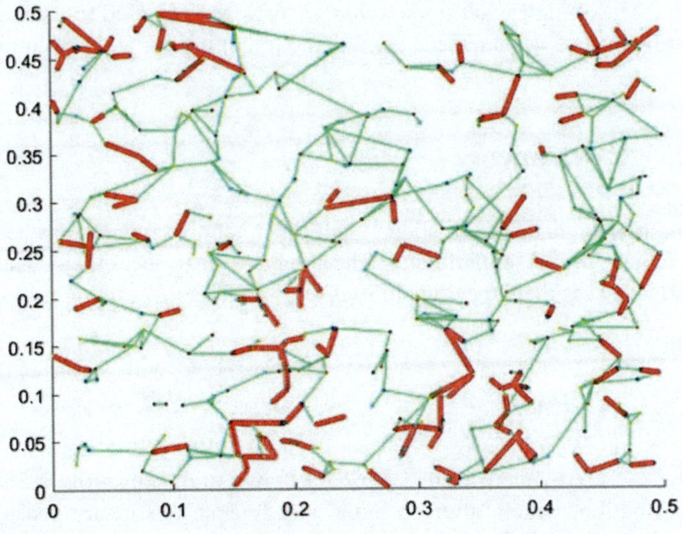

Fig. 7.4 G-EXIN: vertical moving square, stationary phase. Neurons (circles) and their links: edges (green), bridges (red)

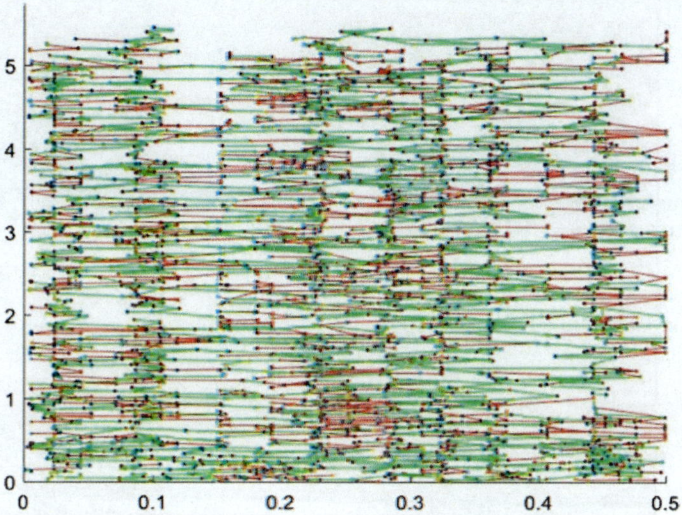

Fig. 7.5 G-EXIN: vertical moving square, non-stationary phase. Neurons (circles) and their links: edges (green), bridges (red)

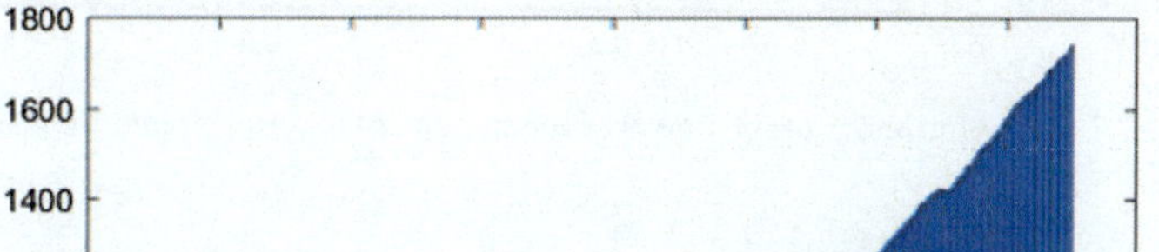

Fig. 7.6 G-EXIN: vertical moving square, number of bridges (Y-axis) over time (X-axis)

to track an abrupt change in the distribution by means of a single, long, bridge. The length of the bridges is proportional to the extent of the distribution change.

Figure 7.7 also shows the G-EXIN graph is able to represent well the borders of the squares because of its anisotropic threshold. On the contrary, this is not possible with a simpler isotropic technique.

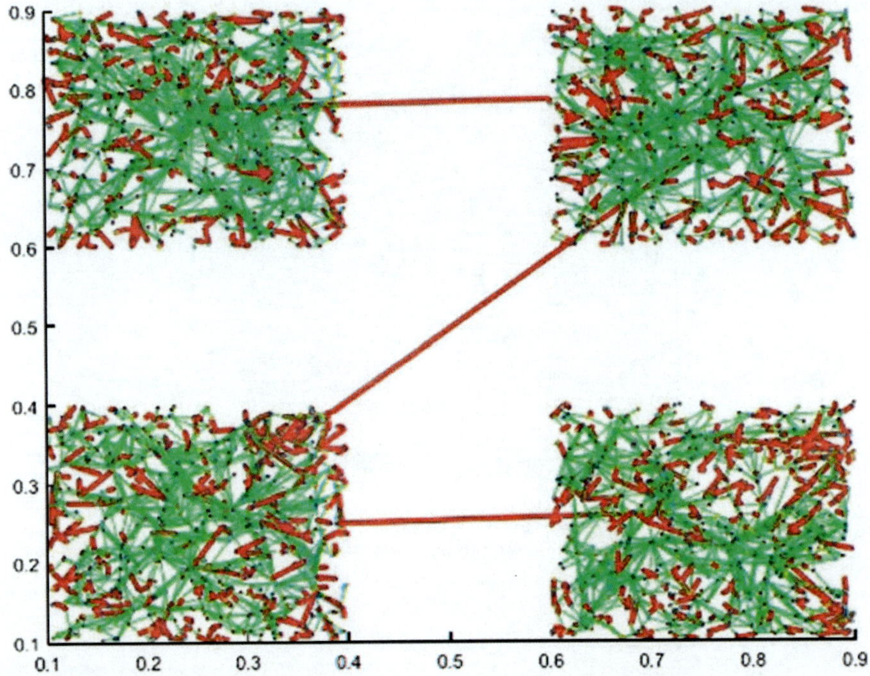

Fig. 7.7 G-EXIN: three jumps moving square. Neurons (circles) and their links: edges (green), bridges (red)

The third experiment deals with a more challenging problem: data drawn from a dataset coming from the bearing failure diagnostic and prognostic platform [30], which provides access to accelerated bearing degradation test. In particular, the test is based on a non-stationary framework that evolves from an initial transient to its healthy state to a double fault. Figure 7.8 shows G-EXIN ($age_{max}= 3$, $\alpha = 0.2$, $\sigma = 0.01$) on the experiment dataset during the complete life of the bearing: the initial transient, the healthy state and the following deterioration (the structure and color legenda are the same as in the previous figures). The transient phase is visible as the small cluster in the bottom left part of the figure. Then, the long vertical bridge signals the onset of the healthy state, which is represented as the central region made neurons connected by green and red edges. Finally, on the right and upper of this region there is the formation of longer and longer bridges which detect the deterioration of the bearing.

Resuming, all these experiments have shown that G-EXIN is able to fully track the non-stationarity by means of bridges, whose length and density carry information on the extent of the non-stationarity of the data distribution.

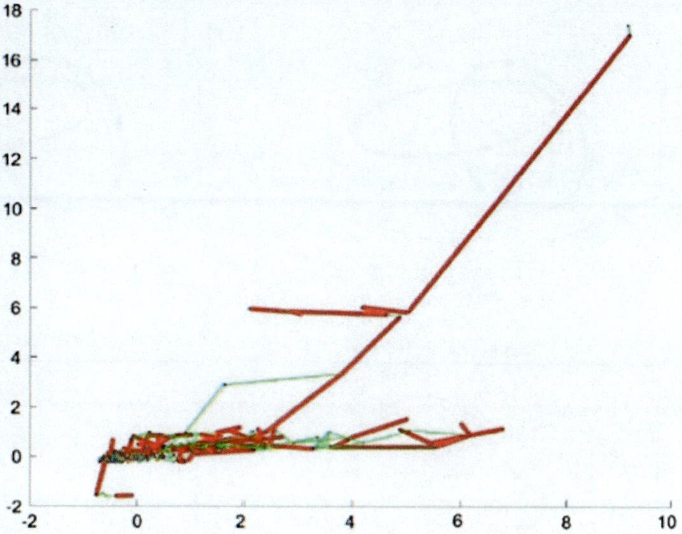

Fig. 7.8 G-EXIN: bearing fault experiment. Neurons (circles) and their links: edges (green), bridges (red)

7.5.2 GCCA

The simulation for GCCA deals with a more challenging synthetic problem: data drawn from a uniform distribution whose domain is given by two interlocked rings (see Fig. 7.9 upper left). Using a batch of 1400 data the projection of the *offline CCA* has been computed, by using a number of *epochs* equal to 10 and λ equal to 1. Figure 7.9 lower left shows that the *offline CCA* correctly unfolds data (the rings are separated). GCCA has then been applied to the same problem. The following parameters have been chosen: $age_{max} = 2$, $\alpha = 1$, $\sigma = 0.03$, $\lambda = 0.05$. Figure 7.9 upper right shows the result of the input space quantization together with the initial dataset. Figure 7.9 lower right yields the GCCA projection. There is a good unfolding (separation) in both the projections; however, it is evident from Fig. 7.9 that GCCA **online projection, based on a single epoch, performs as good as the offline CCA,** which, on the contrary, needs 10 presentations, i.e. *epochs*, of the training set.

In order to check the robustness of GCCA to white noise, an additional experiment has been made, starting from the same training set, but adding a Gaussian noise of zero mean and standard deviation set to 0.1. Figure 7.10 top left shows the resulting noisy distribution. The parameters are the same as in the previous experiment. Figure 7.10 top right yields the X-weight quantization of GCCA. Figure 7.10 bottom left and bottom right show the results of *offline CCA* and GCCA, respectively. It can be observed not only the robustness of GCCA, but also the better accuracy of its projection w.r.t. the *offline CCA*, trained on a batch composed of the same data presented to GCCA.

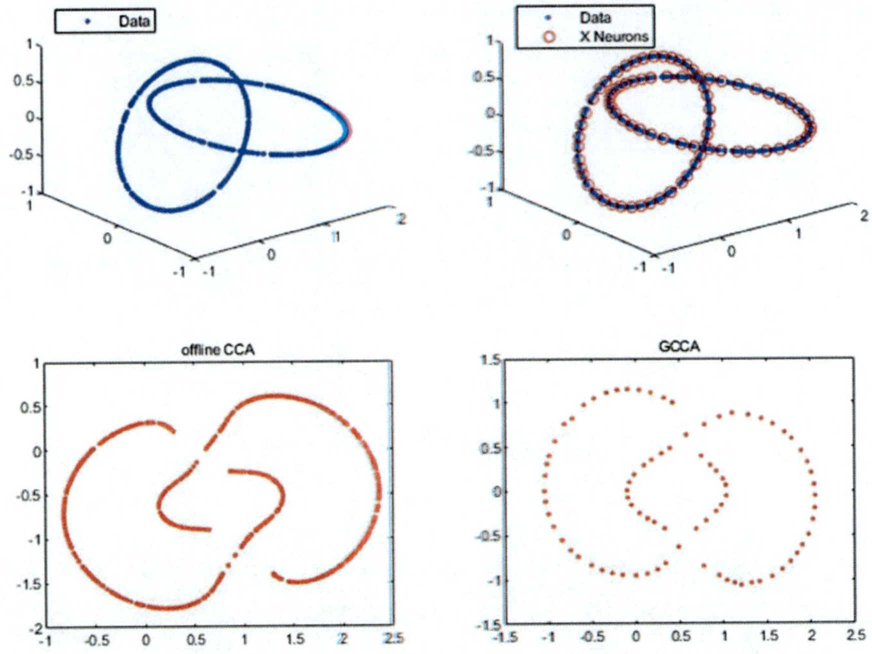

Fig. 7.9 GCCA: interlocked rings—no noise

From the previous simulations and logical considerations, some conclusions about the features of GCCA can be drawn. It retains the same properties of the *offline CCA*, that is the topological preservation of smallest distances and the unfolding of data. The adaptive features allow non-stationary data to be tracked by means of the quantization and the corresponding projection. Finally, GCCA is inherently robust to noise.

7.5.3 GH-EXIN

Considering that GH-EXIN has been conceived for hierarchical clustering, a dataset composed of two Gaussian mixture models has been devised: the first model is made of three Gaussians, the second one of four Gaussians, as shown in Fig. 7.11.

The results, visualized in Fig. 7.12 and Fig. 7.13, clearly show that GH-EXIN ($H_{max} = 0.001$, $H_{perc} = 0.9$, $\boldsymbol{\alpha}_{\gamma 0} = 0.5$, $\boldsymbol{\alpha}_{i0} = 0.05$, $age_{max} = 5$, $min_{card} = 300$) builds the correct hierarchy (the tree is visualized in Fig. 7.14): two nodes in the first layer (level), which represent the two clusters, and as many leaves as Gaussians in the second layer, which represent the mixtures. Neurons are also positioned correctly w.r.t. the centers of the Gaussians.

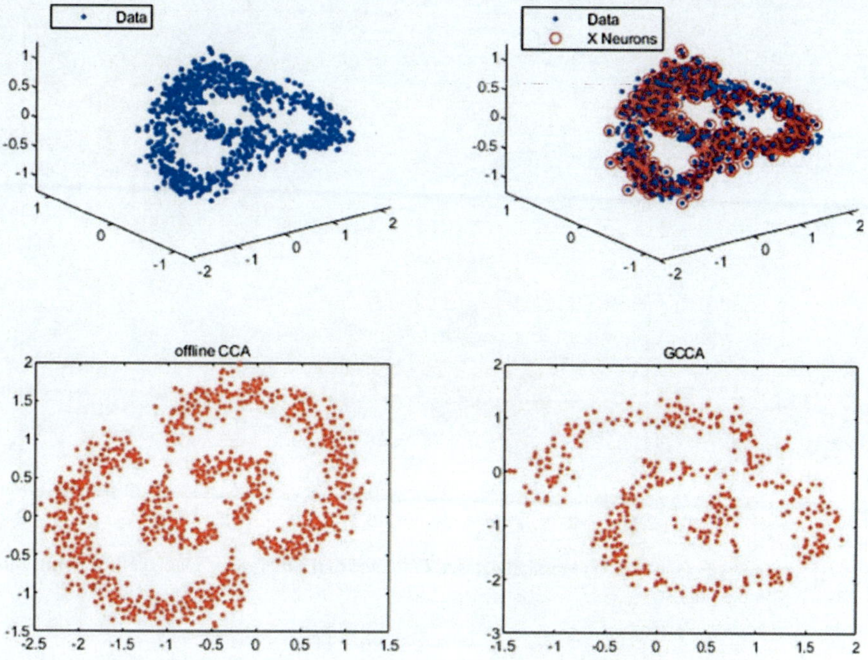

Fig. 7.10 GCCA: interlocked rings—Gaussian noise

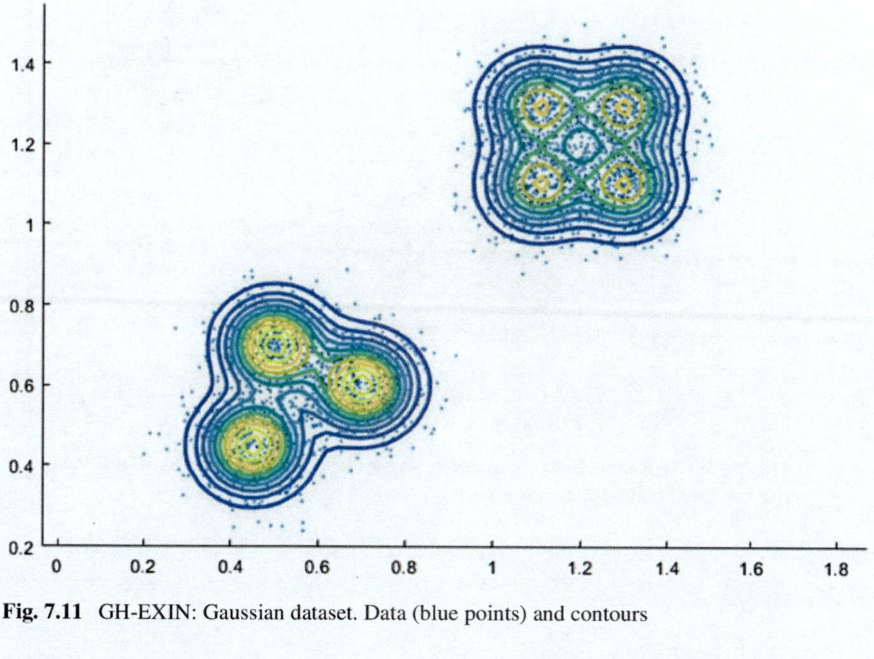

Fig. 7.11 GH-EXIN: Gaussian dataset. Data (blue points) and contours

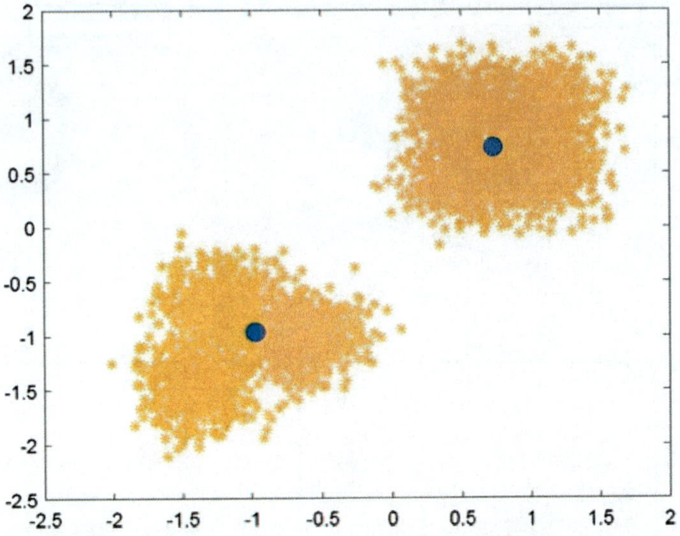

Fig. 7.12 GH-EXIN: Gaussian dataset, first level of the hierarchy. Data (yellow points) and neurons (blue points)

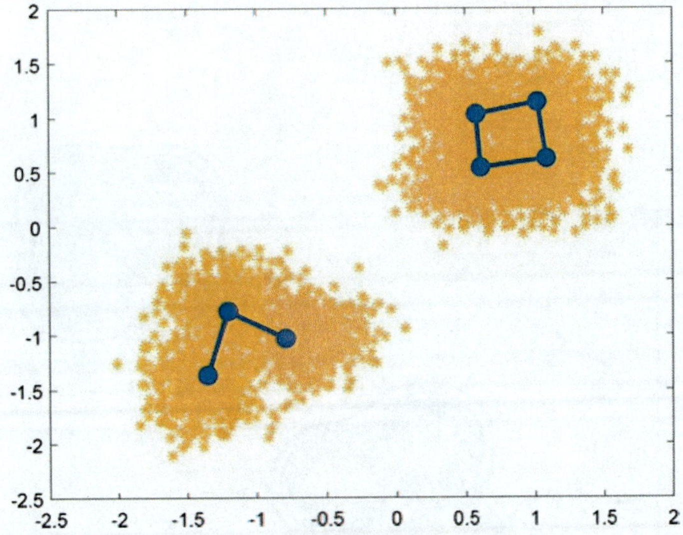

Fig. 7.13 GH-EXIN: Gaussian dataset, second level of the hierarchy. Data (yellow points) and neurons (blue points)

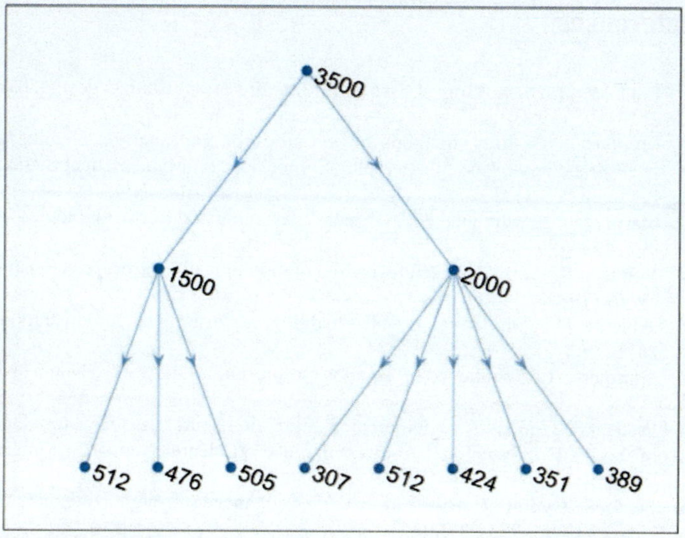

Fig. 7.14 GH-EXIN: Gaussian dataset, final tree and cardinality of nodes and leaves

7.6 Conclusions

This chapter addresses the problem of inferring information from unlabeled data drawn from stationary or non-stationary distributions. At this aim, a family of novel unsupervised neural networks has been introduced. The basic ideas are implemented in the G-EXIN neural network, which is the basic tool of the family. The other neural networks, GCCA and GH-EXIN, are extensions of G-EXIN, for dimensionality reduction and hierarchical clustering, respectively. All these networks exploit new peculiar tools: bridges, which are links for detecting changes in the data distribution; anisotropic threshold for taking into account the shape of the distribution; seed and associated neuron doubling for the *colonization* of new distributions; soft-competitive learning with the use of a Gaussian to represent the winner neighborhood.

The experiments show these neural networks work well both for synthetic and real experiments. In particular, they perform long-life learning, build a quantization of the input space, represent the data topology with edges and the non-stationarity with bridges, perform the CCA non-linear dimensionality reduction with an accuracy comparable to the offline CCA, yield the correct tree in case of hierarchical clustering. These are fast algorithms that require only a few user-dependent parameters.

Future work will deal with the search of new *automatic* variants, which self-calibrate their parameters, and more challenging applications.

References

1. Linde, Y., Buzo, A., Gray, R.: An algorithm for vector quantizer design. IEEE Trans. Commun. **28**, 84–95 (1980)
2. MacQueen, J.: Some methods for classification and analysis of multivariate observations. In: Proceedings of the 5th Berkeley Symposium on Mathematical Statistics and Probability. Berkeley (USA) (1967)
3. Martinetz, T., Schulten, K.: A "neural-gas" network learns topologies. Artif. Neural Netw. 397–402 (1991)
4. Kohonen, T.: Self-organized formation of topologically correct feature maps. Biol. Cybern. **43**, 59–69 (1982)
5. White, R.H.: Competitive hebbian learning: algorithm and demonstrations. In: Neural Netw. **20**(2), 261–275 (1992)
6. Martinetz, T., Schulten, K.: Topology representing networks. Neural Netw. **7**(3), 507–522 (1994)
7. Prudent, Y., Ennaji, A.: An incremental growing neural gas learns topologies. In: Proceedings of the IEEE International Joint Conference on Neural Networks. Motréal, Quebec, Canada (2005)
8. Furao, S., Ogurab, T., Hasegawab, O.: An enhanced self-organizing incremental neural. Neural Netw. **20**, 893–903 (2007)
9. Bouguelia, M.R., Belaïd, Y., Belaïd, A.: An adaptive incremental clustering method based on the growing neural gas algorithm. In: 2nd International Conference on Pattern Recognition Applications and Methods ICPRAM 2013. Barcelona, (Spain) (2013)
10. Bouguelia, M.R., Belaïd, Y., Belaïd, A.: Online unsupervised neural-gas learning method for infinite. In: Pattern Recognition Applications and Methods, pp. 57–70 (2015)
11. Rougier, N.P., Boniface, Y.: Dynamic self-organizing map. Neurocomputing **74**(11), 1840–1847 (2011)
12. Carpenter, G., Grossberg, S.: The ART of adaptive pattern recognition by a self-organizing neural network. IEEE Comput. Soc. **21**, 77–88 (1988)
13. Fritzke, B.: A growing neural gas network learns topologies. In: Advances in Neural Information Processing System, vol. 7, pp. 625–632 (1995)
14. Fritzke, B.: A self-organizing network that can follow non-stationary distributions. In: Proceedings of ICANN 97, International Conference on Artificial Neural Networks. Lausanne, Switzerland (1997)
15. Ghesmoune, M., Lebbah, M., Azzag, H.: State-of-the-art on clustering data streams. In: Big Data Analytics, pp. 1–13 (2016)
16. Ghesmoune, M., Azzag, H., Lebbah, M.: G-stream: growing neural gas over data stream. In: Neural Information Processing, 21st International Conference, ICONIP, Kuching, Malaysia (2014)
17. Zhang, T., Ramakrishnan, R., Livny, M.: BIRCH: an efficient data clustering method for very large databases. In: SIGMOD Conference. New York (1996)
18. Kranen, P., Assent, I., Baldauf, C., & Seidl, T.: The ClusTree indexing microclusters for anytime stream mining. Knowl. Inf. Syst. **29**(2), 249–272 (2011)
19. Aggarwal, C.C., Philip, S.Y., Han, J., Wang, J.: A framework for clustering evolving data streams. In: VLDB2003 Proceedings of the VLDB Endowment. Berlin (2003)
20. Cao, F., Estert, M., Qian, W., Zhou, A.: Density-based clustering over an evolving data stream with noise. In: SIAM International Conference on Data Mining (SDM06). Maryland (2006)
21. Isaksson, C., Dunham, M.H., Hahsler, M.: SOStream: self organizing density-based clustering over data stream. In: 8th International Conference on Machine Learning and Data Mining MLDM 2012. Berlin (2012)
22. Cirrincione, G., Hérault, J., Randazzo, V.: The on-line curvilinear component analysis (onCCA) for real-time data reduction. In: Proceedings of the IEEE International Joint Conference on Neural Networks. Killarney (Ireland) (2015)

23. Cirrincione, G., Randazzo, V., Pasero, E.: Growing curvilinear component analysis (GCCA) for dimensionality reduction of nonstationary data. In: Multidisciplinary Approaches to Neural Computing. Springer International Publishing, pp. 151–160 (2018)
24. Kumar, R.R., Randazzo, V., Cirrincione, G., Cirrincione, M., Pasero, E.: Analysis of stator faults in induction machines using growing curvilinear component analysis. In: International Conference on Electrical Machines and Systems ICEMS2017. Sydney (Australia) (2017)
25. Cirrincione, G., Randazzo, V., Pasero, E.: The Growing curvilinear component analysis (GCCA) neural network. Neural Netw. 108–117 (2018)
26. Cirrincione, G., Randazzo, V., Kumar, R.R., Cirrincione, M., Pasero, E.: Growing curvilinear component analysis (GCCA) for stator fault detection in induction machines. In: Neural Approaches to Dynamics of Signal Exchanges. Springer International Publishing (2019)
27. Randazzo, V., Cirrincione, G., Ciravegna, G., Pasero, E.: Nonstationary topological learning with bridges and convex polytopes: the G-EXIN neural network. In: 2018 International Joint Conference on Neural Networks (IJCNN). Rio de Janeiro (2018)
28. Barbiero, P., Bertotti, A., Ciravegna, G., Cirrincione, G., Pasero, E., Piccolo, E.: Unsupervised gene identification in colorectal cancer. In: Quantifying and Processing Biomedical and Behavioral Signals. Springer International Publishing, pp. 219–227 (2018)
29. Barbiero, P., Bertotti, A., Ciravegna, G., Cirrincione, G., Cirrincione, M., Piccolo, E.: Neural biclustering in gene expression analysis. In: 2017 International Conference on Computational Science and Computational Intelligence (CSCI). Las Vegas (2017)
30. Center, N.A.R.: FEMTO Bearing Data Set, NASA Ames Prognostics Data Repository. http://ti.arc.nasa.gov/project/prognostic-data-repository

Chapter 8
Fall Risk Assessment Using New sEMG-Based Smart Socks

G. Rescio, A. Leone, L. Giampetruzzi, and P. Siciliano

Abstract The electromyography signals (EMG) are widely used for the joint movements and muscles contractions monitoring in several healthcare applications. The recent progresses in surface EMG (sEMG) technologies have allowed for the development of low invasive and reliable sEMG-based wearable devices with this aim. These devices promote long-term monitoring, however they are often very expensive and not easy to be appropriately positioned. Moreover they employ mono-use pre-gelled electrodes that can cause skin redness. To overcome these issues, a prototype of a new smart sock has been realized. It is equipped with reusable stretchable and non-adhesive hybrid polymer electrolytes-based electrodes and can send sEMG data through a low energy wireless transmission connection. The developed device detects EMG signals coming from the Gastrocnemius-Tibialis muscles of the legs and it is suitable for lower-limb related pathology assessment, such as age-related changes in gait, sarcopenia pathology, fall risk, etc. In the paper it has been described, as a case study, the use of the socks to detect the risk of falling. A Machine Learning scheme has been chosen in order to overcome the well-known drawbacks of threshold approaches widely used in pre-fall systems, in which the algorithm parameters have to be set according to the users' specific physical characteristics. The supervised classification phase has been obtained through a low computational cost and a high classification accuracy level Linear Discriminant Analysis. The developed system shows high performance in terms of sensitivity and specificity (about 80%) in controlled conditions, with a mean lead-time before the impact of about 700 ms.

G. Rescio (✉) · A. Leone · L. Giampetruzzi · P. Siciliano
National Research Council of Italy, Institute for Microelectronics and Microsystems, Via Monteroni C/O Campus Ecotekne, Palazzina A3, Lecce, Italy
e-mail: gabriele.rescio@cnr.it

A. Leone
e-mail: alessandro.leone@cnr.it

L. Giampetruzzi
e-mail: lucia.giampetruzzi@le.imm.cnr.it

P. Siciliano
e-mail: pietro.siciliano@le.imm.cnr.it

G. Phillips-Wren et al. (eds.), *Advances in Data Science: Methodologies and Applications*, Intelligent Systems Reference Library 189,
https://doi.org/10.1007/978-3-030-51870-7_8

Keywords Smart wearable device · Surface electromyography · Machine learning scheme

8.1 Introduction

Recently, bio-signal measurements among which electromyography (EMG) and electroencephalography (EEG) have been increasingly demanded. In particular, EMG is a medical procedure that provide the acquisition of the electric potentials produced by the voluntary contraction of the skeletal muscle fibers. These potentials are bio-electric signals, acquired from the human body and then filtered to reduce the noise produced by other electrical activities of the body or inappropriate contact of the sensors, namely artifacts. Than the signals are processed in a control system to acquire information regarding the anatomical and physiological muscles' characteristics and to make a diagnosis. During last years, several works in literature have focused the attention on the use of the EMG signals in medical context [1, 2]. They record and analyze the intramuscular or surface EMG (sEMG) signals in order to study the human body's behaviors under normal and pathological conditions. The sEMG measurement method is safer and less invasive than the intramuscular technique and it presents good performance in the muscle action potentials monitoring. It uses non-invasive, skin surface electrodes, realized with pre-gelled, textile or hydrogel materials, located near the muscles of interest [3]. Application in medicine regarding the use of the electromyography analysis appears relevant for assessment of age-related changes in gait, and for diagnosis in Sarcopenia Pathology (SP), Amyotrophic Lateral Sclerosis (ALS) and Multiple Sclerosis (MS) or other neuropathies, postural anomalies, fall risk, etc. [1, 4]. For the considered diseases, the lower limb muscles are mainly monitored through the medical wired stations or portable and wearable technologies. The last progresses in EMG technology have allowed for the development of low invasive and reliable EMG based wearable devices. They may be used in the monitoring of the elderly during their normal activities for detection of dangerous events in healthcare. In this work the attention has been focused on the leg muscles assessment for the fall risk evaluation.

Fall events represent the second leading cause of accidental death brought about by preventable injury. This rate mostly refers to people over 60 years of age [5]. To date, several automatic integrated wearable devices and ambient sensor devices capable of fall detection have been constructed [6–9]. They present a good performance in terms of sensitivity and specificity and can alert the caregiver allowing a quick medical intervention and the reduction of fall consequences. Although these devices are remarkable, they cannot prevent injuries resulting from the impact on the floor. To overcome this limitation advanced technologies should be developed on the timely recognition of imbalance and fall event; thereby reducing, not only the time of probable medical intervention, but also through the activation of an impact protection system (i.e. an airbag). The current solutions for the assessment of patient physical instability, presented in the literature, primarily monitor the users' body movements and their muscle behaviors [1–10]. While the kinematic analysis of human movements is mainly accomplished through context aware systems, such as motion capture systems, and wearable inertial sensors, implantable and surface electromyography sensors have been used to conduct the analysis on muscle behavior. Considering the wearable devices, they are more invasive than the context aware systems, but they present some important advantages, such as: the re-design of the environments is not required, outdoor operation is possible and ethical issues (e.g. privacy) are always satisfied. For these reasons in this paper the attention has been focused on the fall risk detection systems based on wearable technologies. The majority of the studies presented in the literature for wearable-based fall risk assessment use accelerometer and gyroscopes systems. They measure above all acceleration, velocity and posture of the user's body and appear to be a promising solution to reduce the fall effect. Another strategy to evaluate the human imbalance is provided by the use of the electromyography technique which measures the electrical potentials produced by the lower limb muscles. They mainly describe the changes reaction sequence and muscle contractile force during an imbalance event. These studies suggest that the lack of balance causes a sudden modification on the EMG patterns brought about by reactive/corrective neuromuscular response [11, 12]. This could indicate that imbalance detection systems based on EMG signals may represent a very responsive and effective strategy for fall prevention. In this kind of analysis, wired probes or wireless devices integrating pre-gelled silver/silver chloride (Ag/AgCl) electrodes are mainly used. However, these electrodes are single-use, uncomfortable and unsuitable for a long-term monitoring due to their encumbrance and skin irritation. In Fig. 8.1 some

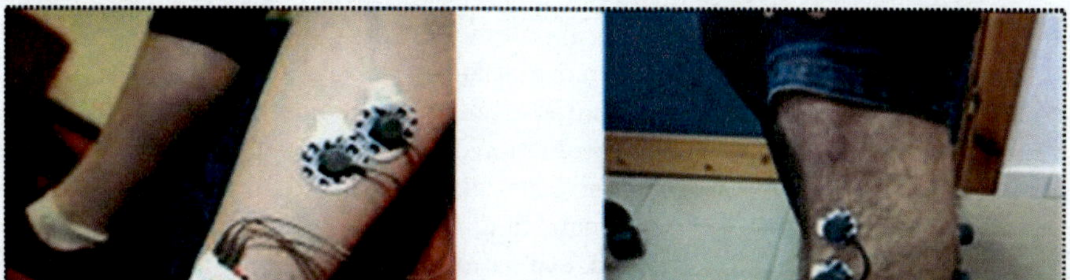

Fig. 8.1 Examples of wearable and wireless sEMG-based devices

examples of wearable sEMG-based devices are reported. Although they are minimally invasive and have a wireless connection, its placement is not very simple and use pre-gelled single-use electrodes.

Recently, new polymer compositions and new materials have emerged in order to address these limitations of traditional electrodes, aiming to adhere seamlessly to the skin [13, 14]. In this regard, many novel polymer materials have been researched to prepare conformable electrodes based on smart textiles and adhesive hydrogel [15, 16]. Considering the polymer made electrodes, for example, cellulose-based composite hydrogels and membranes with conductive compounds show potential applications in the acquisition of bio-signals [17, 18]. In this work, a novel biocompatible polymer electrolyte with good mechanical and conductive performance was prepared using polyvinyl alcohol (PVA), and carboxy-methil-cellulose (CMC) in order to obtain flexible and conductive membrane. Some studies indicated the blending of CMC with PVA in order to increase stability and mechanical properties through preparation of its hybrid materials porosity [19]. In the optical of creating polymers as electrode coatings for EMG, it has been proposed an approach to blending the electrolyte or conducting polymers (CPs) with other polymer forms, such as hydrogels or matrices improving conductivity [20]. Although both synthesis and evaluation on structure–property still remain challenges; these new materials comprised of electrolyte, conducting polymers and co-biopolymers in different forms as matrix hydrogel or textile, are promising in the field of bioactive electrode coatings, useful in smart wearable system research.

In the paper, new wireless and low cost smart socks for the surface EMG signals acquisition were developed to increase the users' level of usability and acceptability. This aim was achieved with the integration into the device of all electronics components and biocompatible hybrid polymer electrolyte (HPe)-based electrodes for the EMG data acquisition and transmission. The hardware was realized customizing commercial devices for that purpose and a low computational cost machine learning paradigm was chosen for the evaluation of imbalance events. The device was designed to monitor the Gastrocnemius Lateralis (GL) and Tibialis Anterior (TA) muscles contractions. The lack of human balance analysis is described by using the realized socks, suitable for the fall risk evaluation.

8.2 Materials and Methods

8.2.1 Hardware Architecture

The hardware architecture of the developed smart sock consist of three main blocks:

- Six hybrid polymer electrolytes (HPe) electrodes integrated in each sock to contact the skin;
- two electronic interface units for each sock to read the signals coming from the electrodes;
- one elaboration and wireless transmission unit for each sock.

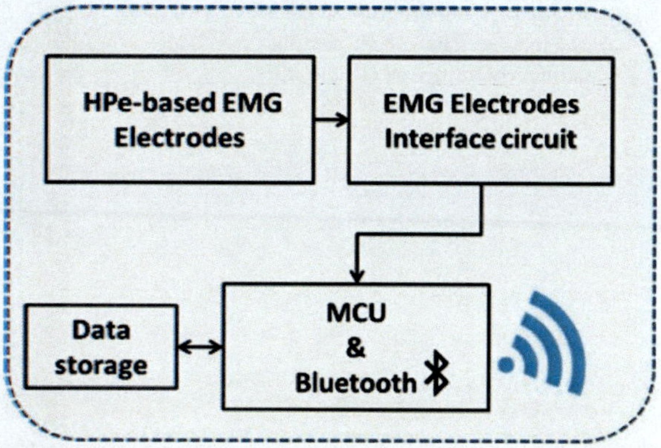

Fig. 8.2 Overview of the socks hardware architecture

In Fig. 8.2 the overview of the hardware architecture of the smart socks is reported. The probes are proper placed in order to monitor the electromyography data coming from the GL and TA muscles. In the following sections it is reported the description of each hardware block.

8.2.1.1 Hybrid Polymer Electrolytes (HPe) Electrodes

The hybrid polymer electrolyte was synthesized, blending a solution of Polyvinyl Alcohol (PVA) into a Carboxymethyl cellulose (CMC) solution. Three different ratio of the two solutions were tested: 20:80, 40:60 and 50:50; while the best ratio resulted the 20:80 (PVA: CMC) as reported in literature [21, 22]. Then 30 wt% of NH_4NO_3 was added and mixed to the resulted polymer solution in order to increase the biopolymer conductivity [23]. The CMC/PVA hybrid (80/20 wt%) electrolytes solution with NH_4NO_3 was placed into a mould containing the socks fabric and the clip. After the drying process, as described in literature [21], the clip and the polymer material resulted embedded into socks (Fig. 8.3).

Recently, a method to crosslink and plasticize the hybrid polymer, in order to maintain the mechanical and physical properties has been evaluated.

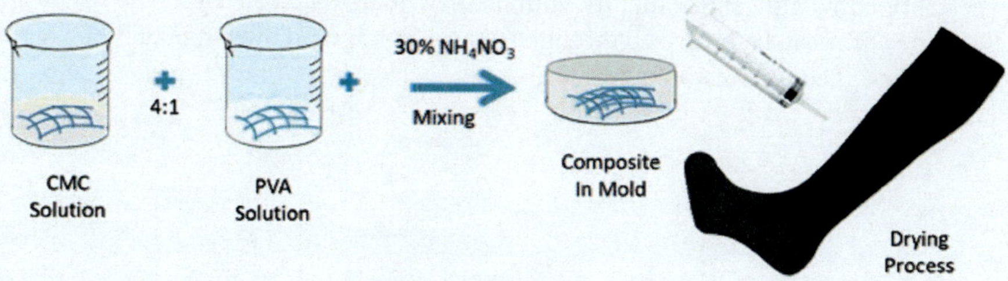

Fig. 8.3 Schema of bio-polymer electrolyte formation

Fig. 8.4 HPe-based electrodes have been casting incorporating the clip, in the site where the Myoware muscle sensor board is placed

The use of citric acid to crosslink the CMC matrix, the glycerol as a plasticizer and the polyethylene glycol as a pore-forming and a front-line curing agent, has been proposed [24, 25]. So the effects about porosity, pH sensitivity and mechanical behaviour are under analysis.

The system: socks-clip-HPe (made of 80:20 CMC/PVA blended) is shown in Fig. 8.4.

8.2.1.2 Electronic Interface Unit

The electronic interface unit has the task of acquiring and to amplifying the signals coming from the HPe electrodes in order to make them suitable for the proper management by the Microcontroller unit. This was obtained through the use of the Myoware Muscle Sensor board interface [26], shown in Fig. 8.5. Myoware Muscle sensors are equipped with three electrodes; two of them must be placed on skin in the measured muscle area, and one on skin, outside the muscle area, which is used as the ground point. Two Myoware devices were sewed on each sock. They normally use disposable pre-gelled electrodes, but through the variable gain of Myoware interface and the pre-elaboration step, described in the following section, it has been possible to obtain a high signal quality with the new realized electrodes. The Myoware Muscle Sensor can be polarized through a single voltage (in the range of 3.3–5 V) and it was designed for wearable devices.

Fig. 8.5 Myoware muscle sensor board EMG signals interface

8.2.1.3 Elaboration and Wireless Transmission Unit

For the data transmission and elaboration unit the Bluno Beetle board [27], shown in Fig. 8.6, was considered. It is lightweight, compact and integrates the low energy Bluetooth 4.0 transmission module.

One board was sewed on each sock and connected to the Myoware device through conductive wires. The whole system was supplied with a rechargeable Lipo battery of 3.7 V of 320 mA with dimension of 26.5 × 25 × 4 mm and weight of 4 gr. It was placed and glued in the rear part of the Beetle board. Figure 8.7 shows the realized prototype. Each electronic component was insulated through an Acrylic resin lacquer; in the future non-invasive packaging will be provided to make the system washable. The total current consumption was measured to evaluate the lifetime of the battery.

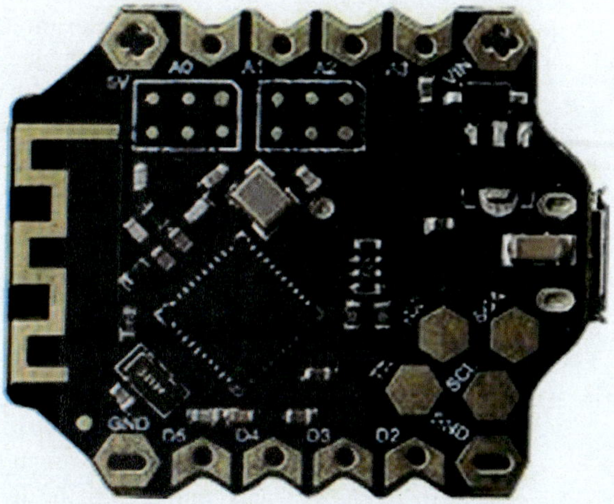

Fig. 8.6 Wireless transmission and elaboration data unit

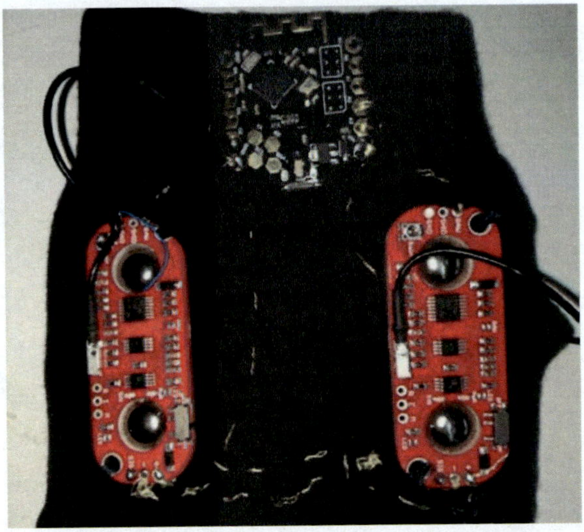

Fig. 8.7 Smart sock prototype

Based on the results the whole system consumes about 40 mA in data transmission mode. So, considering the employed battery, the system is able to monitor the lower limb muscles and to send data to a smartphone/embedded PC for about 8 h. Future improvements should be addressed to increase the system autonomy, optimizing the hardware and their power management logics. The prototype was realized by using an elastic sock to enhance the adhesion between the electrodes and the skin. The sensors were located on the socks in correspondence of the antagonist Gastrocnemius-Tibialis Muscles. The algorithmic framework for the elaboration of the EMG signals, coming from the sensorized socks, was located and tested on an embedded PC, equipped with a Bluetooth module.

In the Fig. 8.8 it is represented an example of application of the realized smart socks. The sEMG data acquired by the device are wirelessly sent to a smartphone or embedded PC for the data elaboration through the low computationally cost software architecture described in the following sections. This architecture must be able to recognize an abnormal condition in order to (a) activate an impact reduction system in a very fast way and (b) work as gateway to contact a relative or to enable a medical service assistance.

8.2.2 Data Acquisition Phase

The data acquisition phase is a relevant step to acquire the data needed for the development and evaluation of the computational system framework. With this aim the electromyography signals coming from the device during the simulation of Activities

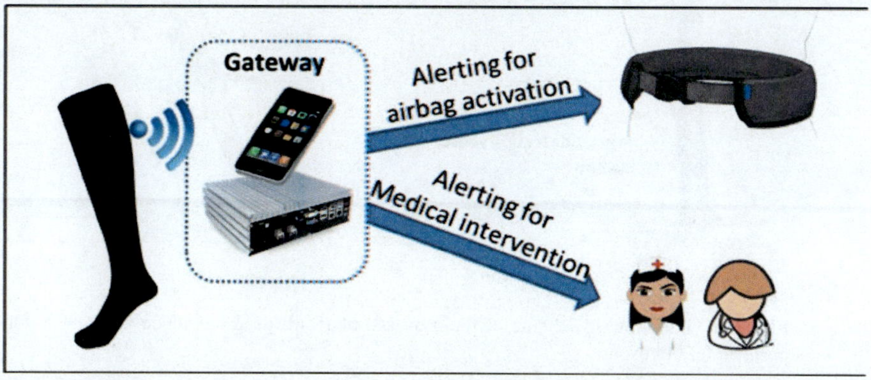

Fig. 8.8 Example of application of the realized smart socks

of Daily Livings (ADLs) and unbalance events (in controlled conditions) performed by six young healthy subjects of different ages (28.9 ± 7.5 years), weight (69.3 ± 8.4 kg), height (1.72 ± 0.3 m) and sex (4 males and 2 females). To acquire data, the socks were located so that the GL and TA muscles can be monitored, as shown in Fig. 8.9. In the zone where the probes were placed, the skin should be shaved and cleaned using an isopropyl alcohol swab to reduce impedance.

In developing and testing the fall risk assessment algorithm, a dataset was created, simulating the following ADLs and fall events:

Fig. 8.9 sEMG sensors mounting setup

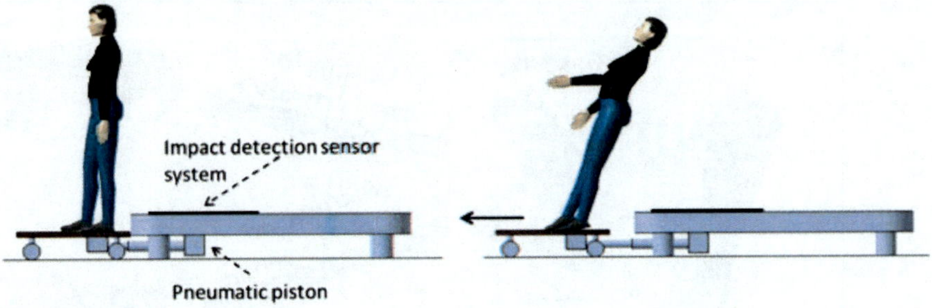

Fig. 8.10 Functioning scheme of the movable platform used to induce imbalance conditions to perform falling events

- Walking;
- Sitting down on a chair;
- Lying down on a mat (height 30 cm);
- Bending;
- Backward, lateral and forward fall events.

Each subject performed about 50 simulated ADLs and 12 falls for a total of about 300 ADLs and 72 falls. The acquired sEMG signals were sent to an embedded PC through the Bluetooth connection, for the data analysis. The imbalance events were simulated with the use of a movable platform designed and built to induce imbalance conditions up and until to the subjects' fall. In Fig. 8.10 the functioning scheme of the platform is shown. The platform consists mainly of a crash mat (height 20 cm and with dimension of about 2 × 2 m) and a carriage of 40 × 40 cm. The volunteer was on the carriage driven by a tunable compressed air piston. Participants wore knee/elbow pad protectors while participating in testing, meeting safety and ethical requirements.

8.2.3 Software Architecture

The data acquired during the campaign described in previous section have been used to develop and to test the computational framework of the system in off-line mode. The study of the system has been developed on Mathworks Matlab. In the primary phase of the data elaboration, the noise caused by movement artifacts were reduced through a band-pass filter within a frequency range of (20,450) Hz. Moreover, for EMG-tension comparison, the signals were processed by generating their full wave rectification and their linear envelope [28]. This was carried out with the use of 10th order low-pass Butterworth filter, with a cut-off frequency of 10 Hz. In Fig. 8.11 it is reported an example of sEMG signals waveform coming from the four sensor channels during a bending action simulation, while in the Fig. 8.12 it is shown the

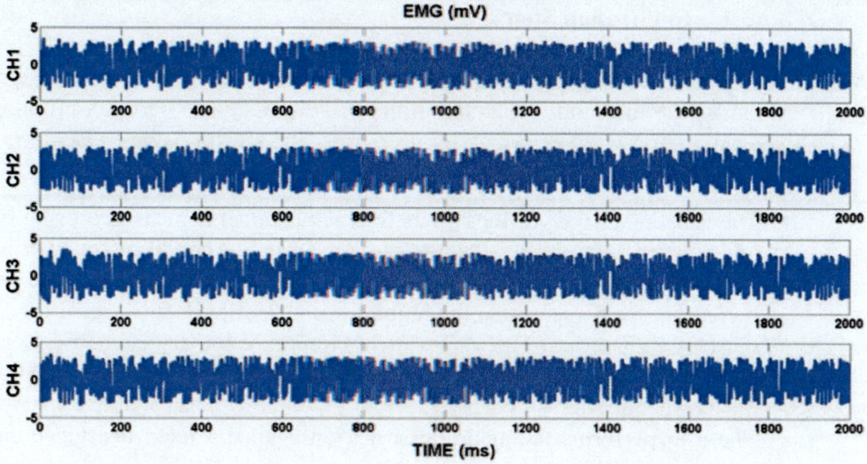

Fig. 8.11 Example of raw signals for the four sEMG channels, obtained during a bending action simulation

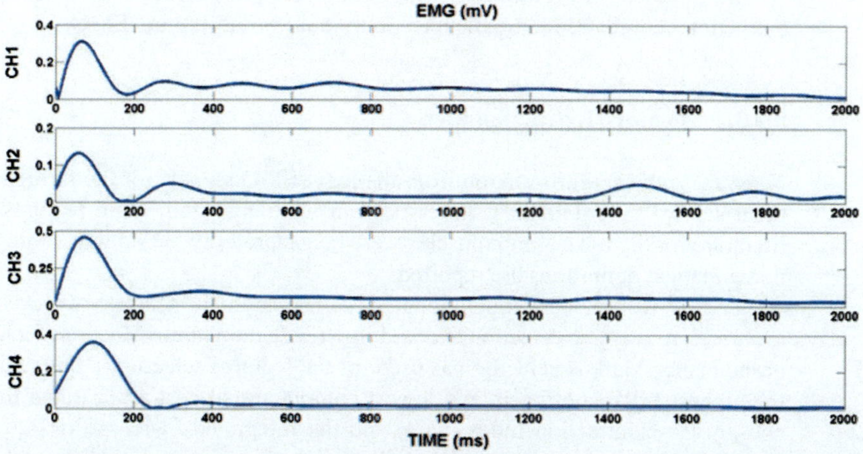

Fig. 8.12 Example of pre-elaborated signals for the four sEMG channels, obtained during a bending action simulation

waveforms after the pre-elaboration phase. It is clear how the noise was eliminated and the peak sEMG value due to the bending action was preserved.

8.2.3.1 Calibration Step

The calibration step was accomplished to measure the baseline of the signals and to reduce the inter-individual variability of sEMG signals between different users. The calibration was performed by users after the sEMG device was placed on the subjects. The calibration process can be divided in three main phases:

- The baseline of the sEMG signals for each channel is measured using the mean of the data acquired when the user remains in an idle condition for a period of 5 s;
- The user performs the ankle plantar flexion against a fixed resistance and holds it constant for 5 s to obtain the highest possible sEMG signal resulting from GL muscles contraction. The values of Maximum Voluntary isometric Contraction (MVC) is calculated taking the mean amplitude of the highest signal portion of the data acquired;
- The user performs the ankle dorsi flexion against a fixed resistance and holds it constant for 5 s to obtain the highest possible sEMG signal resulting from TA muscles contraction. The values of MVC is calculated employing the mean amplitude of the highest signal portion of the data acquired.

The values of MVC are used to normalize the pre-processed data of the feature extraction, thereby reducing the inter-individual variability of sEMG.

8.2.3.2 Feature Extraction Step

To extract relevant information from the legs sEMG signals for the fall risk assessment several time-domain features domain were analyzed. Main features used in literature for the lower-limb muscles were considered [29–31] and in Table 8.1 their mathematical definitions are reported.

For this work low computational cost, time-domain features were chosen to promote a responsive detection. According to [32], the Markov Random Field (MRF) based Fisher-Markov selector was used for the features selection. The features with the highest MRF coefficient and lowest computational cost at the same time were chosen: Co-Contraction Index (CCI) and the Integrated EMG (IEMG). The most significant feature is represented by CCI since it gives an estimation about the simultaneous activation of the Tibialis-Gastrocnemius antagonist muscles. The features were calculated considering a sliding window of 100 ms. The IEMG features were calculated for each muscle of interest, while for CCI the two pair of antagonist muscles were considered. So, the size of the feature vector for the classifier was six. In the Figs. 8.13 and 8.14 some examples of the waveforms for the features are reported.

Table 8.1 Equations of the main considered features

Feature	Mathematical definitions
Integrated EMG (IEMG)	$IEMG = \sum_{i=1}^{N} \lvert EMG_i \rvert$
Co-contraction index (CCI)	$CCI = \sum_{i=1}^{N} \left(\frac{lowEMG_i}{highEMG_i} \right) \times (lowEMG_i + highEMG_i) \times 100$ where $lowEMG_i$ is the EMG signal value for the less activity muscle, while the $highEMG_i$ is the corresponding activity of the higher active muscle
Mean absolute value (MAV)	$MAV = \frac{\sum_{i=1}^{N} EMG_i}{N}$
Root mean square (RMS)	$RMS = \sqrt{\frac{\sum_{i=1}^{N} EMG_i^2}{N}}$
Variance (VAR)	$VAR = \frac{\sum_{i=1}^{N} EMG_i^2}{N-1}$
Waveform length (WL)	$WL = \sum_{i=1}^{N-1} \lvert EMG_{i+1} - EMG_i \rvert$
Zero crossing (ZC)	$ZC = \sum_{i=1}^{N-1} \left[sgn(EMG_i \times EMG_{i+1}) \cap \lvert EMG_i - EMG_{i+1} \rvert \geq 0 \right]$ $sgn(EMG) = \begin{cases} 1 \ if \ EMG \geq thr \\ 0 \ if \ EMG < thr \end{cases}$ where threshold thr = 0.1 mV
Simple square integral (SSI)	$SSI = \sum_{i=1}^{N} \lvert EMG_i \rvert^2$
Slope sign change (SSC)	$SSC = \sum_{i=2}^{N} \left[f\left[(EMG_i - EMG_{i-1}) \times (EMG_i - EMG_{i+1}) \right] \right]$ $f(EMG) = \begin{cases} 1 \ if \ EMG \geq thr \\ 0 \ if \ EMG < thr \end{cases}$ where threshold thr = 0.1 mV
Willison amplitude (WAMP)	$WAMP = \sum_{i=1}^{N} f(\lvert EMG_i - EMG_{i+1} \rvert)$ $f(EMG) = \begin{cases} 1 \ if \ EMG \geq thr \\ 0 \ if \ EMG < thr \end{cases}$ where threshold thr = 0.05 mV

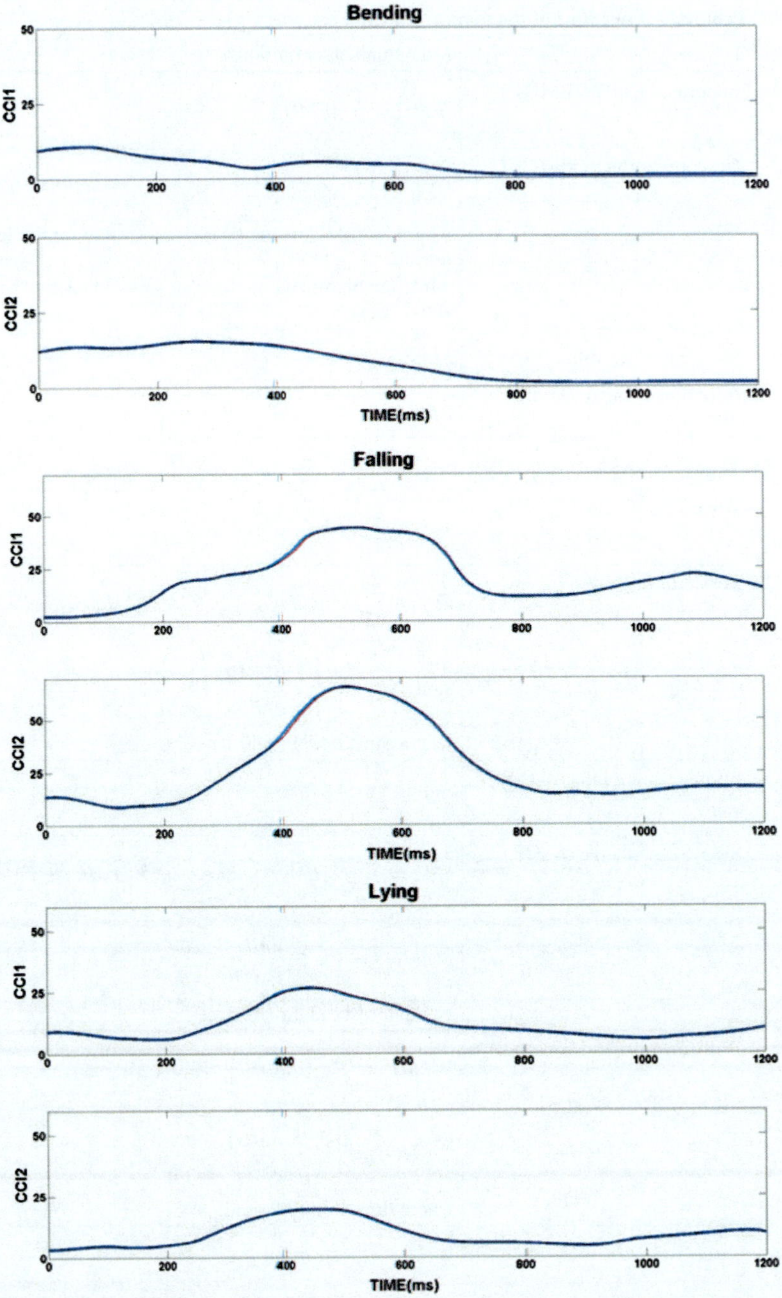

Fig. 8.13 Examples of co-contraction indices features extracted for bending, falling and lying events

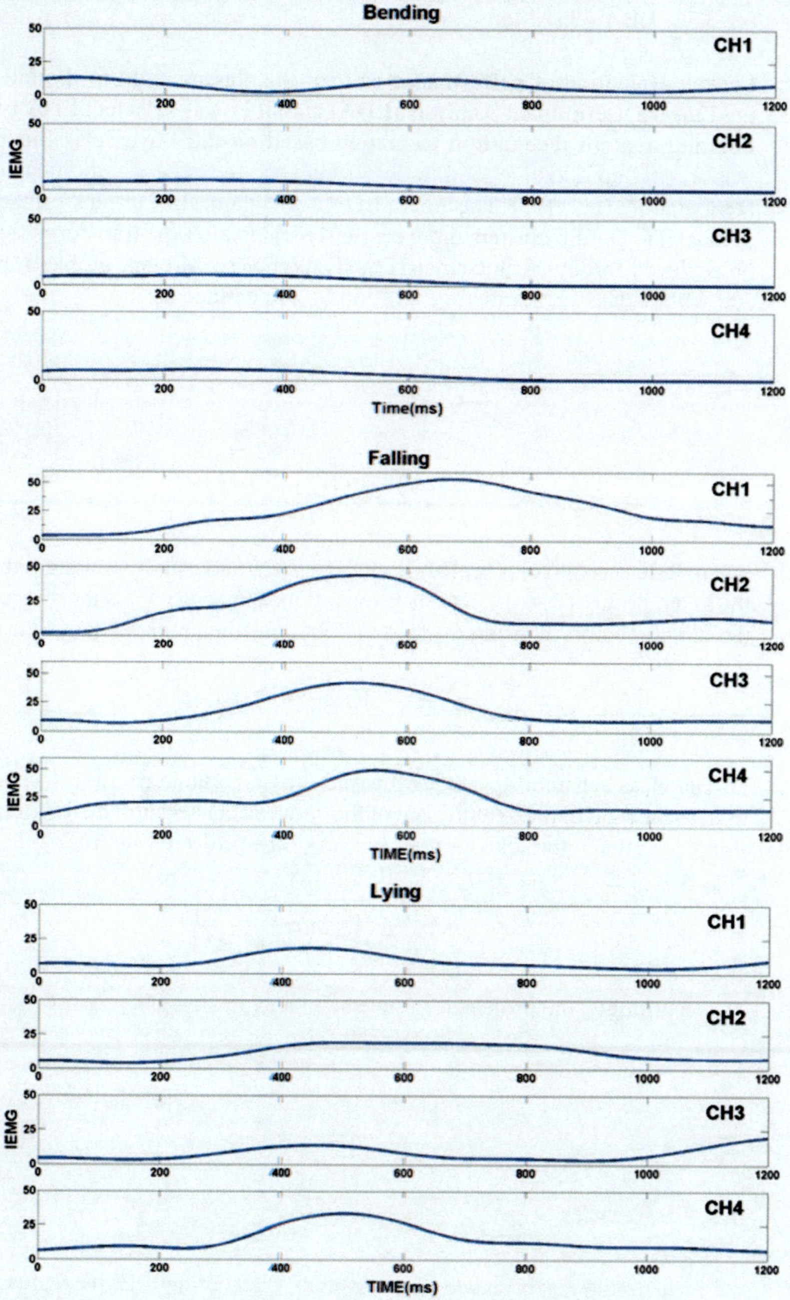

Fig. 8.14 Examples of integrated EMG features extracted for bending, falling and lying events

8.2.3.3 LDA Classifier

For the evaluation of performance and for the classification of the fall risk event, the Linear Discriminant Analysis (LDA) classifier was selected. LDA is a Machine Learning pattern recognition technique based on the Bayes classification rule. It was adopted to obtain a good trade-off between the generalization capabilities and computational cost [33]. The aim of LDA is to obtain a linear transformation in order to make the feature clusters more easily separable after the transformation. This can be achieved through scatter matrix analysis. For an M-class problem, the between and within class scatter matrices S_b and S_w are defined as:

$$S_b = \sum_{i=1}^{M} P\tau(C_i)(\mu_i - \mu)(\mu_i - \mu)^T = \phi_b \phi_b^T$$

$$S_b = \sum_{i=1}^{M} P\tau(C_i)\Sigma_i = \phi_W \phi_W^T$$

where $P\tau(C_i)$ is the prior probability of class (C_i) and usually is assigned to 1/M with the assumption of equal priors; μ is overall mean vector; $\sum_i$ is the average scatter of the sample vectors of different classes (C_i) around their representative mean vector μ_i:

$$\sum_i = \mathrm{E}\left[(x - \mu_i) - (x - \mu_i)^T | C - C_i\right]$$

The class separability can be measured by a certain criterion. (A) is commonly used as the ratio of the determinant of the between class scatter matrix of the projected samples to the within class scatter matrix of the projected samples:

$$(A) = \arg max \frac{|AS_b A^T|}{|AS_w A^T|}$$

Accordingly, the projected sample can be expressed as:

$$x_L = W_{opt}^T X$$

8.2.4 *Results*

To evaluate the performance of the system, the CCI and IEMG features have been calculated for all ADLs and unbalance events, simulated during the aforementioned acquisition campaign. In Table 8.2, the values of the chosen features obtained considering whole dataset.

Table 8.2 Mean and standard deviation of the features for the actions simulated during the data collection phase

Simulated actions		IEMG	CCI
Unbalance	Mean	39.2	32.5
	St. deviation	9.4	9.1
Lying	Mean	28.5	25.8
	St. deviation	8.3	5.7
Sitting	Mean	16.5	15.4
	St. deviation	7.6	5.8
Walking	Mean	11.2	9.8
	St. deviation	3.2	4.2
Bending	Mean	18.2	17.1
	St. deviation	6.1	5.8

The 10-fold cross-validation statistical method has been considered. It allows to give a good estimation about the generalization performance of the algorithm. The data have been portioned into 10 equally sized folds and 10 iterations of training and validations are performed; within each iteration, a different fold of the dataset has been used for the algorithm test and the remain part has been considered to perform the LDA training. The performance was analyzed in terms of sensitivity, specificity and lead of time before the impact [31]:

$$\text{Sensitivity} = \frac{\text{TP}}{\text{TP} + \text{FN}} \times 10$$

$$\text{Specificity} = \frac{\text{TN}}{\text{TN} + \text{FP}} \times 100$$

$$\text{T}_{\text{lead}} = \text{T}_{\text{detection}} - \text{T}_{\text{landing}}$$

where TP (True Positive) indicates that an imbalance event is induced and the algorithm detects it; FP (False Positive) indicates that an imbalance fall event does not occur and the algorithm activates an alarm; TN (True Negative) means that a daily event is performed and the algorithm does not detect it; FN (False negative) implies that an imbalance event occurs but the algorithm does not detect it. Moreover the $\text{T}_{\text{detection}}$ indicates the time when the pre-fall event is detected, $\text{T}_{\text{landing}}$ denotes the moment of the impact on the mat and T_{lead} is the lead-time before the impact. To calculate the period of time between the start of unbalance condition until the impact is sensed on the mat, the data coming from the IMU Xsens MTi10 sensor and the information provided by the time impact detection system integrated in the movable platform were analyzed. Based on the measured results the performance appears high, indeed the values of the specificity and sensitivity are respectively 81.3% $\pm$ 0.7 and 83.8% $\pm$ 0.3. Considering the evaluation of the ability to detect the unbalance event before the impact, the measured lead time is about 700 ms. This demonstrates the effectiveness of the realized wearable EMG-based system to detect fall in very fast way. These outcomes are close to those obtained with a similar analysis in which commercial, not very comfortable and not easy to use devices, were used [32]. Better performance can be obtained improving the adhesion interface between the electrode and skin in order to avoid the cases of signal degradation or loss.

8.3 Conclusion

In the paper new and low invasive surface Electromyography-based smart socks for the monitoring of the antagonist Gastrocnemius-Tibialis muscles is presented. The system is suitable for the evaluation of several diseases related to the lower limb movements and activities such as age-related changes in gait, fall risk, sarcopenia pathology, amyotrophic lateral sclerosis and other peripheral neuropathies. The performance of the developed hardware-software in terms of sensitivity, specificity and lead-time before the impact were high and the level of users' acceptability could be higher, regarding the sEMG/EMG-based wearable systems present in literature and in the market. The realized wearable sEMG-based system may cover a relevant rule in the healthcare applications addressed to monitor the elderly during their normal day-to-day activities in easy and effective way. Moreover, it may be used in the long-term muscular behavior monitoring for fall event recognition and impact protection systems activation. The used Machine Learning scheme is computationally low intensive, however it shows high performance in detection rate and generalization degree by ensuring low detection time.

So it allows for the increase of decision making time before a wearable airbag device activation. This may provide a significant contribution to enhance the effectiveness and reliability of wearable pre-fall systems. Future improvements could be addressed to improve the performance of the hardware system, increasing the lifetime of the battery and the system-level of the impermeability.

References

1. Joyce, N.C., Gregory, G.T.: Electrodiagnosis in persons with amyotrophic lateral sclerosis. PM & R: J. Injury Funct. Rehabil. **5**(5 Suppl), S89–95 (2013)
2. Chowdhury, R.H., Reaz, M.B., Ali, M.A., Bakar, A.A., Chellappan, K., Chang, T.G.: Surface electromyography signal processing and classification techniques. Sensors (Basel). **13**(9), 12431–12466 (2013)
3. Ghasemzadeh, H., Jafari, R., Prabhakaran, B.: A body sensor network with electromyogram and inertial sensors: multimodal interpretation of muscular activities. IEEE Trans. Inf. Technol. Biomed. **14**(2), 198–206 (2010)
4. Leone, A., Rescio, G., Caroppo, A., Siciliano, P.: A wearable EMG-based system pre-fall detector. Procedia Eng. **120**, 455–458 (2015)
5. Chung, T., Prasad, K., Lloyd, T.E.: Peripheral neuropathy: clinical and electrophysiological considerations. Neuroimaging Clin. N. Am. **24**(1), 49–65 (2013)

6. Andò, B., Baglio, S., Marletta, V.: A neurofuzzy approach for fall detection. In: 23rd ICE/IEEE ITMC Conference, Madeira Island, Portugal, 27–29 June 2017
7. Andò, B., Baglio, S., Marletta, V.: A inertial microsensors based wearable solution for the assessment of postural instability. In: ISOCS-MiNaB-ICT-MNBS, Otranto, Lecce, 25–29 June 2016
8. Bagalà, F., Becker, C., Cappello, A., Chiari, L., Aminian, K., Hausdorff, J.M., Zijlstra, W., Klenk, J.: Evaluation of accelerometer-based fall detection algorithms on real-world falls. PLoS ONE **7**, e37062 (2012)
9. Siciliano, P., Leone, A., Diraco, G., Distante, C., Malfatti, M., Gonzo, L., Grassi, M., Lombardi, A., Rescio, G., Malcovati, P.: A networked multisensor system for ambient assisted living application. Advances in sensors and interfaces. In: IWASI, pp. 139–143 (2009)
10. Rescio, G., Leone, A., Siciliano, P.: Supervised expert system for wearable MEMS accelerometer-based fall detector. J. Sens. **2013**, Article ID 254629, 11 (2013)
11. Blenkinsop, G.M., Pain, M.T., Hiley, M.J.: Balance control strategies during perturbed and unperturbed balance in standing and handstand. R. Soc. Open Sci. **4**(7), 161018 (2017)
12. Galeano, D., Brunetti, F., Torricelli, D., Piazza, S., Pons, J.L.: A tool for balance control training using muscle synergies and multimodal interfaces. BioMed Res. Int. 565370 (2014)
13. Park, S., Jayaraman, S.: Smart textiles: wearable electronic systems. MRS Bull. **28**, 585–591 (2013)
14. Matsuhisa, N., Kaltenbrunner, M., Yokota, T., Jinno, H., Kuribara, K., Sekitani, T., Someya, T.: Printable elastic conductors with a high conductivity for electronic textile applications. Nat. Commun. **6**, 7461 (2015)
15. Colyer, S.L., McGuigan, P.M.: Textile electrodes embedded in clothing: a practical alternative to traditional surface electromyography when assessing muscle excitation during functional movements. J. Sports Sci. Med. **17**(1), 101–109 (2018)
16. Posada-Quintero, H., Rood, R., Burnham, K., Pennace, J., Chon, K.: Assessment of carbon/salt/adhesive electrodes for surface electromyography measurements. IEEE J. Transl. Eng. Health Med. **4**, 2100209 (2016)
17. Kim, D., Abidian, M., Martin, D.C.: Conducting polymers grown in hydrogel scaffolds coated on neural prosthetic devices. J. Biomed. Mater. Res. **71A**, 577–585 (2004)
18. Mahmud, H.N., Kassim, A., Zainal, Z., Yunus, W.M.: Fourier transform infrared study of polypyrrole–poly(vinyl alcohol) conducting polymer composite films: evidence of film formation and characterization. J. Appl. Polym. Sci. **100**, 4107–4113 (2006)
19. Li, Y., Zhu, C., Fan, D., Fu, R., Ma, P., Duan, Z., Chi, L.: Construction of porous sponge-like PVA-CMC-PEG hydrogels with pH-sensitivity via phase separation for wound dressing. Int. J. Polym. Mater. Polym. Biomater. 1–11 (2019)
20. Green, R.A., Baek, S., Poole-Warren, L.A., Martens, P.J.: Conducting polymer-hydrogels for medical electrode applications. Sci. Technol. Adv. Mater. **11**(1), 014107 (2010)
21. Dai, W.S., Barbari, T.A.: Hydrogel membranes with mesh size asymmetry based on the gradient crosslinking of poly (vinyl alcohol). J. Membr. Sci. **156**(1), 67–79 (1999)
22. Li, Y., Zhu, C., Fan, D., Fu, R., Ma, P., Duan, Z., Chi, L.: A bi-layer PVA/CMC/PEG hydrogel with gradually changing pore sizes for wound dressing. Macromol. Biosci. 1800424 (2019)
23. Saadiah, M.A., Samsudin, A.S.: Study on ionic conduction of solid bio-polymer hybrid electrolytes based carboxymethyl cellulose (CMC)/polyvinyl alcohol (PVA) doped NH4NO3. In: AIP Conference Proceedings, vol. 2030, no. 1. AIP Publishing (2018)
24. Vieira, M.G.A., da Silva, M.A., dos Santos, L.O., Beppu, M.M.: Natural-based plasticizers and biopolymer films: a review. Eur. Polymer J. **47**(3), 254–263 (2011)
25. Mali, K.K., Dhawale, S.C., Dias, R.J., Dhane, N.S., Ghorpade, V.S.: Citric acid crosslinked carboxymethyl cellulose-based composite hydrogel films for drug delivery. Indian J. Pharm. Sci. **80**(4), 657–667 (2018)
26. http://www.advancertechnologies.com
27. https://www.dfrobot.com
28. De Luca, C.J., Gilmore, L.D., Kuznetsov, M., Roy, S.H.: Filtering the surface EMG signal: movement artifact and baseline noise contamination. J. Biomech. **43**(8), 1573–1579 (2010)

29. Phinyomark, A., Chujit, G., Phukpattaranont, P., Limsakul, C., Huosheng, H.: A preliminary study assessing time-domain EMG features of classifying exercises in preventing falls in the elderly. In: 9th International Conference on Electrical Engineering/Electronics, Computer, Telecommunications and Information Technology (ECTI-CON), pp. 1, 4, 16–18 (2012)
30. Horsak, B., et al.: A. Muscle co-contraction around the knee when walking with unstable shoes. J. Electromyogr. Kinesiol. **25** (2015)
31. Mansor, M.N., Syam, S.H., Rejab, M.N., Syam, A.H.: Automatically infant pain recognition based on LDA classifier. In: 2012 International Symposium on Instrumentation & Measurement, Sensor Network and Automation (IMSNA), Sanya, pp. 380–382 (2012)
32. Rescio, G., Leone, A., Siciliano, P.: Supervised machine learning scheme for electromyography-based pre-fall detection system. Expert Syst. Appl. **100**, 95–105 (2018)
33. Wu, G., Xue, S.: Portable preimpact fall detector with inertial sensors. IEEE Trans. Neural Syst. Rehabil. Eng. **16**(2), 178–183 (2018)

Chapter 9
Describing Smart City Problems with Distributed Vulnerability

Stefano Marrone

Abstract Modern material and immaterial infrastructures have seen a growth of their complexity as well as the criticality of the role played in this interconnected society. Such a growth has brought to a need for protection in particular of vital services (e.g., electricity, water supply, computer networks, etc.). This chapter introduces the problem of defining in mathematical terms a useful definition of vulnerability for distributed and networked systems: this definition is then mapped onto the well-known formalism of Bayesian Networks. A demonstration of the applicability of this theoretical framework is given by describing the distributed plate car recognition problem, one of the possible faces of the smart city model.

9.1 Introduction

The availability of a massive amount of data has enabled the massive application of machine learning and deep learning techniques across the domain of computer-based critical systems. A huge set of automatic learning frameworks are now available and are able to tackle with different kinds of systems, enabling the diffusion of Big Data analysis, cloud computing systems and (Industrial) Internet of Things (IoT). As such applications become more and more widespread, data analysis techniques have shown their capability to identify operational patterns and to predict future behavior for anticipating possible problems. A widespread example is constituted by the installation of cameras inside urban areas; such cameras are used for different purposes that come from traffic to city light management: the images produced by these cameras can be easily and automatically used for other purposes.

On the other hand, models have been extensively used in many computer-intensive activities: one over the others, dependability formal assessment of critical infrastructures (CIs).

S. Marrone (✉)
University of Campania "Luigi Vanvitelli", viale Lincoln, 5, Caserta, Italy
e-mail: stefano.marrone@unicampania.it

G. Phillips-Wren et al. (eds.), *Advances in Data Science: Methodologies and Applications*, Intelligent Systems Reference Library 189,
https://doi.org/10.1007/978-3-030-51870-7_9

One of the main challenges of CIs design and operation management is the quantification of critical aspects such as resilience [6, 7] and security [41] in order to support evidence driven protection mechanisms against several known and unknown threats. Modern infrastructures are demanded to realize more and more critical functions (i.e., to guarantee that this security level fits in the requirements set by customers and/or international standards). These infrastructures are characterized by internal complexities as well as a high degree of inter-dependency among them. This results in an under-specified nature of operations in complex systems that generates potential for unforeseeable failures and cascading effects [8]. Thus, the higher the complexity, more credible that protection systems present exploits and vulnerabilities.

The advantages of the integration between data-driven and explicit knowledge are numerous: (a) to scale up complexity of data analysis allowing reducing size in real world problems; (b) to boost human activities in the supervision of complex system operations; (c) to improve the trustworthiness of the system models built manually; (d) to enhance the accuracy of the results predicted with the analysis; (e) to support the creation of models-at-runtime that is to align models with data logged by the system in operation; (f) to enable automatic validation of models extracted by data mining.

This chapter wants to describe one of these modeling approaches, the *distributed vulnerability* formally defining it.

After recalled and formalized the main concepts of distributed vulnerability, the chapter defines a mapping between this formalism and languages that could ease the capability of analyzing and evaluating the distributed vulnerability. This chapter focuses on Bayesian Networks (BNs) as a tool to easily implement the defined mathematical approach.

The third objective of the chapter is to discuss the application of such framework to a Smart City problem: the problems of image processing and computer vision is License Plate Clone Recognition.

The structure of the chapter is the following: this Section introduces the problem and motivates the chapter. Section 9.2 discusses related works while Sect. 9.3 provides some information needed to understand the chapter easily. Section 9.4 gives a formal definition of the distributed vulnerability concept. Section 9.5 presents the mapping from this language to BNs. Section 9.6 describes the case study with its functions and components: Sect. 9.7 applies the modeling approach to such a problem. Section 9.8 ends the chapter discussing the results and addressing for further research.

9.2 Related Works

This Section provides a first subsection on related scientific works on the theme of Smart City monitoring modeling and analysis by formal methods (Sect. 9.2.1), on the aspects related to critical infrastructures vulnerability assessment (Sect. 9.2.2) and on the topics of the improvement of detection reliability by means of formal modeling (Sect. 9.2.3).

9.2.1 Smart City and Formal Methods

Modeling and trend forecasting as well as early warning systems are prime features in the construction of the future Smart Cities. A good synthesis of the present state of the art and of the future challenge in this topic is reported in [26]: here, authors also highlight the importance of modeling in general. In [24], the importance of modeling approach in managing critical infrastructures is introduced also in the context of resilience, while in [9], the model of critical infrastructure is used to implement a decision support system for system. Another interesting starting point is represented by the paper [1] where Big Data and the supporting modeling approaches are described as one of the enablers of Smart Communities and Cities. A practical example is represented by [8], where Big Data is used to manage smart city critical infrastructure more effectively.

Application of formal modeling and analysis to the structure and the operations of a Smart City is a topic explored in the scientific literature. There are several papers focusing on specific aspects of a Smart City and Bayesian Networks: urban traffic accident management is discussed in [36, 47]; [21, 45] are two similar works where authors use BN modeling and analysis for the security assessment of water networks; BNs have also been applied to predict the future development of the urbanization of an entire area [49]; in [25] Bayesian inference is at the base of early warning systems; BNs have been also applied in the smart management of lifts in Smart Buildings [4].

Smart Cities have been studied also by using Generalized Stochastic Petri Nets (GSPNs) and, more in general, Petri Nets (PNs): urban traffic models have been defined and applied to predict critical blocks in [31] while public transportation have been studied in [11] and in [37] where Stochastic Activity Networks are used to predict the performability of metro systems; energetic aspects of Smart Homes—the elementary cell of a sustainable Smart City—are studied in [22] by means of model-driven and Fluid Stochastic Petri Nets and with PNs in [15]; sustainable waste management systems and their PN models are the center of the work in [14].

9.2.2 Critical Infrastructures Vulnerability

Modeling and evaluation of qualitative and quantitative properties of CIs have attracted the interest of the scientific community with a special focus on CIs interdependencies [35, 40]. A previous definition of distributed vulnerability has been proposed in the field of information security [5] while from a graph theoretic point of view a similar definition are in network evaluation of vulnerability [44].

Various approaches have been taken in the literature to vulnerability model-driven approaches for both information systems and critical infrastructures: UML-CI is a UML profile aiming to define different aspects of an infrastructure organization and behavior [3]; the CORAS method is oriented to model-driven risk analysis of changing systems [32]; UMLsec [27] allows specifying security information during the development of security-critical systems and provides tool-support for formal security verification. A recent research work explores the joint application of two model-driven approaches involving UML Profiles and quantitative formal methods [33]: such approaches are CIP_VAM and SecAM.

More in general, model-based approaches for security evaluation and assessment counts: Defense Trees (an extension of attack trees [34], Attack Response Trees incorporate both attack and response mechanism notations [51], Generalized Stochastic Petri Nets (GPSNs) [18] and BNs [48]. Notwithstanding their flexibility, DBNs have received few attentions from the scientific community [20, 46].

9.2.3 Detection Reliability Improvement

But such systems are also used for homeland security and public trust maintenance purposes. These topics can be framed into the wider context of physical security technologies and advanced surveillance paradigms that, in recent times, creates a research trend called Physical Security Information Management (PSIM) systems; a comprehensive survey of state-of-the-art is provided in [19]. Modern remote surveillance systems for public safety are also discussed in [39]. Technology and market-oriented considerations on PSIM can be also found in [10].

On the other hand the problem of improving the reliability of detection is reported in the scientific literature as classification reliability improvement that is a problem traditionally dealt with Artificial Intelligence techniques. In this research field multi classifier systems have been developed in order to overcome the limitations of traditional classifiers: a comprehensive review of such topic is in [38].

Bayesian Networks and Dynamic Bayesian Networks (DBNs) are a widespread used formalism in Artificial Intelligence and recent research trends apply them in reliability of critical systems such as in [17, 30]. BNs and DBNs have been also used to multi classifier systems to improve classification reliability [23, 43]. Other approaches see BNs (and more in general formal methods) applied also in detection reliability estimation in PSIM [16].

9.3 The Bayesian Network Formalism

BNs [29], also known as belief networks, provide a graphical representation of a joint probability distribution over a set of random variables with a possible mutual causal relationship. The network is a directed acyclic graph (DAG) whose nodes represent random variables and arcs represent casual influences between pair of nodes (i.e., an arc stands for a probabilistic dependence between two random variables). A Conditional Probability Distribution is the function defined for each node in the network which defines how the values of a node are distributed according to the values assumes by parent nodes. A priori probabilities should be provided for the source nodes of the DAG as they have no parents. For discrete random variables, the CPD is often represented by a table (Conditional Probability Table, CPT).

Founded on the Bayes theorem, BNs and their derivatives allow for inferring the posterior conditional probability distribution of an outcome variable based on observed evidence as well as a priori belief in the probability of different hypotheses. Being X an ancestor of Y, there are three different kinds of analysis [29]: (1) Prior belief: $Pr(Y = y)$, the probability that Y has the value y in absence of any observations; (2) Predictive Analysis: $Pr(Y = y|X = x)$, the probability that Y has the value y when the value x is observed for the variable X; (3) Diagnostic Analysis: $Pr(X = x|Y = y)$, the probability that X has the value x when the value y is observed for the variable Y.

9.4 Formalising Distributed Vulnerability

Let us for sake of clarity denote the interval [0, 1] of $\mathbb{R}$ with $\mathbb{R}_{[0,1]}$. In this section a simplified formalization of the notion of the distributed vulnerability is done focusing on the aspects that are strictly related to the case study.

Let a detection system S be the following tuple:

$$S = \langle EV, AS, SE, \mathscr{T}, \mathscr{A}, \mathscr{C} \rangle \tag{9.1}$$

so that:

$$EV = \{e_1, e_2, \ldots, e_{|EV|}\}; \tag{9.2}$$

$$AS = \{rl_1, rl_2, \ldots, rl_{|AS|}\}; \tag{9.3}$$

$$SE = \{d_1, d_2, \ldots, d_{|SE|}\}; \tag{9.4}$$

$$\mathscr{T} \subseteq EV \times SE \times \mathbb{R}_{[0,1]}; \tag{9.5}$$

$$\mathscr{A} \subseteq SE \times AS \times \mathbb{R}_{[0,1]}; \tag{9.6}$$

$$\mathscr{C} : AS \longrightarrow \mathbb{R}_{[0,1]}; \tag{9.7}$$

Equation (9.2) defines a set of events that may occur in the system (EV); Eq. (9.3) defines a set of assessment functions (AS); Eq. (9.4) defines a set of sensor devices (SE); elements of the relation $\mathscr{T}$, see Eq. (9.5), are tuple (a, b, p_a^b) saying that the event a triggers the activation of the sensor b with a probability p_a^b; elements of the

Table 9.1 Discrete semantics of the CI elements

	ok	ko
$a \in EV$	The attack phase is successful	The attack phase fails or it has not been attempted
$r \in AS$	The rule has been activated and The threat detected	The threat has not been detected
$d \in SE$	The sensor has raised a warning	The sensor is not producing any alarm

relation $\mathscr{A}$, see Eq. (9.6), are tuple (a, b, p_a^b) saying that the sensor a activates the assessment function b with a probability p_a^b; the function $\mathscr{C}$, see Eq. (9.7), relates the assessment function a with the probability $\mathscr{C}(a)$ to raise an alarm.

Let $E = EV \cup SE \cup AS$. Moreover, let us assign $\forall e \in E$ a random variable χ_e. Let us also notate with $\mathscr{E}\mathscr{R} \subseteq E \times E$ the relation containing all the elements of the relations already defined: e.g., $(a, b, p_a^b) \in \mathscr{A} \cup \mathscr{T}$.

Let now define the *failure* of a system as a function:

$$f : SE \longrightarrow \mathbb{R}_{[0,1]} \tag{9.8}$$

representing the probability that $d \in SE$ is compromised according to a-priori knowledge about the occurrence of unexpected events; the *distributed failure* is

$$df = \langle f(d_1), f(d_2), \ldots, f(d_{|SE|}) \rangle \in \mathbb{R}_{[0,1]}^{|SE|} \tag{9.9}$$

and it represents the probability of failure of each node of the infrastructure.

Let us suppose that $\forall e \in E, \chi_e \in \mathbb{B} = \{ok, ko\}$. The semantics is summarized in the Table 9.1.

Hence, let us specialise the definitions previously given by clearly stating what we intend for attack and sensing patterns. An attack pattern ap can be as a tuple of $|EV|$ elements of $\mathbb{B}$ while a sensing pattern sp as a tuple of $|SE|$ elements of $\mathbb{B}$:

$$ap \in \mathbb{B}^{|EV|} \tag{9.10}$$

$$sp \in \mathbb{B}^{|SE|} \tag{9.11}$$

In summary, the vulnerability is the probability of successful attack given the occurrence of a threat:

$$V = Pr(\text{not being detected}|\text{occurrence of the threat}) \tag{9.12}$$

so, the vulnerability of the system for the i-th alarm according to the j-th attack pattern is the following:

$$v_{i,j} = Pr(\text{j-th alarm does not raise}|\text{occurrence of the i-th threat}) \tag{9.13}$$

that becomes

$$v_{i,j} = \mathscr{C}_{ap_j}(rl_i); \tag{9.14}$$

Here, the concept of the distributed vulnerability, in response to an attack pattern ap could be defined as following:

$$dv = \langle v_1(ap), v_2(ap), \ldots, v_n(ap)\rangle \in \mathbb{R}_{[0,1]}^{|AS|} \tag{9.15}$$

9.5 Implementing Distributed Vulnerability with Bayesian Networks

This Section has the aim to show how to use Bayesian Networks in to implement functions stated in the previous Section.

Traditional discrete BNs are used to implement the concept of distributed vulnerability.

First, let us give some modelling indications to build a BN model from a formalisation of a CI as shown before. This is given by defining the sets of the node, the arcs and the CPTs. A **BN node** is generated from each element in E: each of these nodes is binary and can assume the values $\{ok, ko\}$. The function bn returns for each element of E such generated node. The links between the nodes are generated according to $\mathscr{ER}$: if $(a, b) \in \mathscr{ER}$ then $bn(a)$ is a parent of $bn(b)$. At this point, a restrictive assumption is made on the model, supposing that $\mathscr{ER}$ does not contain any cycle.

The third aspect is the definition of the CPTs; they are built basing on the relations of CI.

Events

Say a the attack step under consideration, aa the previous attack step conducted with a probability p^a_{aa} and m the counteracting countermeasure stopping a with a probability p^a_m: CPTs are built according to the Table 9.2.

Sensors

Say d the sensor under consideration, s the service (of another infrastructure) triggering d with probability p^d_s, a the attack triggering d with probability p^d_a and rl the assessment rule sensitizing d: CPTs are built according to the Table 9.3.

Table 9.2 CPT of $bn(a)$

	ok	ko
$(aa = ko)$	0	1
$(aa = ok) \wedge (m = ko)$	$p^a_{aa}(1 - p^a_m)$	$1 - p^a_{aa}(1 - p^a_m)$
Otherwise	0	1

Table 9.3 CPT of $bn(d)$

	ok	ko
$(rl = ko)$	0	1
$(rl = ok) \wedge (s = ko) \wedge (a = ko)$	p^d_s	$1 - p^d_s$
$(rl = ok) \wedge (s = ok) \wedge (a = ok)$	p^d_a	$1 - p^d_a$
$(rl = ok) \wedge (s = ko) \wedge (a = ko)$	$1 - (1 - p^d_a)(1 - p^d_s)$	$(1 - p^d_a)(1 - p^d_s)$

Table 9.4 CPT of $bn(rl)$

	ok	ko
$(d = ok)$	1	0
$(d = ko)$	0	1

Assessment Functions

Say rl the assessment rule under consideration and d the sensing device triggering rl: CPTs are built according to the Table9.4.

All of these cases can be extended when there are more than one occurrence per parent type. As example, if we consider that if there is more than one sensor as input to an assessment rule, all must be ok in order to activate the rule.

Computing the posterior probability $P(x = ko \mid a_1 = ok, ..., a_k = ok, a_{k+1} = ko, a_{k+2} = ko ..., a_n = ko)$ on the BN model means to calculate the probability of having a malfunctioning of the component in case of attack. According to the given definitions, it represents the vulnerability function $v(x, \{a_1, a_2, ..., a_k\})$.

BN analysis algorithms allow to evaluate the posterior probability of all the nodes of the model efficiently: thus, this formalism suits to compute the distributed vulnerability function $dv(\{a_1, ..., a_k\})$.

9.6 The Clone Plate Recognition Problem

In this Section an overall description of the studied problem is presented.

The basic idea is to make a communication between different geographically separated towns or cities equipped with an existing traffic monitoring systems. Two constraints must be satisfied by such systems: (1) the presence of digital cameras, (2)

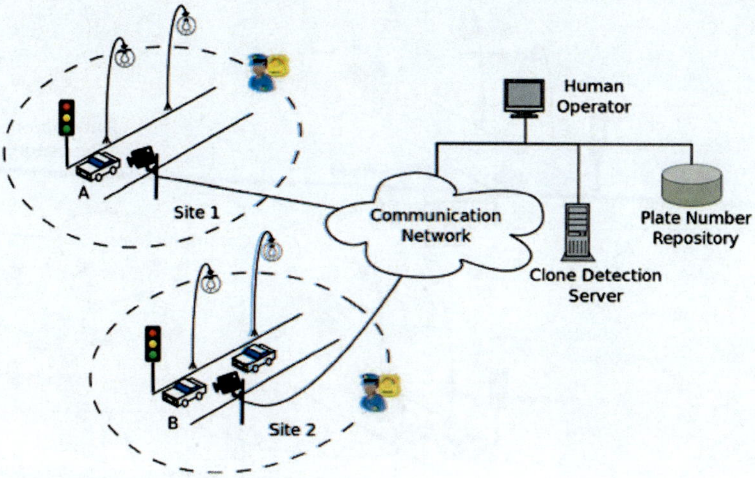

Fig. 9.1 The *naive centralized* architecture

the presence of large bandwidth communication network access. The main objective is to support police in detecting cloned cars and other kind of frauds: when two distinct cars are detected in the same time in different places, the system can raise an alarm.

In order to better clarify this idea, Fig. 9.1 depicts the overall architecture supporting the approach. Let us consider two cars, A and B, that transit in two different sites and let us consider that the license plate number of A has been cloned from the one of B. Cameras in both sites continuously grab car images transmitting them to the *Clone Detection Server*. In this server, a LPR software takes as inputs these images extracting the related licence plate number and storing them into a *Plate Number Repository* with the timestamp and the location of the cameras where it has been detected. On the Clone Detection Server, a specialized software is in charge to correlate the data present in the Plate Number Repository and to determine possible car clones: this software module is called *Plate Matcher*. When a match has been detected, another specialized software running on Clone Detection Server estimates the reliability of the detection in order to minimize both *fpr* and *fnr*: this software module is called *Detector Likelihood Estimator*. When the likelihood of the detection is calculated, the cases that present high likelihood values can be reported to a human operator allowing the assessment of the alarm and then notification to the local police departments.

This architecture is called *Naive Centralized* since it is the most basic architecture supporting our clone detection approach. Obviously, against its simplicity, this approach is very inefficient because sites send to the center grabbed images and because the Clone Detection Server is a performance bottleneck of the system. An easy improvement of this architecture is constituted by the *Centralized* architecture that is depicted in Fig. 9.2. In this model, each site is equipped with a LPR Server

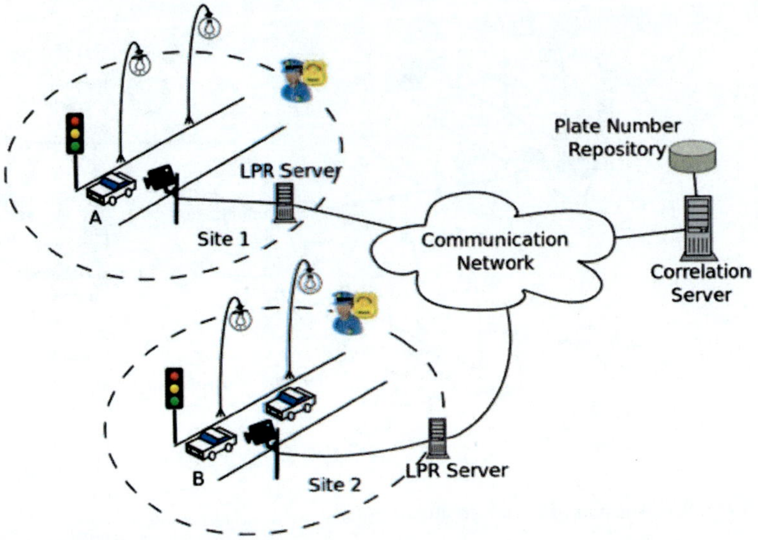

Fig. 9.2 The *centralized* architecture

that is in charge of extracting plate number from the images produced by the cameras of the site: then, recognized plate numbers can be sent to the center where they are stored in the Plate Number Repository. On the Detection Server, they run both Plate Matcher and Detector Likelihood Estimator software.

Figure 9.3 defines a *Decentralized* architecture where the functionalities offered by the Clone Detection Server are distributed over the sites. Each site has a Clone Detection Server that offers the same functionalities of the centralized server in the Naive Centralized model (License Plate Recognition, Plate Matcher and Detector Likelihood Estimator software modules). Each decentralized server is equipped with a Visitor Location Register (VLR). In the architecture there is a single Home Location Register (HLR) that is a repository that stores the site in which each license plate is currently detected. When a car is detected in a site, the local Clone Detection Server inserts its plate number into its VLR and queries the HLR in order to determine if other sites are currently seeing this license plate number in their areas.

If the plate number is not present in the HLR, the site registers itself in the HLR as the "home site" of the plate number and as long the Clone Detection Server sees the car inside its area: the server periodically refreshes such information onto the HLR. At this point if another site detects a plate number another site had already registered in the HLR, the second site sends the data about the detection to the "home site". The responsibility to detect a cloning event is in charge of the "home site". Since all the functionalities of the system are decentralized onto the different sites, this schema is very scalable: the only performance bottleneck is constituted by the HRL that is theoretically queried each car detection: its performance would benefit of caching mechanisms.

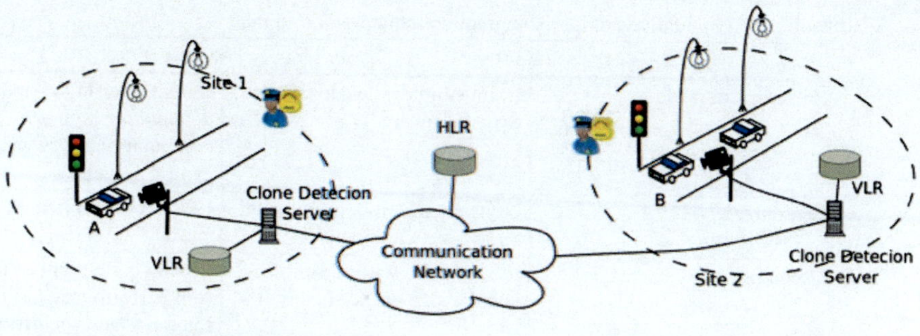

Fig. 9.3 The *decentralized* architecture

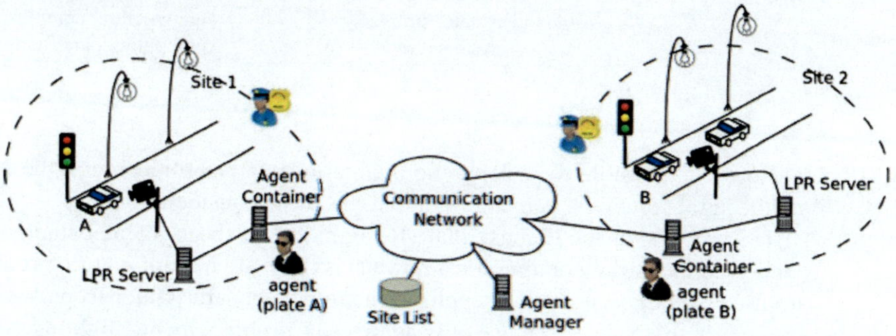

Fig. 9.4 The *distributed* architecture

Figure 9.4 represents a schema where the detection is fully distributed over sites: such result is reached by using mobile agent computing paradigm.[1] With respect to the previous case where the information of the plate recognition sequence are stored into databases that are local to the sites, in this case the state of the detection for a single plate number is in charge of a mobile software agent that can move across the network. When a site detects a plate number, a software agent is created inside the local agent container in order to manage this plate number. Then it clones itself and starts to move one replica across the sites in order to find other mobile agents dealing with the same plate number. If found, the two agents merge to each other and make a decision about the cloning of the plate number. According to the mobile agents research, all the non functional properties of this software system such as persistence, consistency, security (in both the senses of data integrity and privacy) can be guaranteed by adopting the proper architectural facilities. A further discussion of these architectural elements are out of the scope of this chapter. Two things are worth to note: (1) according to the computing paradigm, when moving mobile agents bring application code and data; (2) the only centralized elements is the list of the sites that changes very slowly and thus can be easily cached.

[1] We suppose the reader is acquainted to this computing paradigm: for further details see [13].

Table 9.5 Qualitative comparison among architectural schemes

	Pros	Cons
Naive centralized	Extra simple; sites do not need extra hardware	Demand for high bandwidth; demand for high computational power of the central server
Centralized	Quite simple; it does not need large bandwidth network	LPR server replicated on each site
Decentralized	It does scale with the number and the size of the sites	Still a single point of failure (not performance but fault tolerance and security) is present (HLR); quite complex
Distributed	Fully scales with growth of the system; simple system architecture	Complex software architecture; complex computing paradigm

Now it is possible to make some qualitative comparisons among the proposed architectural solutions. Table 9.5 summarizes these considerations.

The problem of the License Plate Recognition has been already studied by academic and industrial communities: nowadays there are a lot of mature products that are used everyday in a lot of application from road safety tutor to parking billing systems. In this Subsection, we only address the problem by highlighting the critical aspects of this phase. First applications of LPR systems go back to the 1979: a more recent survey on technologies and methods for LPR is in [50] while recent research improvements are in [12, 28]. Hw/sw techniques had made big improvements in this field allowing more than 90% of successful recognition in different climate and enlightenment conditions. Several image features affect the confidence of the recognition: quality of the image that can be expressed in terms of resolution of the digital camera; the camera positioning, i.e. the angle at which the camera is positioned, the distance of the object, the level of enlightenment, the quantity of rain as well the quantity of mist, etc.

The LPR module has the responsibility to recognize the car plate number from the acquired image and to estimate the likelihood associated to such recognition. In order to accomplish to this objective, LPR module requires as input not only the image with the car plate to detect but also some meta-data as modelled in Fig. 9.5.

Figure 9.6 depicts the order in which the four phases of the LPR process are accomplished: Reliability Estimation, Car Plate Detection and Number Extraction.

Car Plate Detection and *Number Extraction* phases are traditionally based on pattern recognition and artificial intelligence methods; the scientific literature is rich of approaches and algorithm for both the problems: thus they will not further been studied in this chapter. *Reliability Estimation* has a twofold scope: it solves the problem of choosing the best algorithm for the detection and it allows a quick estimation of the reliability of the detection. Some scientific works have analyzed different recognition algorithm trying to classify their effectiveness under image features [2, 42].

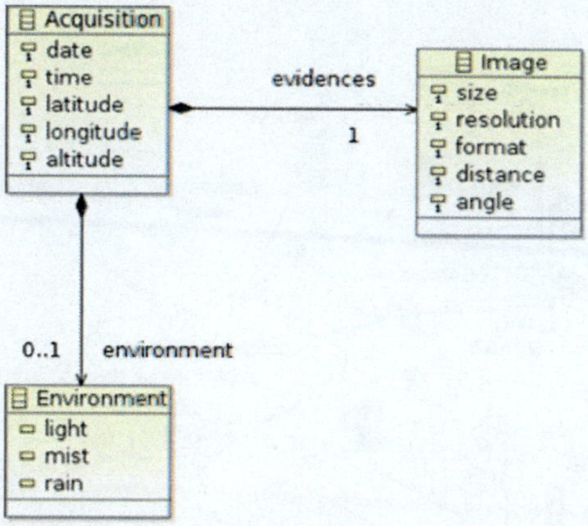

Fig. 9.5 *Acquisition* domain model

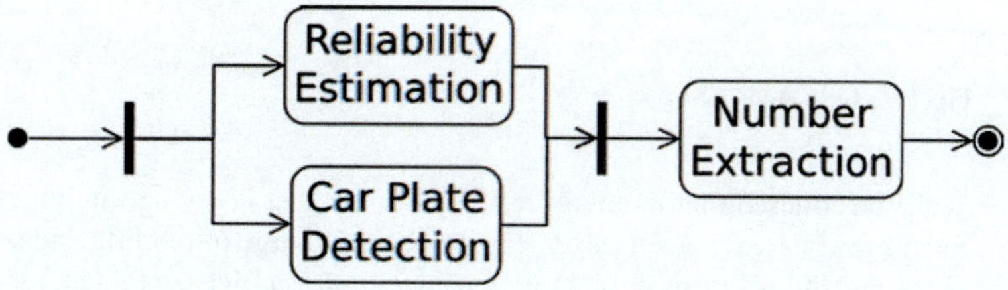

Fig. 9.6 The LPR process schema

Here, some of these affecting features are considered: image angle, object distance, enlightenment as well as weather conditions.

9.7 Applying Distributed Vulnerability Concepts

This section has the aim to apply the formalization of the distributed vulnerability to the clone license plate recognition system. The final objective is to show how such formalism can boost the possibility to have a quantification of the effectiveness of such a system: such quantification, on the other hand, is hard to obtain simply by testing the applications for a short time. Furthermore, by means of a formal model, organization in charge of operating such a system could also tune recognition parameters to maximize efficiency (in terms of minimize false positive and false negative events). Since the formalization is made at application level and a discussion of the performance/scalability issues of the recognition application is not in the scope of this chapter, we consider, for sake of the simplicity, a centralized approach.

This situation is summarized in Fig. 9.7.

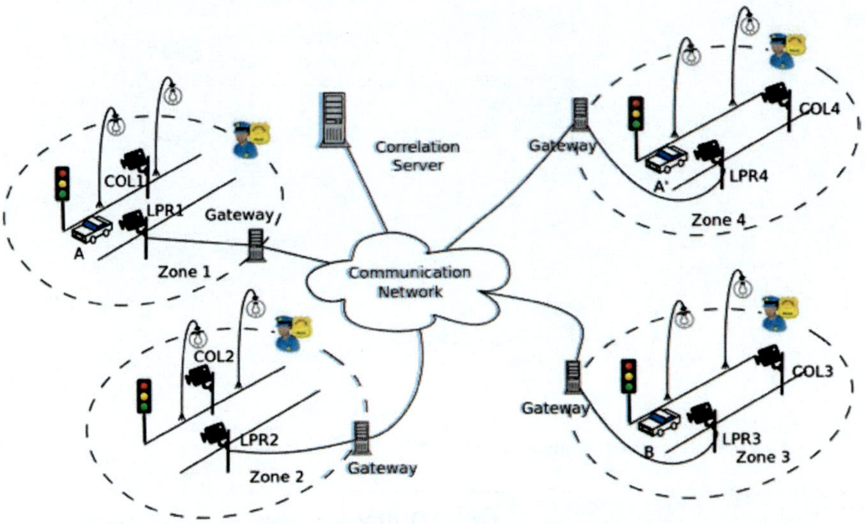

Fig. 9.7 Case study

To be concrete, let us imagine a Smart City that is divided in zones or districts.[2] Each zone has a set of smart sensing devices in charge of reading and recognizing the plates and the colors of the cars. We want to model a License Plate Clone Recognition (LPCR) application running on the correlation server. Furthermore, let us suppose three cars: A, B and A' where A and B have different license plates and A' is a clone of A (with a different color).

Now, in conformance with the notions formalized in Sect. 9.4, we may say that our LPCR application is modelled by:

$$LPCR = \langle VE, DEV, CORR, \mathscr{T}, \mathscr{A}, \mathscr{C} \rangle \tag{9.16}$$

where:

$$VE = PLATES \cup COLORS \tag{9.17}$$

is the set of the possible events, parted in two subsets:

$$PLATES = \{PLATE\_A\_1, \ldots, , PLATE\_A\_4, PLATE\_B\_1, \ldots\} \tag{9.18}$$

representing the possible plate recognition events and:

$$COLORS = \{RED\_1, \ldots, RED\_4, BLUE\_1, BLUE\_2, \ldots\} \tag{9.19}$$

[2]Smaller the zones, finer the grain of the detection.

representing all the possible color recognition events. Furthermore we have:

$$DEV = \{LPR\_1, \ldots, LPR\_4, COL\_1, \ldots, COL\_4\} \tag{9.20}$$

that is the set of the devices present in the system. We can now define our correlation logic rules saying that a car/vehicle cannot be present in two different zones within the same time interval δ_T ($RULE^A_{i,j,\delta_T}$) and that the same car cannot have two different colors within the same time interval δ_C ($RULE^B_{i,\delta_C}$)

$$RULE^A_{i,j,\delta_T} : (LPR\_i(t) = LPR\_j(t')) \wedge (COL\_i(t) \neq COL\_j(t')) \\ \text{with } |t - t'| \leq \delta_T \wedge i \neq j \tag{9.21}$$

$$RULE^B_{i,j,\delta_C} : (LPR\_i(t) = LPR\_j(t')) \wedge (COL\_i(t) \neq COL\_j(t')) \\ \text{with } |t - t'| \leq \delta_C \tag{9.22}$$

Consequently, the set of all the assessment rules of the system is parametrized in both δ_T and δ_C:

$$DEV_{\delta_T,\delta_C} = \bigcup_{i=1}^{4} \bigcup_{j=1 \wedge i \neq j}^{4} \left\{RULE^A_{i,j,\delta_T}\right\} \cup \bigcup_{i=1}^{4} \bigcup_{j=1}^{4} \left\{RULE^B_{i,j,\delta_C}\right\} \tag{9.23}$$

Let us suppose that the LPR devices are all of the same kind—i.e., have the same performance: the same with COL sensors. For what concerns the $\mathscr{T}$ relation, this set has three kind of elements:

$$(PLATE\_X\_i, LPR\_i, p_x) \forall x \in \{A, B\}, \forall i \in \{1, \ldots, 4\} \tag{9.24}$$

where p_x is the probability to detect a plate x, or

$$(RED\_i, COL\_i, p_{red}) \forall i \in \{1, \ldots, 4\} \tag{9.25}$$

where p_{red} is the probability to detect the RED color, or

$$(BLUE\_i, COL\_i, p_{blu}) \forall i \in \{1, \ldots, 4\} \tag{9.26}$$

where p_{blu} is the probability to detect the BLUE color. Furthermore, elements of $\mathscr{A}$ are:

$$(s, r, p_s) \tag{9.27}$$

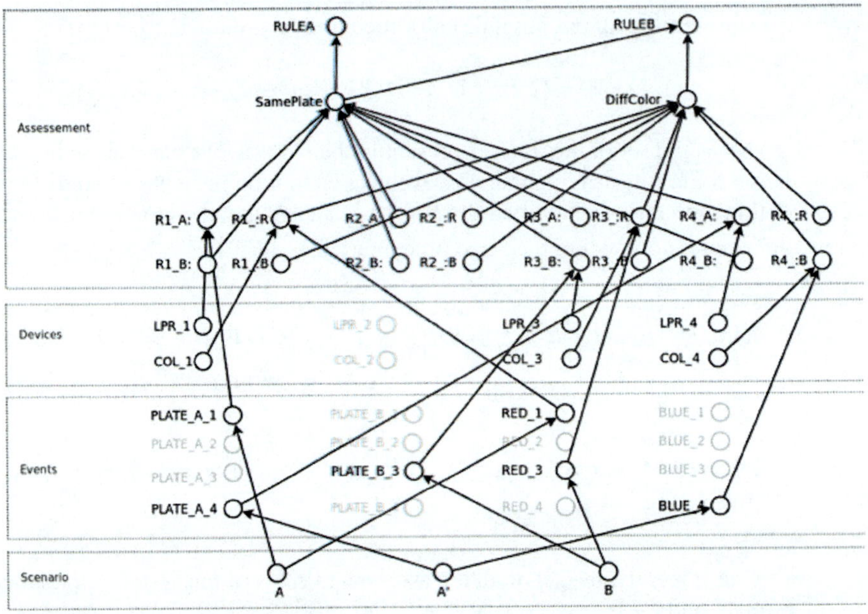

Fig. 9.8 BN model of the case study

with s a generic sensor, r a generic rule and p_s the probability that the sensor s is working. Let us suppose for simplicity that the rules are deterministic, i.e., all the rules have probability 1 to succeed when preconditions are met.

$$\mathscr{C}(r) = 1 \forall r \in CORR \tag{9.28}$$

According to this formalization, it is possible to generate a BN model as depicted in Fig. 9.8 where gray nodes are present but related arcs are not report to make the draw readable. Up to now, there is no tool in charge of automating such translation process: implementing it is straightforward task and, as future research work, an automatic translation and analysis tool will be provided and made publicly available.

Furthermore, this is a high level formalization: a finer grain specification method must be available making concrete the specification (e.g., by means of a formalized grammar and a proper parser) also enabling the specification of further details needed by a complete and comprehensive approach: some of these parameters are the probability of failures of the devices, the rates of the confusion matrices and model-specific parameters (e.g., $\boldsymbol{\delta}_C$ and $\boldsymbol{\delta}_T$ for this domain).

This notwithstanding, the structure of the BN model drives the translation process: the nodes present in the layers *Events*, *Sensor* are generated from the elements respectively in VE, DEV.

Table 9.6 CPT of Rz_x

LPR_z	x	ok	ko
ko	ko	0	1
ko	ok	0	1
ok	ko	0	1
ok	ok	p_x	$1 - p_x$

Another node layer is present—i.e., the *scenario* layer—representing our "test cases": the nodes in this layer are out of the scope of the formalization and transformational approach and are not strictly needed in our approach.

For what concerns the assessment layer, generating just the nodes corresponding to the rules enumerated in the $CORR$ set is possible but it generate CPTs that are hard to understand. To overcome to this problem, it is possible to break the rules in smaller chunks. In particular, there are nodes that are related to the assessment of basic events: $Rz\_x:$—i.e. a plate x is recognized in the zone z) and $Rz\_:y$—i.e. a color y is recognized in the zone z). These nodes can be parents for second level nodes: `SamePlate`, that is `OK` when the same plates is recognized in two different zones, and `DiffColor` that is `OK` when the same color is seen in different zone.

The last two nodes contribute to top level alarm nodes: `RULEA` and `RULEB`.

In the following some of the most interesting CPTs of the model are reported. The CPT of $Rz\_x:$ is reported in Table 9.6: it means that when the device is broken ($LPR_z = ko$) or the plate is not present ($x = ko$), the recognition is false; otherwise, the sensor recognizes the plates with a probability of p_x.

The CPT of $SamePlate$ is reported in Table 9.7[3]: this CPT is deterministic, in the sense that values are only 1 and 0. n particular, it is `ok` just when the same plate is detected in two different zones.

The CPT of $DiffColor$ is reported in Table 9.8[4] and it is very similar to the one in Table 9.7; the difference is, of course, in the logic: the value is `ok` when there is no difference in the detected color.

The CPT of `RULEB` is reported in Table 9.9: it represents a simple logical and between the parent events `SamePlate` and `DiffColor`.

Another advantage of the approach, but not explored in the chapter, is in its hierarchical approach that allow a finer grain, as an example, in introducing the possibility to have different confusion rates according to the different values. In fact it is easier to have confusion between `RED` and `ORANGE` rather than between `RED` and `GREEN`.

---

[3]Limited to two zones for sake of the simplicity.

[4]Limited to two zones for sake of the simplicity.

Table 9.7 CPT of *SamePlate*

$R1\_A$:	$R1\_B$:	$R2\_A$:	$R2\_B$:	ok	ko
ko	*ko*	*ko*	*ko*	0	1
ko	*ko*	*ko*	*ok*	0	1
ko	*ko*	*ok*	*ko*	0	1
ko	*ko*	*ok*	*ok*	1	0
ko	*ok*	*ko*	*ko*	0	1
ko	*ok*	*ko*	*ok*	1	0
ko	*ok*	*ok*	*ko*	1	0
ko	*ok*	*ok*	*ok*	1	0
ok	*ko*	*ko*	*ko*	0	1
ok	*ko*	*ko*	*ok*	1	0
ok	*ko*	*ok*	*ko*	1	0
ok	*ko*	*ok*	*ok*	1	0
ok	*ok*	*ko*	*ko*	1	0
ok	*ok*	*ko*	*ok*	1	0
ok	*ok*	*ok*	*ko*	1	0
ok	*ok*	*ok*	*ok*	1	0

Table 9.8 CPT of *DiffColor*

$R1\_:R$	$R1\_:B$	$R2\_:R$	$R2\_:B$	ok	ko
ko	*ko*	*ko*	*ko*	0	1
ko	*ko*	*ko*	*ok*	0	1
ko	*ko*	*ok*	*ko*	0	1
ko	*ko*	*ok*	*ok*	1	0
ko	*ok*	*ko*	*ko*	0	1
ko	*ok*	*ko*	*ok*	0	1
ko	*ok*	*ok*	*ko*	1	0
ko	*ok*	*ok*	*ok*	1	0
ok	*ko*	*ko*	*ko*	0	1
ok	*ko*	*ko*	*ok*	1	0
ok	*ko*	*ok*	*ko*	0	1
ok	*ko*	*ok*	*ok*	1	0
ok	*ok*	*ko*	*ko*	1	0
ok	*ok*	*ko*	*ok*	1	0
ok	*ok*	*ok*	*ko*	1	0
ok	*ok*	*ok*	*ok*	1	0

Table 9.9 CPT of RULEB

SamePlate	*DiffColor*	ok	ko
ko	*ko*	0	1
ko	*ok*	0	1
ok	*ko*	0	1
ok	*ok*	1	0

9.8 Conclusions

This chapter has discussed the feasibility to apply data-aware formal modeling approaches in the quantitative evaluation of the trustworthiness of Smart City applications where data coming from sensors and IoT devices must be framed into a formalized knowledge to exploit the best of both the worlds.

In particular, this chapter focused on the notion of distributed vulnerability and related formalisms as a mean to describe the interactions between sensors, possible events and correlation schemes by means of a probabilistic approach. Such a formalism can be then translated into Bayesian Network to exploit the solution tools available for such a notation.

Concluding, the approach is able to hide many low-level details of the BNs delegating the construction of a large error prone model to algorithms.

Future research will first focus on the construction of such a tool to refine the approach. Then, continuous variables (e.g., temperature, humidity, people density, etc.) will be considered and the formalism will be extended.

Another important advancement to run after is the introduction of time-aware formalism to overcome with the limitations of state-less combinatorial formalisms as BNs. In fact, an important consideration is needed on the occurring time of the events. When considering $RULE^{A}_{i,j,\delta_T}$ and $RULE^{B}_{i,j,\delta_C}$ as in Sect. 9.7, we must correlate events that are time-related with a combinatorial formalism: a first solution is to define an oblivion mechanism to forget an event after a while to avoid the interference of old events in present correlation schemes. A more powerful formalism (e.g., Petri Nets) could have memory also of the sequence of the event occurrences.

References

1. Allam, Z., Dhunny, Z.A.: On big data, artificial intelligence and smart cities. Cities **89**, 80–91 (2019)
2. Anagnostopoulos, C.-N.E., Anagnostopoulos, I.E., Psoroulas, I.D., Loumos, V., Kayafas, E.: License plate recognition from still images and video sequences: a survey. IEEE Trans. Intell. Transp. Syst. **9**(3), 377–391 (2008)
3. Bagheri, E., Ghorbani, A.A.: UML-CI: a reference model for profiling critical infrastructure systems. Inf. Syst. Front. **12**(2), 115–139 (2010)

4. Bapin, Y., Zarikas, V.: Smart building's elevator with intelligent control algorithm based on bayesian networks. Int. J. Adv. Comput. Sci. Appl. **10**(2), 16–24 (2019)
5. Barrere, M., Badonnel, R., Festor, O.: Towards the assessment of distributed vulnerabilities in autonomic networks and systems. In: 2012 IEEE Network Operations and Management Symposium (NOMS), pp. 335–342 (2012)
6. Bellini, E., Ceravolo, P., Nesi, P.: Quantify resilience enhancement of UTS through exploiting connected community and internet of everything emerging technologies. ACM Trans. Internet Technol. **18**(1) (2017)
7. Bellini, E., Coconea, L., Nesi, P.: A functional resonance analysis method driven resilience quantification for socio-technical systems. IEEE Syst. J. 1–11 (2019)
8. Bellini, E., Nesi, P., Coconea, L., Gaitanidou, E., Ferreira, P., Simoes, A., Candelieri, A.: Towards resilience operationalization in urban transport system: the resolute project approach. In: Proceedings of the 26th European Safety and Reliability Conference on Risk, Reliability and Safety: Innovating Theory and Practice, ESREL 2016, p. 345 (2017)
9. Bellini, E., Nesi, P., Pantaleo, G., Venturi, A.: Functional resonance analysis method based-decision support tool for urban transport system resilience management. In: IEEE 2nd International Smart Cities Conference: Improving the Citizens Quality of Life, ISC2 2016, Proceedings (2016)
10. Bobbio, A., Ciancamerla, E., Franceschinis, G., Gaeta, R., Minichino, M., Portinale, L.: Sequential application of heterogeneous models for the safety analysis of a control system: a case study. Reliab. Eng. Syst. Saf. **81**(3), 269–280 (2003)
11. Boreiko, O., Teslyuk, V.: Model of a controller for registering passenger flow of public transport for the "smart" city system. In: 2017 14th International Conference The Experience of Designing and Application of CAD Systems in Microelectronics, CADSM 2017, Proceedings, pp. 207–209 (2017)
12. Chang, S.-L., Chen, L.-S., Chung, Y.-C., Chen, S.-W.: Automatic license plate recognition. IEEE Trans. Intell. Transp. Syst. **5**(1), 42–53 (2004)
13. Chen, B., Cheng, H.H.: A review of the applications of agent technology in traffic and transportation systems. IEEE Trans. Intell. Transp. Syst. **11**(2), 485–497 (2010)
14. Dolinina, O., Pechenkin, V., Gubin, N., Kushnikov, V.: A petri net model for the waste disposal process system in the "smart clean city" project. In: ACM International Conference Proceeding Series (2018)
15. Fanti, M.P., Mangini, A.M., Roccotelli, M.: A petri net model for a building energy management system based on a demand response approach. In: 2014 22nd Mediterranean Conference on Control and Automation, MED 2014, pp. 816–821 (2014)
16. Flammini, F., Marrone, S., Mazzocca, N., Pappalardo, A., Pragliola, C., Vittorini, V.: Trustworthiness evaluation of multi-sensor situation recognition in transit surveillance scenarios. In: Proceedings of SECIHD Conference. LNCS, vol. 8128 (2013)
17. Flammini, F., Marrone, S., Mazzocca, N., Vittorini, V.: A new modeling approach to the safety evaluation of n-modular redundant computer systems in presence of imperfect maintenance. Reliab. Eng. Syst. Saf. **94**(9), 1422–1432 (2009)
18. Flammini, F., Marrone, S., Mazzocca, N., Vittorini, V.: Petri net modelling of physical vulnerability. Critical Information Infrastructure Security. LNCS, vol. 6983, pp. 128–139. Springer (2013)
19. Flammini, F., Vittorini, V., Pappalardo, A.: Challenges and emerging paradigms for augmented surveillance. Effective Surveillance for Homeland Security. Chapman and Hall/CRC (2013)
20. Frigault, M., Wang, L., Singhal, A., Jajodia, S.: Measuring network security using dynamic Bayesian network. In: Proceedings of the 4th ACM Workshop on Quality of Protection, QoP '08, New York, NY, USA, pp. 23–30. ACM (2008)
21. Gentile, U., Marrone, S., De Paola, F., Nardone, R., Mazzocca, N., Giugni, M.: Model-based water quality assurance in ground and surface provisioning systems. In: Proceedings—2015 10th International Conference on P2P, Parallel, Grid, Cloud and Internet Computing, 3PGCIC 2015, pp. 527–532

22. Gentile, U., Marrone, S., Mazzocca, N., Nardone, R.: Cost-energy modelling and profiling of smart domestic grids. Int. J. Grid Utility Comput. **7**(4), 257–271 (2016)
23. Ghahramani, Z., Ghahramani, Z., Kim, H.C.: Bayesian classifier combination (2003)
24. Häring, I., Sansavini, G., Bellini, E., Martyn, N., Kovalenko, T., Kitsak, M., Vogelbacher, G., Ross, K., Bergerhausen, U., Barker, K., Linkov, I.: Towards a generic resilience management, quantification and development process: general definitions, requirements, methods, techniques and measures, and case studies. NATO Science Peace Secur. Ser. C Environ. Secur. Part **F1**, 21–80 (2017)
25. Huang, C., Wu, X., Wang, D.: Crowdsourcing-based urban anomaly prediction system for smart cities. In: International Conference on Information and Knowledge Management, Proceedings, 24–28-Oct 2016, pp. 1969–1972 (2016)
26. Ismagilova, E., Hughes, L., Dwivedi, Y.K., Raman, K.R.: Smart cities: advances in research—an information systems perspective. Int. J. Inf. Manag. **47**, 88–100 (2019)
27. Jürjens, J.: UMLsec: extending UML for secure systems development. In: Proceedings of the 5th International Conference on The Unified Modeling Language, UML '02, London, UK, UK, pp. 412–425. Springer(2002)
28. Kasaei, S.H.M., Kasaei, S.M.M.: Extraction and recognition of the vehicle license plate for passing under outside environment. In: 2011 European Intelligence and Security Informatics Conference (EISIC), pp. 234–237 (2011)
29. Korb, K.B., Nicholson, A.E.: Bayesian Artificial Intelligence, 2nd edn. CRC Press Inc., Boca Raton, FL, USA (2010)
30. Langseth, H., Portinale, L.: Bayesian networks in reliability. Reliab. Eng. Syst. Saf. **92**(1), 92–108 (2007)
31. Latorre-Biel, J.-I., Faulin, J., Jiménez, E., Juan, A.A.: Simulation model of traffic in smart cities for decision-making support: case study in Tudela (Navarre, Spain). Lecture Notes in Computer Science (including subseries Lecture Notes in Artificial Intelligence and Lecture Notes in Bioinformatics). LNCS, vol. 10268, pp. 144–153 (2017)
32. Lund, M.S., Solhaug, B., Stølen, K.: Risk analysis of changing and evolving systems using CORAS. In: Aldini, A., Gorrieri, R. (eds.) Foundations of Security Analysis and Design VI, pp. 231–274. Springer, Berlin, Heidelberg (2011)
33. Marrone, S., Rodríguez, R.J., Nardone, R., Flammini, F., Vittorini, V.: On synergies of cyber and physical security modelling in vulnerability assessment of railway systems. Comput. Electr. Eng. **47**, 275–285 (2015)
34. Mauw, S., Oostdijk, M.: Foundations of attack trees. In: 8th International Conference on Information Security and Cryptology—ICISC 2005, Seoul, Korea, 1–2 Dec 2005, pp. 186–198. Revised Selected Papers (2005)
35. Pederson, P., Dudenhoeffer, D., Hartley, S., Permann, M.: Critical infrastructure interdependency modeling: a survey of U.S. and international research. Technical Report, Idaho National Laboratory (2006)
36. Pettet, G., Nannapaneni, S., Stadnick, B., Dubey, A., Biswas, G.: Incident analysis and prediction using clustering and Bayesian network. In: 2017 IEEE SmartWorld Ubiquitous Intelligence and Computing, Advanced and Trusted Computed, Scalable Computing and Communications, Cloud and Big Data Computing, Internet of People and Smart City Innovation, SmartWorld/SCALCOM/UIC/ATC/CBDCom/IOP/SCI 2017—Conference Proceedings, pp. 1–8 (2018)
37. Quaglietta, E., D'Acierno, L., Punzo, V., Nardone, R., Mazzocca, N.: A simulation framework for supporting design and real-time decisional phases in railway systems. In: IEEE Conference on Intelligent Transportation Systems, Proceedings, ITSC, pp. 846–851 (2011)
38. Ranawana, R., Palade, V.: Multi-classifier systems: review and a roadmap for developers. Int. J. Hybrid Intell. Syst. **3**(1) (2006)
39. Räty, T.D.: Survey on contemporary remote surveillance systems for public safety. Trans. Syst. Man Cyber Part C **40**(5), 493–515 (2010)
40. Rinaldi, S.M., Peerenboom, J.P., Kelly, T.K.: Identifying, understanding, and analyzing critical infrastructure interdependencies. IEEE Control Syst. Mag. **21**(6), 11–25 (2001)

41. Sha, L., Gopalakrishnan, S., Liu, X., Wang, Q.: Cyber-physical systems: a new frontier. In: IEEE International Conference on Sensor Networks, Ubiquitous and Trustworthy Computing, 2008, SUTC '08, pp. 1–9 (2008)
42. Sharifi, H., Shahbahrami, A.: A comparative study on different license plate recognition algorithms. In: Cherifi, H., Zain, J.M., El-Qawasmeh, E. (eds.) Digital Information and Communication Technology and Its Applications. Communications in Computer and Information Science, vol. 167, pp. 686–691. Springer, Berlin, Heidelberg (2011)
43. Simpson, E., Roberts, S., Psorakis, I., Smith, A.: Dynamic Bayesian combination of multiple imperfect classifiers. In: Guy, T.V., Karny, M., Wolpert, D. (eds.) Decision Making and Imperfection. Studies in Computational Intelligence, vol. 474. Springer (2013)
44. Skinner, S.C., Stracener, J.T.: A graph theoretic approach to modeling subsystem dependencies within complex systems. In: WMSCI 2007, ISAS 2007, Proceedings, vol. 3, pp. 41–46 (2007)
45. Sun, F., Wu, C., Sheng, D.: Bayesian networks for intrusion dependency analysis in water controlling systems. J. Inf. Sci. Eng. **33**(4), 1069–1083 (2017)
46. Tang, K., Zhou, M.-T., Wang, W.-Y.: Insider cyber threat situational awareness framework using dynamic Bayesian networks. In: Proceedings of the 4th International Conference on Computer Science Education (ICCSE), July 2009, pp. 1146–1150
47. Vaniš, M., Urbaniec, K.: Employing Bayesian networks and conditional probability functions for determining dependences in road traffic accidents data. In: 2017 Smart Cities Symposium Prague, SCSP 2017—IEEE Proceedings (2017)
48. Xie, P., Li, J.H., Ou, X., Liu, P., Levy, R.: Using Bayesian networks for cyber security analysis. In: 2010 IEEE/IFIP International Conference on Dependable Systems and Networks (DSN), June 2010, pp. 211–220
49. Yousef, R., Liginlal, D., Fass, S., Aoun, C.: Combining morphological analysis and Bayesian belief networks: a DSS for safer construction of a smart city. In: 2015 Americas Conference on Information Systems, AMCIS 2015 (2015)
50. Zhao, J., Ma, S., Han, W., Yang, Y., Wang, X.: Research and implementation of license plate recognition technology. In: 2012 24th Chinese Control and Decision Conference (CCDC), pp. 3768–3773 (2012)
51. Zonouz, S.A., Khurana, H., Sanders, W.H., Yardley, T.M.: RRE: a game-theoretic intrusion response and recovery engine. IEEE Trans. Parallel Distrib. Syst. **25**(2), 395–406 (2014)

Chapter 10
Feature Set Ensembles for Sentiment Analysis of Tweets

D. Griol, C. Kanagal-Balakrishna, and Z. Callejas

Abstract In recent years, sentiment analysis has attracted a lot of research attention due to the explosive growth of online social media usage and the abundant user data they generate. Twitter is one of the most popular online social networks and a microblogging platform where users share their thoughts and opinions on various topics. Twitter enforces a character limit on tweets, which makes users find creative ways to express themselves using acronyms, abbreviations, emoticons, etc. Additionally, communication on Twitter does not always follow standard grammar or spelling rules. These peculiarities can be used as features for performing sentiment classification of tweets. In this chapter, we propose a Maximum Entropy classifier that uses an ensemble of feature sets that encompass opinion lexicons, n-grams and word clusters to boost the performance of the sentiment classifier. We also demonstrate that using several opinion lexicons as feature sets provides a better performance than using just one, at the same time as adding word cluster information enriches the feature space.

10.1 Introduction

Due to the explosive growth of online social media in the last few years, people are increasingly turning to social media platforms such as Facebook, Twitter, Instagram, Tumblr, LinkedIn, etc., to share their thoughts, views and opinions on products, services, politics, celebrities, events, and companies. This has resulted in a massive amount of user-generated data [24].

D. Griol (✉) · Z. Callejas
University of Granada, Granada, Spain
e-mail: dgriol@ugr.es

Z. Callejas
e-mail: zoraida@ugr.es

C. Kanagal-Balakrishna
Universidad Carlos III de Madrid, Getafe, Spain
e-mail: 100353591@alumnos.uc3m.es

G. Phillips-Wren et al. (eds.), *Advances in Data Science: Methodologies and Applications*, Intelligent Systems Reference Library 189,
https://doi.org/10.1007/978-3-030-51870-7_10

As the usage of online social media has grown, so has the interest in the field of sentiment analysis [17, 25, 27]. For the scientific community, sentiment analysis is a challenging and complex field of study with applications in multiple disciplines and has become one of the most active research areas in Natural Language Processing, data mining, web mining and management sciences. For industry, the massive amount of user-generated data is fertile ground for extracting consumer opinion and sentiment towards their brands. In recent years, we have seen how social media has helped reshape businesses and sway public opinion and sentiment, sometimes with a single viral post or tweet. Therefore, monitoring public sentiment towards their products and services enables them to cater to their customers better.

In the last few years, Twitter has become a hugely popular microblogging platform with over 500 million tweets a day. However, Twitter only allows short messages of up to 140 characters which results in users using abbreviations, acronyms, emoticons, etc., to better express themselves. The field of sentiment analysis in Twitter therefore includes the various complexities brought by this form of communication using short informal text. The main motivation for studying sentiment analysis in Twitter is due to the immense academic as well as commercial value that it provides [1, 3, 26].

Besides its commercial applications, the number of application-oriented research papers published on sentiment analysis has been steadily increasing. For example, several researchers have used sentiment information to predict movie success and box-office revenue. Mishne and Glance showed that positive sentiment is a better predictor of movie success than simple buzz count [15]. Researchers have also analyzed sentiments of public opinions in the context of electoral politics. For example, in [20], a sentiment score was computed based simply on counting positive and negative sentiment words, which was shown to correlate well with presidential approval, political election polls, and consumer confidence surveys. Market prediction is also another popular research area for sentiment analysis [13].

The main research question that we want to ask in this chapter is: *Can we combine different feature extraction methods to boost the performance of sentiment classification of tweets?*

Raw data cannot be fed directly to the algorithms themselves as most of the algorithms expect numerical feature vectors with a fixed size rather than the raw text documents with variable length. Feature extraction is the process of transforming text documents into numerical feature vectors. There are many standard feature extraction methods for sentiment analysis of text data such as Bag of Words representation, tokenization, etc. Since feature extraction usually results in high dimensionality of features, it is important to use features that provide useful information to the machine learning algorithm.

Sub-question 1: Does extracting features using opinion lexicons add value to the feature space?

Opinion Lexicons refers to a list of opinion words such as good, excellent, poor, bad, etc., which are used to indicate positive and negative sentiment. The positive and negative sentiment scores of each tweet can be extracted as features using Opinion Lexicons. We investigate if Opinion Lexicons boost the performance of sentiment classification of tweets.

Sub-question 2: Does using word clusters as features add value to the feature space

Word clustering is a technique for partitioning sets of words into subsets of semantically similar words, for example, Monday, Tuesday, Wednesday, etc., would be included in a word cluster together. Word clusters can be used as features themselves. Thus, word clustering has a potential to reduce sparsity of the feature space. We investigate if using word clusters as features improves the performance of sentiment classification of tweets.

The remainder of the chapter is as follows. Section 10.2 describes the motivation of our proposal and related work. Section 10.3 summarizes the basic terminology, levels and approaches for sentiment analysis. Section 10.4 describes the main data sources used in our research. Section 10.5 presents the experimental process that we have followed, the feature sets and results of the evaluation. Finally, Sect. 10.6 presents the conclusions and suggests some future work guidelines.

10.2 State of the Art

Sentiment Analysis can be defined as a field of study consisting of a series of methods, techniques, and tools about detecting and extracting subjective information of people's opinions, sentiments, evaluations, appraisals, attitudes, and emotions towards entities such as products, services, organizations, individuals, issues, events, topics, and their attributes expressed in written text [13, 23]. Though there are some nuances in the definition of the terms as well as their applications, for our study, we will treat sentiment analysis, opinion mining, subjectivity analysis, opinion analysis, review mining, opinion extraction, etc., interchangeably.

Traditionally, the desired practical outcome of performing sentiment analysis on text is to classify the polarity of the opinion. Opinion polarity can be classified into 3 categories, i.e., if the opinion expressed in the text is positive, negative or neutral towards the entity.

An important part of our information-gathering behavior has always been to find out what other people think. With the growing availability and popularity of opinion-rich resources such as online review sites and personal blogs, new opportunities and challenges arise as people now can, and do, actively use information technologies to seek out and understand the opinions of others [21]. It is not always feasible for potential customers to go to a physical store to examine the features and performance of various products. It is also difficult to predict how the products will hold up over time. The general trend now before selecting a product and making a purchase is to read the reviews, blog posts, etc., written by other customers about their experiences with the product to better gauge if it will be a good fit in accordance to their product requirements.

Factors that further advanced sentiment analysis during the last decade are:

- The rise of machine learning methods in natural language processing and information retrieval;
- The availability of datasets for machine learning algorithms to be trained on, due to the World Wide Web and, specifically, the development of review-aggregation web-sites;
- Realization of the intellectual challenges and commercial and intelligence applications that the area offers [21].
- Evolution of the web from Web 1.0 to Web 2.0. Web 2.0 is an evolution from passive viewing of information to interactive creation of user generated data by the collaboration of users on the Web. The evolution of Web from Web 1.0 to Web 2.0 was enabled by the rise of read/write platforms such as blogging, social networks, and free image and video sharing sites. These platforms have jointly allowed exceptionally effortless content creation and sharing by anyone [10].

With the proliferation of Web 2.0 applications, research field of sentiment analysis has been progressing rapidly due to the vast amounts of data generated by such applications. Blogs, review sites, forums, microblogging sites, wikis and social networks have all provided different dimensions to the data used for sentiment analysis.

10.3 Basic Terminology, Levels and Approaches of Sentiment Analysis

Formally, Sentiment Analysis is the computational study of opinions, sentiments and emotions expressed in text. The goal of sentiment analysis is to detect subjective information contained in various sources and determine the mind-set of an author towards an issue or the overall disposition of a document. The analysis is done on user generated content on the Web which contains opinions, sentiments or views. An opinionated document can be a product review, a forum post, a blog or a tweet, that evaluates an object. The opinions indicated can be about anything or anybody, for e.g. products, issues, people, organizations or a service [10].

Mathematically, Liu defines an opinion as a quintuple, (e, a, s, h, t), where e is the target entity; also known as object, a is the target aspect of entity e on which the opinion has been given; also known as feature of the object, s is the sentiment of the opinion on aspect a of entity e, h is the opinion holder, and t is the opinion posting time [13].

- Object: An entity which can be a product, person, event, organization, or topic. The object can have attributes, features or components associated with it. Further on the components can have subcomponents and attributes.
- Feature: An attribute (or a part) of the object with respect to which evaluation is made.

- Opinion orientation or polarity: The orientation of an opinion on a feature indicates whether the opinion is positive, negative or neutral. It can also be a rating (e.g., 1–5 stars). Most work has been done on binary classification i.e. into positive or negative. But opinions can vary in intensity from very strong to weak. For example a positive sentiment can range from content to happy to ecstatic. Thus, strength of opinion can be scaled and depending on the application the number of levels can be decided.
- Opinion holder: The holder of an opinion is the person or organization that expresses the opinion [10].

Sentiment Analysis can be performed at different structural levels, ranging from individual words to entire documents. Depending on the granularity required, Sentiment Analysis Research has been mainly carried out at three levels namely: Document Level, Sentence Level and Aspect Level.

Document level Sentiment Analysis is the simplest form of classification. The whole document is considered as a basic unit of information. The task at the document level is to classify whether the whole document expresses a positive, negative or neutral sentiment. However, there are two assumptions to be made. Firstly, this level of analysis assumes that the entire document expresses opinions on a single entity (film, book, hotel, etc.). Secondly, it is assumed that the opinions are from a single opinion holder. Thus, document level Sentiment Analysis is not applicable to documents that evaluate or compare opinions on multiple entities [13].

Sentence level Sentiment Analysis aims to go to the sentences and determine whether each sentence expresses a positive, negative or neutral opinion. Neutral usually means no opinion. Sentence level classification assumes that the sentence expresses only one opinion, which is not true in many cases. Sentence level classification is closely related to subjectivity classification which distinguishes sentences which provide factual information from sentences that express subjective opinions. The former is called an objective sentence, while the latter is called a subjective sentence [12, 13, 23]. Therefore, the first task at this level is to determine if the sentence is opinionated or not, i.e., subjective or objective. The second task is to determine the polarity of the sentence, i.e., positive, negative or neutral.

Aspect level sentiment analysis is based on the idea that an opinion consists of a sentiment, i.e., positive, negative or neutral, as well as a target of the opinion, aspect. Aspect level sentiment analysis performs a finer-grained analysis compared to document level and sentence level sentiment analysis. The goal of this level of analysis is to discover sentiments on entities and/or their aspects. Thus, aspect level sentiment analysis is a better representation when it comes to texts such as product reviews which usually involve opinions on multiple aspects.

There are two well-established approaches to carrying out sentiment analysis. One is the lexicon-based approach where the classification process relies on the rules and heuristics obtained from linguistic knowledge. The other is the machine-learning approach where algorithms learn underlying information from previously annotated data which allows them to classify new unlabeled data. There have also been a growing number of studies which have successfully implemented a hybrid approach by combining lexicon-based approach and machine-learning approach.

The lexicon-based approach depends on finding the opinion lexicon which can be used to analyze the text. There are two methods in this approach; dictionary-based approach and corpus based approach. The dictionary based approach depends on finding opinion seed words, and then searching the dictionary for their synonyms and antonyms. On the other hand, the corpus based approach begins with a seed list of opinion words, and then finds other opinion words in a large corpus to help in finding opinion words with context specific orientations. This can be accomplished using statistical or semantic methods [14].

In the dictionary-based approach, a small set of opinion words is collected manually with known prior polarity or sentiment orientations. Then, this seed set is expanded by searching in a well-known corpora such as WordNet or a thesaurus for their synonyms and antonyms. The newly found words are added to the seed list then the next iteration starts. The iterative process stops when no new words are found. After the process is completed, manual inspection can be carried out to remove or correct errors [14, 23].

Corpus based methods rely on syntactic or statistical techniques like co-occurrence of word with another word whose polarity is known. For this approach, [8] used a corpus and some seed adjective sentiment words to find additional sentiment adjectives in the corpus. Their technique exploited a set of linguistic rules or conventions on connectives to identify more adjective sentiment words and their orientations from the corpus.

Using the corpus-based approach alone is not as effective as the dictionary-based approach because it is hard to prepare a huge corpus which covers all English words. However, the advantage of corpus-based approach is that it can help to find domain and context specific opinion words and their orientations using a domain corpus [14]. But it is important to note that having a sentiment lexicon (even with domain specific orientations), does not mean that a word in the lexicon always expresses an opinion/sentiment in a specific sentence. For example, in "I am looking for a good car to buy", "good" here does not express either a positive or negative opinion on any particular car. Due to contributions of many researchers, several general-purpose subjectivity, sentiment, and emotion lexicons have been constructed and are also publicly available [4, 14].

The text classification methods using Machine Learning approach can be roughly divided into supervised and unsupervised learning methods. The supervised methods make use of a large number of labeled training documents. The unsupervised methods are used when it is difficult to find these labeled training documents. Machine learning approach relies on Machine Learning algorithms to solve the problem of sentiment classification. To achieve this, the machine learning approach treats sentiment analysis as a regular text classification problem, where instead of classifying documents of different topics (e.g., politics, sciences, and sports), we estimate positive, negative, and neutral classes [22].

The goal of the supervised machine learning approach is to predict and classify the sentiment of a given text based on information learned from past examples. The supervised learning methods, therefore, depend on the existence of labeled training documents. To build the classification model, training data with annotated sentiment is applied to the chosen supervised machine learning classifier. Then, the unlabeled testing data which is not used for training is applied to the trained classifier model. With the results obtained, sentiment polarity of the test data is predicted. Typical classifiers used in this approach include: probabilistic classifiers, linear classifiers, decision trees classifiers, and rule-based classifiers.

Probabilistic classifiers are among the most popular classifiers used in the machine learning community and increasingly in many applications. These classifiers are derived from generative probability models which provide a principled way to the study of statistical classification in complex domains such as natural language and visual processing. Probabilistic classification is the study of approximating a joint distribution with a product distribution. Bayes rule is used to estimate the conditional probability of a class label, and then assumptions are made on the model, to decompose this probability into a product of conditional probabilities [7]. Three of the most famous probabilistic classifiers are Naive Bayes classifiers, Bayesian Network and Maximum Entropy classifiers.

There are many kinds of linear classifiers, among which Support Vector Machines is popularly used for text data. These classifiers are supervised machine learning models used for binary classification and regression analysis. However, research studies have proposed various approaches to handle multiclass classification using SVM. Support vector machines (SVMs) are highly effective for traditional text categorization, and can outperform Naive Bayes [21].

Decision trees are based on a hierarchical decomposition of the training data, in which a condition on the attribute value is used in order to divide the data space hierarchically. The division of the data space is performed recursively in the decision tree, until the leaf nodes contain a certain minimum number of records, or some conditions on class purity. The majority class label in the leaf node is used for the purposes of classification. For a given test instance, the sequence of predicates is applied at the nodes, in order to traverse a path of the tree in top-down fashion and determine the relevant leaf node.

In rule-based classifiers, the data space is modeled with a set of rules, in which the left hand side is a condition on the underlying feature set, and the right hand side is the class label. The rule set is essentially the model which is generated from the training data. For a given test instance, we determine the set of rules for which the test instance satisfies the condition on the left hand side of the rule. We determine the predicted class label as a function of the class labels of the rules which are satisfied by the test instance. Rule-based Classifiers are related to the decision tree classifiers because both encode rules on the feature space. The main difference is that the decision tree classifier uses the hierarchical approach, whereas the rule-based classifier allows for overlap in the decision space [2]. In these classifiers, the training phase generates the rules based on different criteria. Two of the most common conditions which are used for rule generation are those of support and confidence.

Classifier ensemble have been also proposed to combine different classifiers in conjunction with a voting mechanism in order to perform the classification. The basis is that since different classifiers are susceptible to different kinds of overtraining and errors, a combination classifier is likely to yield much more robust results. This technique is also sometimes referred to as stacking or classifier committee construction. Ensemble learning has been used quite frequently in text categorization. Most methods simply use weighted combinations of classifier outputs (either in terms of scores or ranks) in order to provide the final classification result. The major challenge in ensemble learning is to provide the appropriate combination of classifiers for a particular scenario. This combination can significantly vary with different scenarios and data sets [2, 3].

10.4 Data Sources

The dataset chosen to build a classifier for sentiment analysis can have a significant impact on the performance of the classifier when implemented on the test data. Several important factors need to be considered before choosing a dataset. When it comes to analyzing tweets, we need to consider the effect of the domain focused tweets, data structure as well as the objective of the classification.

Twitter Sentiment Analysis SemEval Task B Dataset was chosen for experimentation using various classification methods. To remedy the lack of datasets which is hindering sentiment analysis research, Nakov et al. [18] released a twitter training dataset to the research community to be used for evaluation and comparison between approaches. The SemEval Tweet corpus contains tweets with sentiment with sentiment expressions annotated with overall message-level polarity. The tweets that were gathered express sentiment about popular topics. The collection of tweets span over a one-year period from January 2012 to January 2013. Public streaming Twitter API was used to download tweets.

The dataset was annotated for sentiment on Mechanical Turk, a crowdsourcing marketplace that enables individuals or businesses to use human intelligence to perform tasks that computers are currently unable to do such as image recognition, audio transcription, machine learning algorithm training, sentiment analysis, data normalization, surveys, etc., in exchange for a reward.[1] Each sentence was annotated by five Mechanical Turk workers. They had to indicate the overall polarity of the sentence as positive, negative or neutral as well as the polarity of a subjective word or phrase. However, the dataset used to build our classifier only contains annotations of overall message-level polarity. The final polarity of the entire sentence was determined based on the majority of the labels [18].

SemEval Twitter Corpus consists of 13,541 tweets (or instances) collected between January 2012 and January 2013. The domain of the tweets is not indicated in [18]. Each instance in the corpus contains values for two attributes namely; Content and Class. The instances of the content attribute contain the tweets themselves containing data in a string format. The instances of the class attribute contain

[1] Amazon Mechanical Turk, https://www.mturk.com/.

Table 10.1 Examples from SemEval Twitter Corpus

Class	Count	Example
Positive	5,232	Gas by my house hit $3.39!!!! I am going to Chapel Hill on Sat :)
Negative	6,242	Theo Walcott is still shit, watch Rafa and Johnny deal with him on Saturday
Neutral	2,067	Fact of the day; Halloween night is Papa John's second busiest night of the year behind Super Bowl Sunday

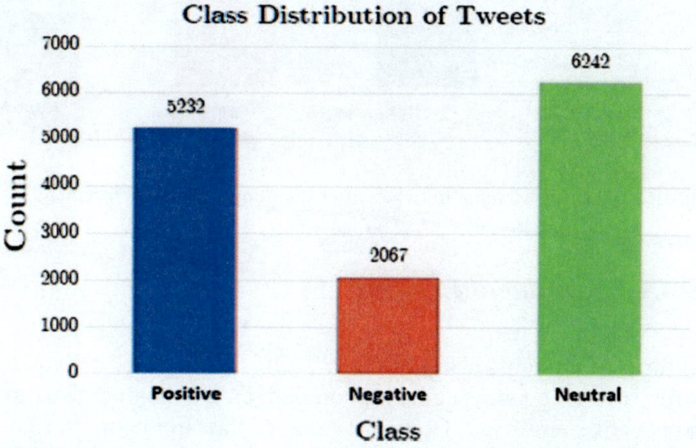

Fig. 10.1 Class distribution of tweets

three nominal values (classes) namely positive, negative and neutral. It should be noted that, the turkers were instructed to choose the stronger sentiment in messages conveying both positive and negative sentiments. Table 10.1 illustrates the distribution of tweets from the corpus as well as an example tweet and its class as labeled by the turkers.

We see from Fig. 10.1 that the class distribution is not balanced. For model training and classification, balanced class distribution is very important to ensure the prior probabilities are not biased caused by the imbalanced class distribution.

There are many methods to address the class imbalance problem such as collecting more data, changing the performance metric, resampling the dataset, generating synthetic samples, penalized models, etc. In order to balance the dataset, we are going to implement resampling of the dataset. Since the class with the lowest number of instances (Negative) still has a considerable number of instances which can be used to train the classifiers, we are going to perform random sampling without replacement so that instances from the over-represented classes are removed from the dataset. Figure 10.2 illustrates the class distribution by tweets after sampling without replacement.

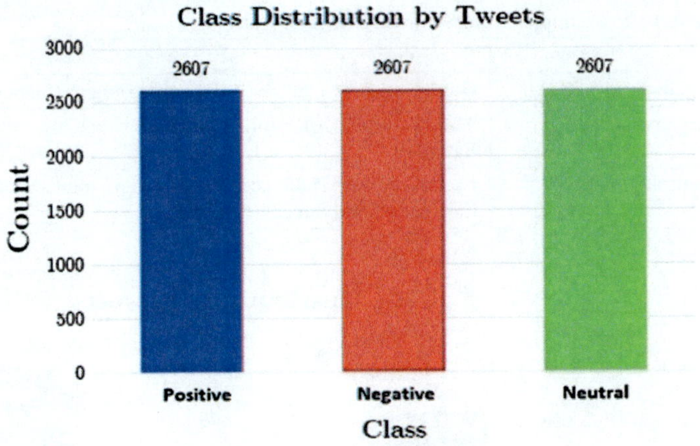

Fig. 10.2 Class distribution of tweets after sampling without replacement

10.4.1 Sentiment Lexicons

Sentiment Lexicons, also known as Opinion Lexicons, refers to a list of opinion words such as good, excellent, poor, bad, etc., which are used to indicate positive and negative sentiment. Opinion Lexicons play an important role in extracting two very important features; positive and negative sentiment scores. Extraction of these features could enhance the accuracy of the classification system and the frequency of these sentiment words directly maps to overall sentiment of a tweet. Therefore, we can enrich the feature space with opinion lexicon information, where each tweet (or instance) as the associated positive and negative sentiment score.

The AFINN lexicon is based on the Affective Norms for English Words lexicon (ANEW) proposed in [5]. ANEW provides emotional ratings for a large number of English words. These ratings are calculated according to the psychological reaction of a person to a specific word, being the valence the most useful value for sentiment analysis. Valence ranges in the scale pleasant-unpleasant. This lexicon was released before the rise of microblogging and therefore does not contain the common slang words used on microblogging platforms such as Twitter. Nielsen created the AFINN lexicon [19], which is more focused on the language used in microblogging platforms. The word list includes slang and obscene words as well as acronyms and web jargon. Positive words are scored from 1 to 5 and negative words from -1 to -5, reason why this lexicon is useful for strength estimation. The lexicon includes 2,477 English words [6].

The AFINN lexicon extracts two features from each tweet (or instance). AFINN Positivity Score and AFINN Negativity Score, that are the sum of the ratings of positive and negative words of the tweet that matches the AFINN lexicon, respectively.

The Bing Liu Opinion lexicon is one of the most widely used sentiment lexicons for sentiment analysis. Hu and Liu [9] proposed a lexicon-based algorithm for aspect level sentiment classification, but the method can determine the sentiment orientation of a sentence as well. It was based on a sentiment lexicon generated using a bootstrapping strategy with some given positive and negative sentiment word seeds and the synonyms and antonyms relations in WordNet. The sentiment orientation of a sentence was determined by summing up the orientation scores of all sentiment words in the sentence. A positive word was given the sentiment score of +1 and a negative word was given the sentiment score of −1. Negation words and contrary words (e.g., but and however) were also considered [13]. The Lexicon includes 6,800 English words.

The Bing Liu Opinion lexicon extracts two features from the tweets (or instances). Bing Liu Positivity Score and Bing Liu Negativity Score, that are the sum of the orientation scores of positive and negative sentiment words in the tweet that matches the Bing Liu lexicon, respectively.

The NRC Word-Emotion Association Lexicon is a lexicon that includes a large set of human-provided words with their emotional tags. By conducting a tagging process in the crowdsourcing Amazon Mechanical Turk platform, Mohammad and Turney [16] created a word lexicon that contains more than 14,000 distinct English words annotated according to the Plutchik's wheel of emotions. The wheel is composed by four pairs of opposite emotion states: joy-trust, sadness-anger, surprise-fear, and anticipation-disgust. These words can be tagged to multiple categories. Additionally, NRC words are tagged according to polarity classes positive and negative [6]. The NRC Word-Emotion Association lexicon extracts ten features from the tweets (or instances) namely; NRC Joy, NRC Trust, NRC Sadness, NRC Anger, NRC Surprise, NRC Fear, NRC Anticipation, NRC Positive and NRC Negative.

NRC Word-Emotion Association Lexicon did not include expressions such as hashtags, slang words, misspelled words, etc., that are commonly seen on social media (i.e. twitter, facebook, etc.). The NRC-10 Expanded Lexicon was created to address this issue. The NRC-10 Expanded lexicon extracts ten features from the tweets (or instances): NRC Joy, NRC Trust, NRC Sadness, NRC Anger, NRC Surprise, NRC Fear, NRC Anticipation, NRC Positive and NRC Negative.

The NRC Hashtag Emotion Lexicon consists of an association of words with eight emotions (anger, fear, anticipation, trust, surprise, sadness, joy, and disgust) generated automatically from tweets with emotion-word hashtags such as #happy and #angry. It contains 16832 distinct English words. NRC Hashtag Emotion Lexicon extracts 8 features from the tweets (or instances) namely; NRC Joy, NRC Trust, NRC Sadness, NRC Anger, NRC Surprise, NRC Fear, NRC Anticipation.

The NRC Hashtag Sentiment Lexicon consists of an association of words with positive and negative sentiment generated automatically from tweets with sentiment-word hashtags such as #amazing and #terrible. It consists of 54,129 unigrams (words), 316,531 bigrams and 308,808 pairs. NRC Hashtag Sentiment Lexicon extracts two features from the tweets (or instances) namely; NRC Positive and NRC Negative.[2]

[2]NRC Emotion and Sentiment Lexicons, http://saifmohammad.com/WebPages/AccessResource.htm.

10.5 Experimental Procedure

In this section, we describe the experimentation performed on the SemEval Twitter Corpus. As previously described, we build our baseline classifiers using the sub-feature sets from the three feature sets defined. The preceding steps such as pre-processing and feature extraction are performed on the classifiers. Feature selection will be performed only on feature set 2 and feature set 3. The proposed classifier will be trained using the feature set PFS where we combine various models from feature set 1, feature set 2 and feature set 3. All the models will be trained using the classification algorithms; Maximum Entropy and Support Vector Machines. The model is evaluated as described in Sect. 6.6. Finally, we compare the performance metrics of our baseline classifiers with that of the proposed classifier(s).

10.5.1 Feature Sets

We have defined three feature sets that will be tested for our baseline classifier models. These feature sets are further sub-divided into classifier models that use specific feature extraction and feature selection methods. All the models will be trained using two classification algorithms; Maximum Entropy and Support Vector machines.

In feature set 1, we make use of six Sentiment Lexicons; AFINN, Bing Liu Lexicon, NRC-10 Word Emotion Association Lexicon, NRC-10 Expanded Lexicon, NRC Hashtag Emotion Lexicon, NRC Hashtag Sentiment Lexicon and Negation to extract their respective features. The Lexicons are employed in various combinations. For data preprocessing, we reduce length of elongated words, convert to lower case and replace user mentions and URLs with generic tokens.

In feature set 2, we use a combination of word N-grams such as Unigrams, Unigrams and Bigrams, Unigrams, Bigrams and Trigrams, for feature extraction. In feature set 3, we use a combination of cluster N-grams such as Unigrams, Unigrams and Bigrams, Unigrams, Bigrams and Trigrams, for feature extraction. Additionally, we also use the Twokenize tokenizer from CMU Tweet NLP tool, binary frequency of terms as well as weighted frequency as feature extraction methods in all the models. For data preprocessing, we use negation handling, reduce length of elongated words and convert words to lower case. Table 10.2 shows the feature extraction method and types of features used for the different models defined for each set.

10.5.2 Results of the Evaluation

Performance of classifiers is commonly measured with reference to a baseline classifier. Some of the most used baseline classifiers for text classification and sentiment

Table 10.2 Models defined for the feature sets 1, 2 and 3

Feature set	Feature extraction method
FS1A	AFINN Lexicon
FS1B	Bing Liu Lexicon
FS1C	NRC-10 Word Emotion Association Lexicon, NRC-10 Expanded Lexicon, NRC Hashtag Emotion Lexico and NRC Hashtag Sentiment Lexicon, Negation
FS1D	AFINN Lexicon, Bing Liu Lexicon, NRC-10 Word Emotion Association Lexicon, NRC-10 Expanded Lexicon, NRC Hashtag Emotion Lexicon and NRC Hashtag Sentiment Lexicon, Negation
FS2A	Word Unigrams, Twokenize, Binary frequency, Frequency weighting
FS2B	Word Unigrams + Bigrams, Twokenize, Binary frequency, Frequency weighting
FS2C	Word Unigrams + Bigrams + Trigrams, Twokenize, Binary Frequency, Frequency weighting
FS3A	Cluster Unigrams, Twokenize, Binary frequency, Frequency weighting
FS3B	Cluster Unigrams-Bigrams, Twokenize, Binary frequency, Frequency weighting
FS3C	Cluster Unigrams-Bigrams-Trigrams, Twokenize, Binary frequency, Frequency weighting

analysis include Support Vector Machines, Maximum Entropy, Naive Bayes, Decision Trees and Random Forest. For the purpose of performance comparison, we consider the classifiers built using the feature set 2 model, FS2A as our primary baseline classifiers. The FS2A classifiers are built using standard data preprocessing steps such as lowering case, reducing the length of elongated words, etc. It also uses Unigrams for feature extraction which is standard for classification of tweets. The classification accuracy of models built using the feature set 1, feature set 2, feature set 3, as well as that of the proposed feature set is illustrated in Fig. 10.3.

The classification accuracy of the baseline classifier, FS2A, which uses Maximum Entropy algorithm is 73.63%, whereas the LibLinear SVM algorithm provides an accuracy of 73.13%. While Maximum Entropy performs slightly better, the difference is not significant. When we compare the baseline classifiers which models from feature set 1, we see that none of the classifiers perform as well as the baseline classifiers for both the algorithms. Feature set 1, which uses various combinations of opinion lexicons, provides the highest classification accuracy when we combine the opinion lexicons; AFINN, Bing Liu Lexicon, NRC-10 Word Emotion Association Lexicon, NRC-10 Expanded Lexicon, NRC Hashtag Emotion Lexicon and NRC Hashtag Sentiment Lexicon with accuracies of 67.58% and 70.15% for Maximum Entropy and LibLinear SVM respectively. LibLinear SVM consistently outperforms Maximum Entropy in feature set 1.

Feature set 2 includes models built using various word n-gram combinations. FS2C Maximum Entropy classifier achieves the highest overall accuracy with 79.64%. We observe that the classification accuracy rises when we include Bigrams, Bigrams and Trigrams to the baseline classifier which only uses Unigrams. While this is true of both Maximum Entropy and LibLinear SVM, the performance improvement

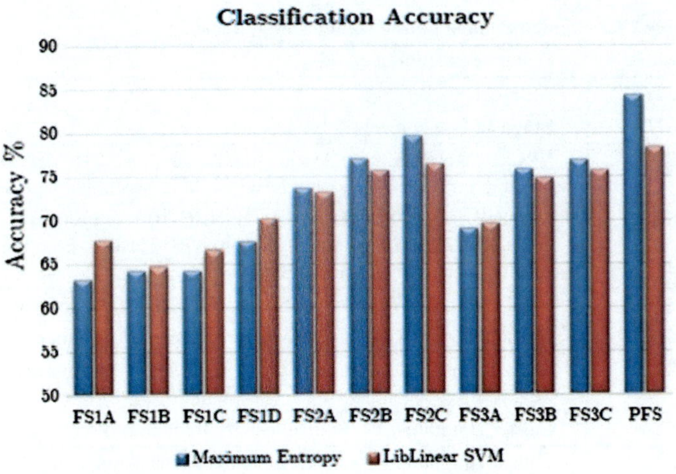

Fig. 10.3 Classification accuracy obtained for the set of models

is more apparent with Maximum Entropy which shows a significant improvement over the baseline when the n-gram combination of unigrams, bigrams and trigrams. While LibLinear SVM shows an improvement over the unigram model, the difference between the unigram-bigram and unigram-bigram-trigram model is not significant.

Feature set 3 includes models built using various cluster n-gram combinations. FS3C Maximum Entropy classifier achieves the highest overall accuracy with 76.87%. We observe that the classification accuracy rises when we include Bigrams, Bigrams and Trigrams to the baseline classifier which only uses Unigrams. This is the case for both Maximum Entropy and LibLinear SVM, although the performance improvement is more apparent with Maximum Entropy which shows a significant improvement over the baseline when the n-gram combination of unigrams, bigrams and trigrams. While both the algorithms show an improvement over the unigram model, the difference between the unigram-bigram and unigram-bigram-trigram model is not large.

The proposed feature set uses the best performing model among the 3 feature sets. Therefore, we combine the models FS1D, FS2C and FS3C to generate the proposed classifier model. The LibLinear SVM model achieves an accuracy of 78.32% which is better than the performance of all the other LibLinear SVM classifiers built using the 3 feature sets. However, Maximum Entropy shows a significant improvement in performance. It achieves the highest classification accuracy of 84.3% as well as the highest overall classification accuracy of all the models used. The Kappa statistic of models built using the feature sets 1, 2, and 3, as well as that of the proposed feature set is illustrated in Fig. 10.4.

By following the guidelines of Landis and Koch [11] to interpret the Kappa statistic measures, we observe that the baseline model, FS2A, are in the 0.41 and 0.60 range which indicates a moderate strength of agreement. With feature set 1, LibLinear SVM

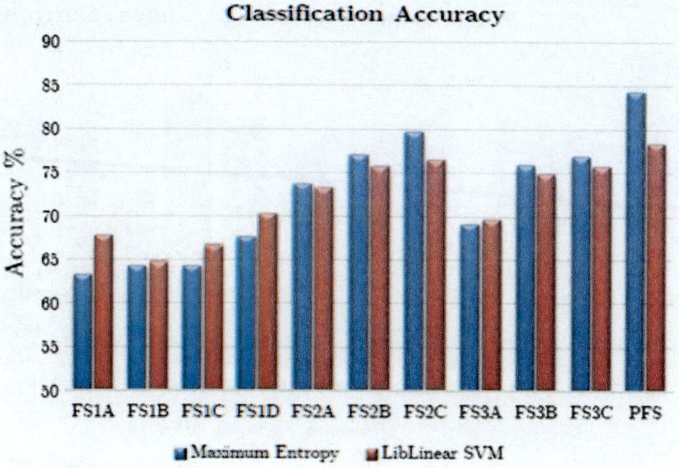

Fig. 10.4 Kappa statistic values obtained for the set of models

performs better than Maximum entropy in all cases except FS1B, where Maximum Entropy and Liblinear SVM perform at the same level. FS1D performs better among all the models in feature set 1 and performs moderately well, being in the 0.41–0.6 range.

With feature set 2, we observe that the kappa statistic improves consistently when higher order word n-gram combinations are used for both Maximum Entropy and LibLinear SVM, with Maximum Entropy achieving the highest overall kappa measure of 0.6947 which falls in the 0.61–0.80 range. We can thus infer that the strength of agreement is substantial.

With feature set 3, we observe that the Kappa statistic increases with higher order cluster n-grams. Maximum Entropy outperforms LibLinear SVM, but only slightly, with a kappa statistic of 0.6531 indicating a substantial strength of agreement. The LibLinear SVM has a Kappa statistic of 0.6355 which also indicates a substantial strength of agreement.

Overall, the highest kappa statistic measure is obtained by FS2C, which includes features extracted using word unigram=bigram-trigram combination.

Figure 10.5 indicates the performance metrics of precision, recall and F-score for feature sets 1, 2, 3 and the proposed feature set for Maximum Entropy classifier.

For Maximum Entropy, the precision, recall and the F-score of the baseline model, FS2A, is 0.75, 0.738 and 0.739 respectively, thus having a slightly better precision compared to recall. For feature set 1, the precision ranges from 0.67 to 0.681, recall ranges from 0.632 to 0.676 and F-score ranges from 0.614 to 0.675. Thus, none of the models perform as well as the baseline model in terms of these metrics. FS1D achieves the highest precision, recall and accuracy among the feature set 1 models. For feature set 2, FS2C performs the best in terms of accuracy precision and recall achieving values of 0.803, 0.796 and 0.798 respectively. For feature set 3, FS3C performs better

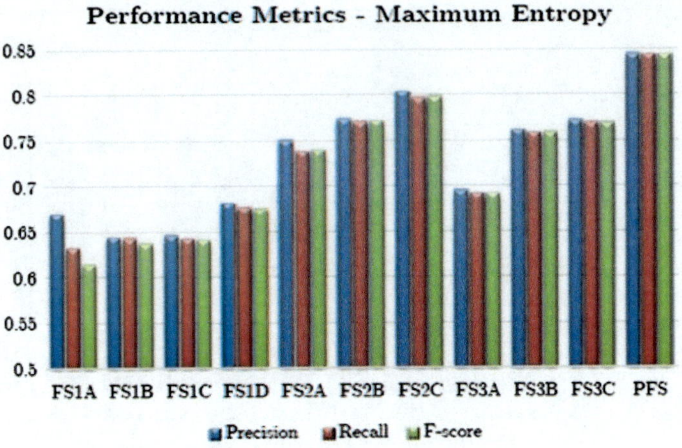

Fig. 10.5 Performance metrics of Maximum Entropy models

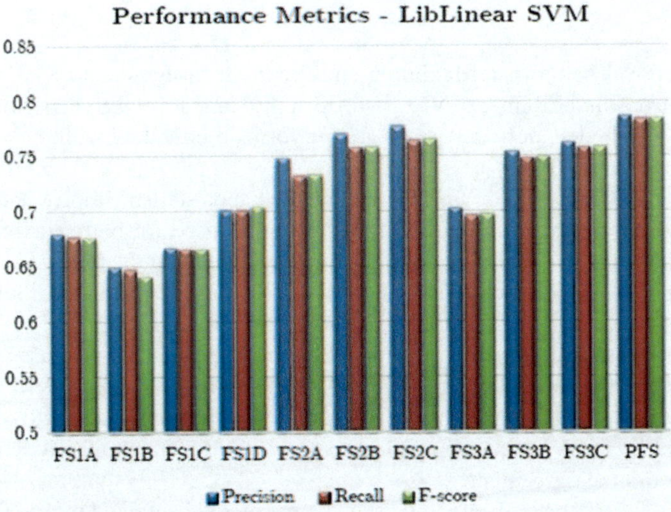

Fig. 10.6 Performance metrics of LibLinear SVM models

than the baseline values achieving 0.772, 0.769 and 0.769. PFS, model from the proposed feature set which includes cluster unigram-bigram-trigram combination, word unigram-bigram-trigram combination achieves the highest overall performance metrics compared to the baseline model with precision, recall and F-score values of 0.844, 0.843 and 0.843.

Figure 10.6 indicates the performance metrics of precision, recall and F-score for feature sets 1, 2, 3 and the proposed feature set for LibLinear SVM classifier.

For LibLinear SVM, the precision, recall and the F-score of the baseline model, FS2A, is 0.748, 0.732 and 0.733 respectively, thus having a slightly better precision compared to recall. For feature set 1, the precision ranges from 0.68 to 0.701, recall ranges from 0.677 to 0.701 and F-score ranges from 0.676 to 0.704. Thus, none of the models perform as well as the baseline model in terms of these metrics. FS1D achieves the highest precision, recall and accuracy among the feature set 1 models. For feature set 2, FS2C performs the best in terms of accuracy precision and recall achieving values of 0.777, 0.764 and 0.765 respectively. For feature set 3, FS3C performs the better than the baseline values achieving 0.762, 0757 and 0.758. However, we do not see a significant improvement in the metrics for the proposed feature set model which uses LibLinear SVM compared to the other high-performing LibLinear models such as FS2C.

From our discussion, it appears that using Opinion Lexicons alone as features to train machine learning algorithms such as Maximum Entropy and Support Vector Machines does not raise classification accuracy significantly. However, using multiple Opinion Lexicons to generate features seems to provide a better performance than using them individually. Though using a standard word n-gram iteration such as unigrams to train machine learning algorithms provides a better performance than using Opinion Lexicons, adding higher order word n-grams as features significantly improves performance. However, it was observed during our experimentation that this effect only carries until trigrams.

Generating features with word n-grams of higher order than trigrams does not improve the performance and is computationally expensive since it generates a large number of features and increases sparsity. When cluster n-grams are uses ad features by themselves, they too provide a better performance with higher order n-grams. As with the word n-grams, higher order cluster n-grams provided better performance than cluster unigrams alone. And similar to word n-grams, this effect was only noticed until we reached trigrams. Using cluster n-grams of higher order not only increased the time taken for feature extraction, feature selection and model training, it also did not keep the pattern of increased performance seen with the addition of cluster bigrams and trigrams. When Opinion Lexicons, word n-grams and cluster n-grams were combined from all the high performing models of the three feature sets, Maximum Entropy classifier showed a marked improvement in performance while LibLinear SVM did not show any significant improvement.

From the different experiments, it can be concluded that a combination of word unigrams-bigrams-trigrams, cluster unigrams-bigrams-trigrams as well as a combination of six opinion lexicons used a features and then ranked using Information Gain alogorithm and the Ranker Search method provided the best performance in terms of accuracy, precision, recall, F-score and Kappa statistic when used with the Maximum Entropy Classifier with the conjugate gradient descent method.

10.6 Conclusions and Future Work

In this chapter we have presented an approach that yields improved sentiment classification of Twitter data. Sentiment classification of tweets poses a unique challenge compared to text classification performed in other mediums.

For our research, we used the SemEval Twitter Corpus which contained a large number of tweets in the neutral class compared to that of positive and negative classes. In order to reduce bias, we balanced the dataset by reducing the number of neutral tweets to that of positive and negative tweets. We explored various feature extraction methods which could enrich the feature model space such that problems of sparsity commonly associated with datasets that have a large number of attributes, such as Twitter data, is addressed.

Our major contributions are four-folds. We extensively study various feature extraction methods individually and combined using a supervised machine learning approach. First, we demonstrated that using a combination of opinion lexicons to extract features improves the sentiment classification accuracy than using an individual opinion lexicon by itself. Second, we demonstrated that using unigram-bigram-trigram Bag of Words feature improves the sentiment classification accuracy than using lower order n-gram features alone. Third, we demonstrate that when using brown word clusters as features by themselves, unigram-bigram-trigram clusters provide an improvement in performance than lower order cluster n-grams. And fourth, we proposed a classifier model which significantly raises the classification accuracy by combining various feature extraction methods. We demonstrated that by taking the external knowledge of a word cluster into account while classifying sentiment of tweets improves the performance of the classifier using a machine-based learning algorithm.

The proposed classifier uses a combination of six mainstream opinion lexicons, unigram-bigram-trigram Bag of Words and unigram-bigram-trigram clusters as features. The dimensionality of the features was reduced by feature extraction methods such as information gain algorithm and the ranker search method. Using the Multinomial Logistic Regression algorithm (Maximum Entropy) with conjugate gradient descent with the proposed set of features not only improved the accuracy over the baseline Unigram Bag of Words model by 10.67%, but still maintained a comparable training time.

As future work, additional studies need to be undertaken to determine if the results obtained can be generalized to other domains which use short informal text for communication such as Tumblr, SMS, Plurk, etc.

Acknowledgements This research has received funding from the European Union's Horizon 2020 research and innovation programme under grant agreement No 823907 (MENHIR: Mental health monitoring through interactive conversations https://menhir-project.eu).

References

1. Abid, F., Alam, M., Yasir, M., Li, C.: Sentiment analysis through recurrent variants latterly on convolutional neural network of twitter. Future Gener. Comput. Syst. **95**, 292–308 (2019)
2. Aggarwal, C., Zhai, C.: Mining Text Data. Springer Science and Business Media (2012)
3. An ensemble classification system for twitter sentiment analysis: Ankit, Saleena, N. Procedia Comput. Sci. **132**, 937–946 (2018)
4. Balazs, J.A., Velásquez, J.D.: Opinion mining and information fusion: a survey. Inf. Fusion **27**, 95–110 (2016)
5. Bradley, M., Lang, P.: Affective norms for English words (ANEW): instruction manual and affective ratings. Technical Report, Center for Research in Psychophysiology, University of Florida (1999)
6. Bravo-Marquez, F., Mendoza, M., Poblete, B.: Combining strengths, emotions and polarities for boosting twitter sentiment analysis. In: Proceedings of Second International Workshop on Issues of Sentiment Discovery and Opinion Mining, pp. 1–9. Chicago, USA (2013)
7. Garg, A., Roth, D.: Understanding probabilistic classifiers. machine learning. In: Proceedings of 12th European Conference on Machine Learning (ECML'01), pp. 179–191. Freiburg, Germany (2001)
8. Hatzivassiloglou, V., McKeown, K.: Predicting the semantic orientation of adjectives. In: Proceedings of ACL'98, pp. 174–181 (1998)
9. Hu, M., Liu, B.: Mining and summarizing customer reviews. In: Proceedings of 10th ACM SIGKDD International Conference on Knowledge Discovery and Data Mining (KDD'04), pp. 168–177. Seattle, WA, USA (2004)
10. Kumar, A., Sebastian, T.: Sentiment analysis: a perspective on its past, present and futures. Int. J. Intell. Syst. Appl. **4**(10), 1–14 (2012)
11. Landis, J., Koch, G.: The measurement of observer agreement for categorical data. Biometrics **33**(1), 159–174 (1977)
12. Liu, B.: Sentiment Analysis and Opinion Mining. Synthesis Digital Library of Engineering and Computer Science. Morgan & Claypool (2012)
13. Liu, B.: Sentiment Analysis: Mining Opinions, Sentiments, and Emotions. Cambridge University Press (2016)
14. Medhat, W., Hassan, A., Korashy, H.: Sentiment analysis algorithms and applications: a survey. Ain Shams Eng. J. **5**(4), 1093–1113 (2014)
15. Mishne, G., Glance, N.: Predicting movie sales from blogger sentiments. In: Proceedings of Computational Approaches to Analyzing Weblogs, Papers from the 2006 AAAI Spring Symposium, pp. 1–4. Stanford, California, USA (2006)
16. Mohammad, S., Turney, P.: Crowdsourcing a word şemotion association lexicon. Comput. Intell. **29**(3), 436–465 (2013)
17. Montoyo, A., Martínez-Barco, P., Balahur, A.: Subjectivity and sentiment analysis: an overview of the current state of the area and envisaged developments. Decis. Support Syst. **53**(4), 675–679 (2012)
18. Nakov, P., Kozareva, Z., Ritte, A., Rosenthal, S., Stoyanov, V., Wilson, T.: SemEval-2013 Task 2: sentiment analysis in Twitter. In: Proceedings of 7th International Workshop on Semantic Evaluation (SemEval'13), pp. 312–320. Atlanta, Georgia, USA (2013)
19. Nielsen, F.: A new anew: evaluation of a word list for sentiment analysis in microblogs. In: Proceedings of the ESWC2011 Workshop on 'Making Sense of Microposts': Big Things Come in Small Packages, pp. 93–98. Crete, Greece (2011)
20. O'Connor, B., Balasubramanyan, R., Routledge, B., Smith, N.: From tweets to polls: Linking text sentiment to public opinion time series. In: Proceedings of AAAI Conference on Weblogs and Social Media, pp. 122–129. Stanford, California, USA (2010)
21. Pang, B., Lee, L.: Opinion Mining and Sentiment Analysis. Now Publishers (2008)
22. Pozzi, F., Fersini, E., Messina, E., Liu, B.: Sentiment Analysis in Social Networks. Morgan Kaufmann (2017)

23. Schuller, B., Batliner, A.: Computational Paralinguistics: Emotion, Affect and Personality in Speech and Language Processing. Wiley (2013)
24. Thai, M.T., Wu, W., Xiong, H.: Big Data in Complex and Social Networks. Chapman and Hall/CRC (2016)
25. Wang, D., Zhu, S., Li, T.: SumView: a Web-based engine for summarizing product reviews and customer opinions. Expert Syst. Appl. **40**(1), 27–33 (2013)
26. Xiong, S., Lv, H., Zhao, W., Ji, D.: Towards twitter sentiment classification by multi-level sentiment-enriched word embeddings. Neurocomputing **275**, 2459–2466 (2018)
27. Yu, L.C., Wu, J.L., Chang, P.C., Chu, H.S.: Using a contextual entropy model to expand emotion words and their intensity for the sentiment classification of stock market news. Knowl.-Based Syst. **41**, 89–97 (2013)

Chapter 11
Supporting Data Science in Automotive and Robotics Applications with Advanced Visual Big Data Analytics

Marco Xaver Bornschlegl and Matthias L. Hemmje

Abstract Handling Big Data requires new techniques with regard to data access, integration, analysis, information visualization, perception, interaction, and insight within innovative and successful information strategies supporting informed decision making. After deriving and qualitatively evaluating the conceptual IVIS4BigData Reference Model as well as defining a Service-Oriented Architecture, two prototypical reference applications for demonstrations and hands-on exercises for previous identified e-Science user stereotypes with special attention to the overall user experience to meet the users' expectation and way-of-working will be outlined within this book chapter. In this way and based on the requirements as well as data know-how and other expert know-how of an international leading automotive original equipment manufacturer and a leading international player in industrial automation, two specific industrial Big Data analysis application scenarios (**anomaly detection on car-to-cloud data** and (**predictive maintenance analysis on robotic sensor data**) will be utilized to demonstrate the practical applicability of the IVIS4BigData Reference Model and proof this applicability through a comprehensive evaluation. By instantiation of an IVIS4BigData infrastructure and its exemplary prototypical proof-of-concept reference implementation, both application scenarios aim at performing anomaly detection on real-world data that empowers different end user stereotypes in the automotive and robotics application domain to gain insight from car-to-cloud as well as from robotic sensor data.

11.1 Introduction and Motivation

The availability of data has changed dramatically over the past ten years. The wide distribution of web-enabled mobile devices and the evolution of web 2.0 and **Internet**

M. X. Bornschlegl (✉) · M. L. Hemmje
Faculty of Mathematics and Computer Science, University of Hagen, 58085 Hagen, Germany
e-mail: marco-xaver.bornschlegl@fernuni-hagen.de

M. L. Hemmje
e-mail: matthias.hemmje@fernuni-hagen.de

G. Phillips-Wren et al. (eds.), *Advances in Data Science: Methodologies and Applications*, Intelligent Systems Reference Library 189,
https://doi.org/10.1007/978-3-030-51870-7_11

of Things (IoT) technologies are contributing to a large amount of data (so-called Big Data) [33]. Due to the fact that *"we live in the Information Age"* [77] cognitive efficient perception and interpretation of knowledge and information to uncover hidden patterns, unknown correlations, and other useful information within the huge amount of data (of a variety of types) [55] is a big challenge [8]. This challenge will become one of the key factors in competition, underpinning new waves of productivity growth, innovation, and consumer surplus [47]. *"The revolutionary potential"* of the benefits of Big Data technologies [74] and the use of scientific methods in business, as e.g. operational data analysis and problem solving for managing scientific or industrial enterprise operations in order to stay innovative and competitive and at the same time being able to provide advanced customer-centric service delivery, has also been recognized by industry [46].

Nevertheless, usable access to complex and large amounts of data poses an immense challenge for current solutions in business analytics [8]. Handling the complexity of relevant data (generated through information deluge and being targeted with Big Data technologies) requires new techniques with regard to data access, visualization, perception, and interaction supporting innovative and successful information strategies [8]. These challenges emerge at the border between automated data analysis and decision-making [32]. As a consequence, academic research communities as well as industrial ones but especially research teams at small universities and in **Small and Medium-sized Enterprises (SMEs)** will be facing enormous challenges because these changes in data processing technologies have increased the demand for new types of specialists with strong technical background and deep knowledge of the so-called **Data Intensive Technologies (DITs)** [35].

After deriving and qualitatively evaluating the conceptual IVIS4BigData Reference Model as well as defining a Service-Oriented Architecture, two prototypical reference applications for demonstrations and hands-on exercises for previous identified e-Science user stereotypes with special attention to the overall user experience to meet users' expectation and way-of-working will be outlined within this chapter.

In this way and based on the requirements as well as data know-how and other expert know-how of an international leading automotive original equipment manufacturer and a leading international player in industrial automation, two specific industrial Big Data analysis application scenarios (**anomaly detection on car-to-cloud data** and (**predictive maintenance analysis on robotic sensor data**) will be utilized to demonstrate the practical applicability of the IVIS4BigData Reference Model and proof this applicability through a comprehensive evaluation. By instantiation of an IVIS4BigData infrastructure and its exemplary prototypical proof-of-concept reference implementation, both application scenarios aim at performing anomaly detection on real-world data that empowers different end user stereotypes in the automotive and robotics application domain to gain insight from car-to-cloud as well as from robotic sensor data.

11.2 State of the Art in Science and Technology

11.2.1 Information Visualization and Visual Analytics

Information Visualization (IVIS) has emerged *"from research in Human-Computer Interaction, Computer Science, graphics, visual design, psychology, and business methods"* [67]. Nevertheless, IVIS can also be seen as a result of the question for interchanging ideas and information between humans keeping with Rainer Kuhlen [44] because of the missing direct way.

Hutchins [36] describes that the human cognitive process takes place both inside and outside the minds of people. Furthermore, he states *"unfortunately, [...] in a mind that is profoundly disconnected from its environment, it is necessary to invent internal representations of the environment that is outside the head"*.

Shneiderman [63] suggests that *"exploring information collections becomes increasingly difficult as the volume grows. A page of information is easy to explore, but when the information becomes the size of a book, or library, or even larger, it may be difficult to locate known items or to browse to gain an overview"*. Moreover, he also indicates that *"a picture is often cited to be worth a thousand words and, for some tasks, it is clear that a visual presentation—such as a map or photograph—is dramatically easier to use than a textual description or a spoken report"*.

In consequence of both arguments and of the way the human mind processes information, *"it is faster to grasp the meaning of many data points when they are displayed in charts and graphs rather than poring over piles of spreadsheets or reading pages of reports"* [62]. *"Even when data volumes are very large, patterns can be spotted quickly and easily"* [62]. Nevertheless, IVIS is not only a computational domain. In 1644, a graphic by Michael Florent v. Langren (Fig. 11.1), a Flemish astronomer to the court of Spain, believed to be the first visual representation of statistical data [70].

Notable in this basic example (which shows all 12 known estimates of the difference in longitude between Toledo and Rome, and the name of the astronomer Mercator, Tycho Brahe, Ptolemy, etc. who provided each observation) is that Langren could have presented this information in various tables (e.g. ordered by author to show provenance, by date to show priority, or by distance). However, only a visualization displays the wide variation in the estimates [34] because Information Visualization presumes that *"visual representations and interaction techniques take*

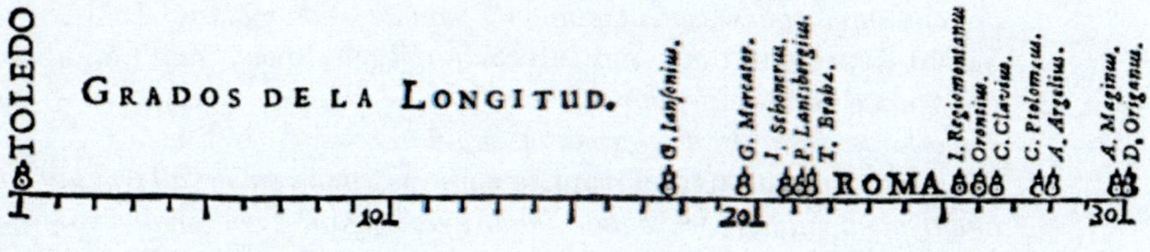

Fig. 11.1 Langren's graph of determinations of the distance from toledo to Rome [34, 70]

advantage of the human eye's broad bandwidth pathway into the mind to allow users to see, explore, and understand large amounts of information at once" [67]. Information Visualization focuses on the creation of approaches for conveying abstract information and sharing ideas with others in intuitive ways and a universal manner [62, 67]. The most precise and common definition of IVIS as *"the use of computer-supported, interactive, visual representations of abstract data to amplify cognition"* stems from Card et al. [17].

Whereas the purpose of IVIS is insight [17, 63], the purpose of **Visual Analytics (VA)** is to enable and discover information and knowledge that supports insight [18, 68]. To be more precise, Wong et al. [76] define VA as *"a contemporary and proven approach to combine the art of human intuition and the science of mathematical deduction to directly perceive patterns and derive knowledge and insight from them. [...] VA is an outgrowth of the fields of Scientific and Information Visualization but includes technologies from many other fields, including KM, statistical analysis, cognitive science, decision science, and many more."*

Keim et al. [41] describe VA as *"an iterative process, which has historically evolved out of the fields of Information- and Scientific Visualization and involves collecting information, data preprocessing, knowledge representation, interaction, and decision making."* Furthermore, they characterize the overarching driving vision of VA as turning the information overload into an opportunity: *"Just as Information Visualization has changed the view on databases, the goal of VA is to make the way of processing data and information transparent for an analytic discourse. VA (whose complete scope is outlined in Fig. 11.2) will foster the constructive evaluation, correction, and rapid improvement of our processes and models—and ultimately—the improvement of our knowledge and our decisions"* [40].

On a grand scale, VA solutions provide technology that combines the advantages of machines with the strengths of humans. While methods from statistics and mathematics are the driving force on the automatic analysis side, capabilities to perceive, relate, and conclude turn VA into a very promising field of research [32, 41, 42].

According to the definitions above, Thomas and Cook [67] describe Visual Analytics as a multidisciplinary field as well, *"where Visual Analytics tools and techniques are used to synthesize information and derive insight from massive, dynamic, ambiguous, and often conflicting data; detect the expected and discover the unexpected; provide timely, defensible, and understandable assessments; and communicate assessment effectively for action"* [67]. Furthermore, they diversify Visual Analytics in four focus areas:

1. **Analytical Reasoning Techniques** *"that let users obtain deep insights that directly support assessment, planning, and decision making"* [67].
2. **Visual Representations and Interaction Techniques** *"that exploit the human eye's broad bandwidth pathway into the mind to let users see, explore, and understand large amounts of information simultaneously"* [67].
3. **Data representations and transformations** *"that convert all types of conflicting and dynamic data in ways that support visualization and analysis"* [67].

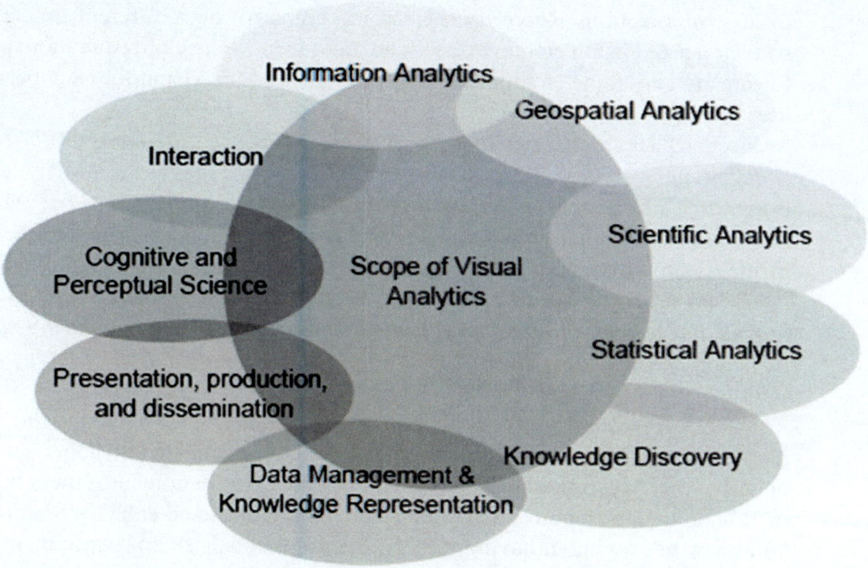

Fig. 11.2 Scope of visual analytics [41]

4. **Techniques to Support Production, Presentation, and Dissemination of Analytical Results** *"to communicate information in the appropriate context to a variety of audiences"* [67].

Summarizing the definitions in this section, it can be concluded that the purpose of Information Visualization is insight [17, 63], whereas the purpose of Visual Analytics is to enable and discover information and knowledge that supports insight [18, 68]. In this way, Visual Analytics can be defined as an outgrowth of techniques of the fields of Scientific- and Information Visualization that are used to synthesize information and derive insight from massive, dynamic, ambiguous, and often conflicting data [67].

11.2.2 End User Empowerment and Meta Design

Current development of Information and Communication Technology leads to a continuous growth of both computer systems and end user population [19]. Thus, designing visual interfaces for HCI supporting VA requires a critical decision which of the involved parties—the user or the software—will control the interaction.

User-friendly interfaces, which are often more intuitive, are focusing on providing users only basic information and less interoperability. This type of implementation is more suitable for naive users without deep technical understanding [50]. Nevertheless, in situations where users need more control over different aspects of the software, user empowered interfaces can provide more specialized or more powerful features to enrich the environment with the fruits of the vision for each person who uses them [37, 50].

Fischer [30] emphasizes that *"people and tasks are different, [...] they become engaged and excited about personally meaningful ideas, they strive for self-expression, and they want to work and learn in a self-directed way."* Moreover, he explains that humans start from a partial specification of a task, and refine it incrementally, on the basis of the feedback that they get from their environment. Thus, *"users must be able to articulate incrementally the task at hand. The information provided in response to these* ***problem-solving activities based on partial specifications and constructions must assist users to refine the definition of their problem"*** [30].

"Users are increasingly willing and, indeed, determined to shape the software they use to tailor it to their own needs" [5]. To turn computers into convivial tools, to underpin the evolution of end users from passive information consumers into information producers, requires that people can use, change, and enhance their tools and build new ones without having to become professional-level programmers [5, 28]. Thus, empowering individuals requires conceptual frameworks and computational environments which extend the traditional notion of system design beyond the original development of a system to include an ongoing process in which the users of the system become co-designers that will give domain workers more independence from computer specialists [28, 29].

For enabling end users *"to articulate incrementally the task at hand"* [30], *"the information provided in response to their problem-solving activities based on partial* ***specifications and constructions must assist users to refine the definition of their*** *problem"* [30]. To realize this interaction, Fig. 11.3 outlines the different elements of the specification and construction process in context of the users' problem-solving activities.

In this user empowerment architecture model, five central elements (specification, construction, argumentation base, catalog base, and semantics base) can be identified, ***"that assist end users to refine the definition of their problem in their problem-solving activities based on partial specifications and constructions"*** [30] [10].

Derived from an end users' configuration perspective, a domain independent problem-solving, i.e., User Interface configuration process can be divided in three layers [10]. The design creation layer contains the construction and specification components, that represent the interactive part of this process utilizing the three static components argumentation base, catalog base, and semantics base within the lowest domain knowledge layer [10]. Moreover, the feedback layer in the middle of this architecture represents the interactive user actions (critics, case-based reasoning, and simulation), that are initiated during the specification or construction process [10]. In addition to this architectural illustration and to emphasize the importance of the construction and specification elements, Fischer and Nakakoji defined a process-based illustration of the whole design process, that is outlined in Fig. 11.4 [10].

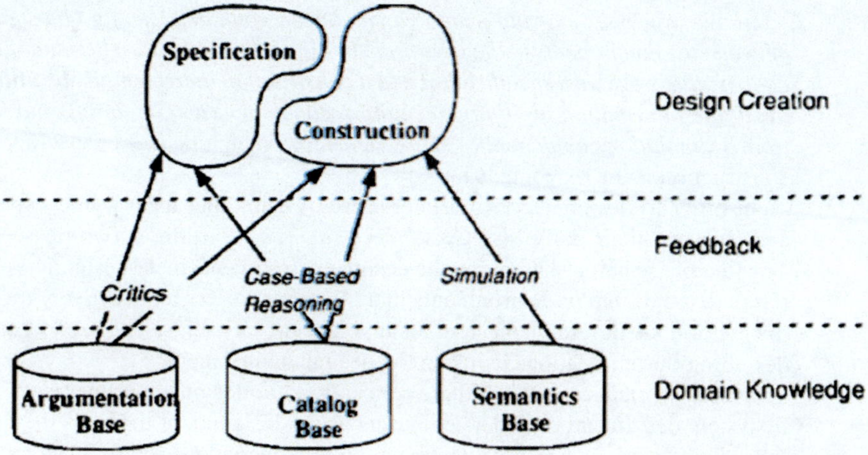

Fig. 11.3 Elements of a multifaceted architecture [30]

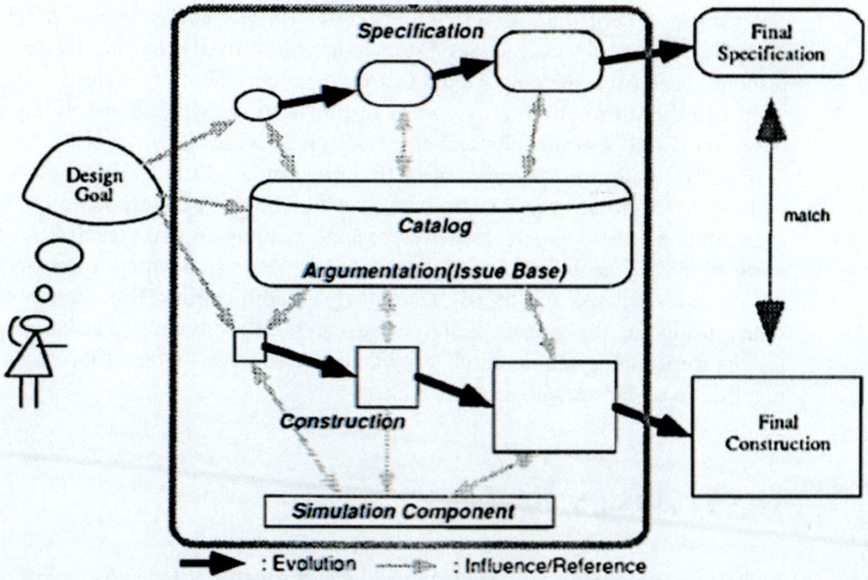

Fig. 11.4 Co-evolution of construction and specification of design in multifaceted architecture [30]

In this process, *"starting with a vague design goal, designers go back and forth between the components in the environment"* [30]. Thus, *"a designer and the system cooperatively evolve a specification and a construction incrementally by utilizing the available information in an argumentation component and a catalog and feedback from a simulation component"* [30]. As a result, a matching pair of specification and construction is the outcome [9].

In order to design successful interactive systems that meet users' expectations and improve their daily life, Costabile et al. [19] consider a two-phase process. The first phase being sharpening the design environment (meta-design phase) which refers to the design of environments that allows end users to be actively involved in the continuous development, use, and evolution of systems. The second one being designing the applications by using the design environment.

Discussing and concluding these concepts, meta-design underlines a novel vision of system design and considers end users as co-designers of the tools they will use. All stakeholders of an interactive system, including end users, are *"owners"* [19] of a part of the problem: Software engineers know the technology, end users know the application domain, Human-Computer Interaction experts know human factors, etc.; *"they must all contribute to system design by bringing in their own expertise"* [19]. Thus, implementing interactive systems supported by the end user empowerment principle, end users are empowered to articulate incrementally their task at hand and to utilize the information provided in response to their problem-solving activities based on partial specification and construction activities [30].

Finally, to address the observation that modern Big Data analysis infrastructures and software frameworks have to consider a high degree of interoperability by adopting common existing open standards for access, analysis, and visualization for realizing an ubiquitous collaborative workspace for end users which is able to facilitate the research process and its Big Data analysis applications, this section continues with introducing the concept of Virtual Research Environments (that serves as a basis for the resulting system architecture where end users in different locations to work together in real time without restrictions).

11.2.3 IVIS4BigData

In 2016, Bornschlegl et al. systematically performed the Road Mapping of Infrastructures for Advanced Visual Interfaces Supporting Big Data workshop [14], where academic and industrial researchers and practitioners working in the area of Big Data, Visual Analytics, and Information Visualization were invited to discuss and validate future visions of Advanced Visual Interface infrastructures supporting Big Data applications. Within that context, the conceptual IVIS4BigData reference model (c.f. Fig. 11.5) was derived, presented [11], and qualitatively evaluated [7] within the workshop's road mapping and validation activities [13]. Afterwards, a set of conceptual end user empowering use cases that serve as a base for a functional, i.e. conceptual as well as technical IVIS4BigData system specification supporting end users, domain experts, as well as for software architects in utilizing IVIS4BigData have been modeled and published [9].

In IVIS4BigData, the IVIS pipeline is segmented in a series of data transformations [10]. Furthermore, due to the direct manipulative interaction between different user stereotypes within the single process stages and their adjustments and configurations of the respective transformations by means of user-operated controls, each segment in the IVIS4BigData pipeline needs to support an interactive user empowerment, i.e., system configuration workflow allowing to configure the transformations and visualizations in the different phases [10].

For this reason, the IVIS4BigData pipeline has been divided into four consecutive process stages that empower end users to define, configure, simulate, optimize, and run each user empowered phase of the pipeline in an interactive way. Arranged in the sequence of the IVIS4BigData pipeline, each process stage contains all actions of its interactive and user empowered transformation configuration workflow between the two IVIS phases [10]. Starting from raw data on the left side, the four consecutive IVIS4BigData process stages, where each stage represents a certain IVIS4BigData transformation data integration, data transformation, visual mapping, and view transformation), are defined the following way:

- **Data Collection, Management, and Curation**: Harmonization and Semantic Integration of individual, distributed, and heterogeneous raw data sources into a uniform schema (data integration from local source schemata into a global integrated mediator schema) [8].
- **Analytics**: Distributed and cloud-based Big Data analysis of the integrated raw data (data transformation) [8].
- **Visualization**: Definition and creation of a visualization based on the structured data (visual mapping) [8].
- **Perception and Effectuation**: Facilitation of the interaction with appropriate views of the generated visual structures to enable suitable interpretations of the Big Data analysis results (view transformation) [8].

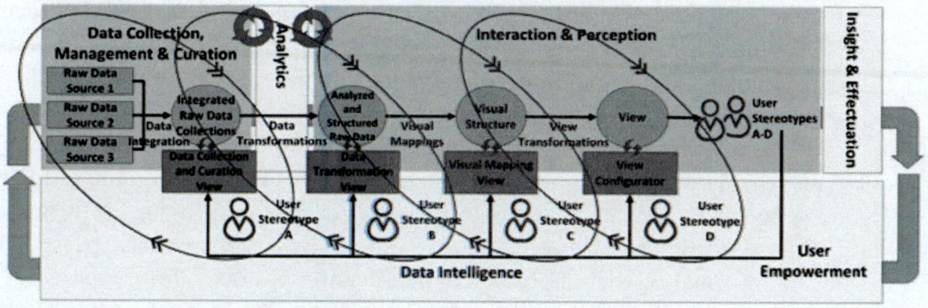

Fig. 11.5 IVIS4BigData reference model [11]

11.2.3.1 Conceptual IVIS4BigData End User Empowering Use Cases

"Users are increasingly willing and, indeed, determined to shape a software they use to tailor it to their own needs" [5]. For turning computers into convivial tools, to underpin the evolution of end users from passive information consumers into information producers, requires that people can use, change, and enhance their tools and build new ones without having to become professional-level programmers [5, 28]. After deriving the IVIS4BigData process structure including the description of their interactive configuration workflow objectives between each two IVIS4BigData phases and the definition of the interactive configuration use case framework, the gap between the architectural and the functional mapping of the process stages still exists [8]. To close this gap and to derive a set of conceptual user empowerment use case for each IVIS4BigData process stage, each process stage is considered from a functional perspective [8]. These conceptual configuration use cases, that contain all configuration activities of their respective process stage, serve as a base and a functional system description for end users, domain experts, as well as for software architects utilizing IVIS4BigData [8].

Configuration of the Data Collection, Management, and Curation Phase. The first conceptual configuration use case data collection, management, and curation in the sequence of the IVIS4BigData process stages describes functions that can configure how to integrate distributed and heterogeneous raw data sources into a uniform raw data collection, by means of Semantic Integration [10]. Thus, as illustrated in Fig. 11.6, several functions are provided to facilitate the main configuration functionality of this informal use case and the first IVIS4BigData transformation data integration [10].

Starting from raw, already processed, or stored data of previous IVIS4BigData process iterations, the configuration functions of the application layer represent the mediator-wrapper functionality of the concept of Semantic Integration. Thus, they empower end users to select individual data sources as well as the semantic representation of the entire data of the connected data sources to design, configure, and finally create and manage integrated data sets as an intermediate result and preliminary input for the next consecutive use case in the next IVIS4BigData process phase [10]. Within the application layer, beginning with the distributed and heterogeneous data sources, the data source management function in the integration and analytics area enables domain experts to connect data sources from a technical perspective [10]. For this purpose, this function makes use of the data instance description, the data schema description, and the data model description within the domain knowledge layer [10]. Whereas the data instance description provides information about technical attributes (like, e.g., data type, data host address, data port, data source log-in information, or supported communication protocol) for the physical connection, the data schema description contains information about the data structure (like, e.g., table names, columns, or property lists) and the data model description contains information about data content (like, e.g., data model, representation, syntactical relationships, or constraints) for logical connections of data sources [10]. The data

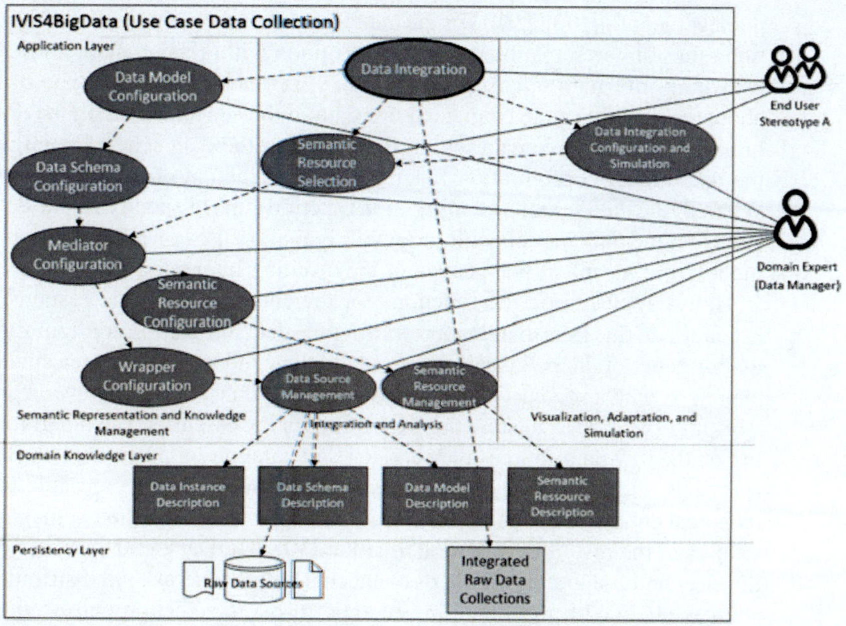

Fig. 11.6 Configuration support in IVIS4BigData use case data collection, management, and curation [10]

source management function is located at the lowest level within this use case to emphasize that there is no relation to raw data within the data sources at this level of abstraction although this function accesses information of the domain knowledge layer and provides it to the subsequent functions. Based on this base functionality, hereafter the process of the data integration is segmented in a technical and a logical path [10].

From a technical perspective, the wrapper configuration function, located in the semantic representation and knowledge management area, provides access to the data of the data sources by exporting some relevant information about their schema, data, and query processing capabilities [9]. Moreover, the mediator configuration function represents the second step of the technical path of the data integration. In addition to the functionality of the mediator configuration, the defined mediator combines both paths to a resulting logical path [10].

To *"exploit encoded knowledge about certain sets or subsets of data to create information for a higher layer of applications"* [75], and to store the data provided by the wrappers in a unified view of all available data with central data dictionary by the utilization of Semantic Integration, the mediator relies on the information of the logical function semantic resource configuration [10]. It configures the data sources from a semantic perspective with focus on their logical content based on the available semantic resources, which can be configured within the semantic resource management function [10]. For the management of the semantic resources, this function relies on the semantic resource description within the domain knowledge layer, providing information (like, e.g., full text information or other type of meta-data) about the content of the connected raw data sources [10]. Based on the unified views of all available source data within the mediator, the data schema configuration and the data model configuration function consider the data from a target perspective. Whereas the data schema configuration function aims of specifying the data structure and type, the data model configuration function focuses on the definition of the data model from a content perspective of the resulting integrated data [10].

Before focusing on the functions for the end users, that are responsible for the execution of this IVIS4BigData workflow, the data integration configuration and simulation function in the visualization, adaptation, and simulation area enables domain experts as well as end users to configure and simulate individual raw data integration workflows as well as to store the workflows for the essential data integration depending on the raw data sources data and the data integration purpose [10]. Based on the functions with enhanced capabilities, where domain experts are able to configure technical details of the connected data sources as well as of their semantic representation and the resulting data model within IVIS4BigData, end users are empowered to select and integrate the data of connected heterogeneous and distributed raw data sources [10]. With the function semantic resource selection end users are able to select data sources with their semantic representation of their respective data [10]. Finally, the function data integration, that represents the first transformation of the IVIS4BigData pipeline, integrates the data by utilizing the configured and stored data integration workflows and provides the resulting integrated raw data set to the integrated raw data collection process phase in the persistency layer [10].

Configuration of the Analytics Phase. The conceptual configuration use case for configuring the analytics phase of IVIS4BigData, that is illustrated in Fig. 11.7 and is located at the second position in the sequence of the IVIS4BigData process stages, describes functions for end users as well as for domain experts to facilitate the essential technical Big Data analysis [10]. This main functionality represents the second IVIS4BigData transformation and transforms the integrated and unstructured raw data in analyzed structured data [10].

Starting from the integrated raw data of the heterogeneous and distributed raw data sources, the functions of the application layer empower end users to select unstructured raw data sets, configure and simulate Big Data analysis workflows, execute the configured workflows, and export the resulting analyzed and structured data for the consecutive use case [10]. Before focusing on the functions for the end users, that are responsible for the execution of this IVIS4BigData workflow, two central configuration functions for domain experts are considered within the application layer at first. Starting with the Big Data analysis method configuration function in the semantic representation and knowledge management area, that resorts to the Big Data analysis method catalog within the domain knowledge layer, this function enables domain experts to configure Big Data analysis methods (like, e.g., Hadoop [3], Spark [4], or R [66]) and for the utilization in IVIS4BigData [10]. Afterwards, these methods can

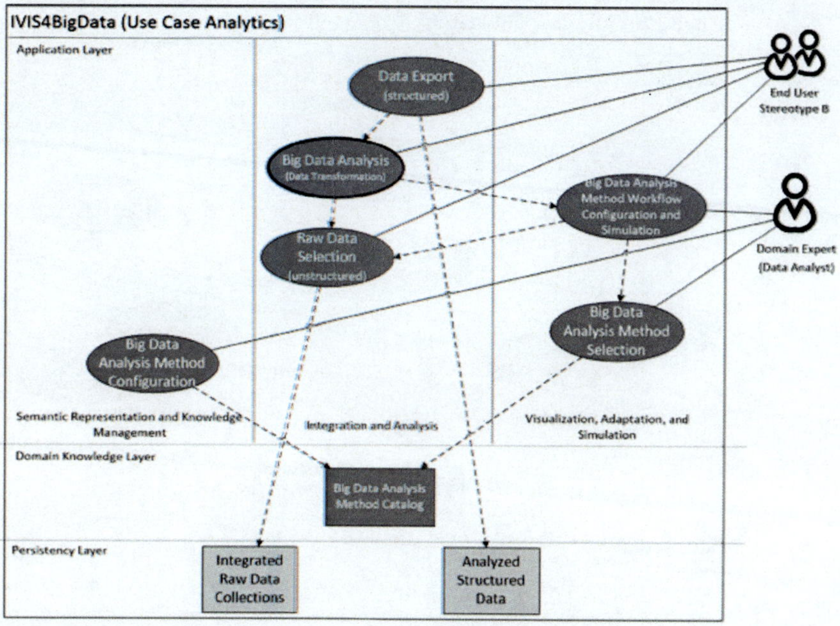

Fig. 11.7 Configuration support in IVIS4BigData use case analytics [10]

be selected by the Big Data analysis method selection function in the visualization, adaptation, and simulation area [10]. The last function in the visualization, adaptation, and simulation area (Big Data analysis method workflow configuration and simulation) enables domain experts as well as end users to configure and simulate individual Big Data analysis workflows as well as to store the workflows for the essential analysis depending on the source data and the analysis purpose [10].

Thus, after configuration and simulation of analysis algorithms and methods, the end users are empowered to perform their Big Data analysis with the aid of three functions in the integration and analysis area [10]. First of all, the function raw data selection enables the selection of the integrated but unstructured data of the heterogeneous and distributed raw data sources. Afterwards, the main function Big Data analysis, that represents the second transformation of the IVIS4BigData pipeline, utilizes the configured and stored analysis workflows to transform the unstructured data to structured data [10]. Finally, the data export function provides the resulting structured data to the analyzed and structured data process phase in the persistency layer for the consecutive use case [10].

Configuration of the Visualization Phase. The third conceptual configuration use case in the sequence of the IVIS4BigData process stages describes functions to transform analyzed and structured data into visual structures [10]. As illustrated in Fig. 11.8, several functions are provided to end users as well as to domain experts to

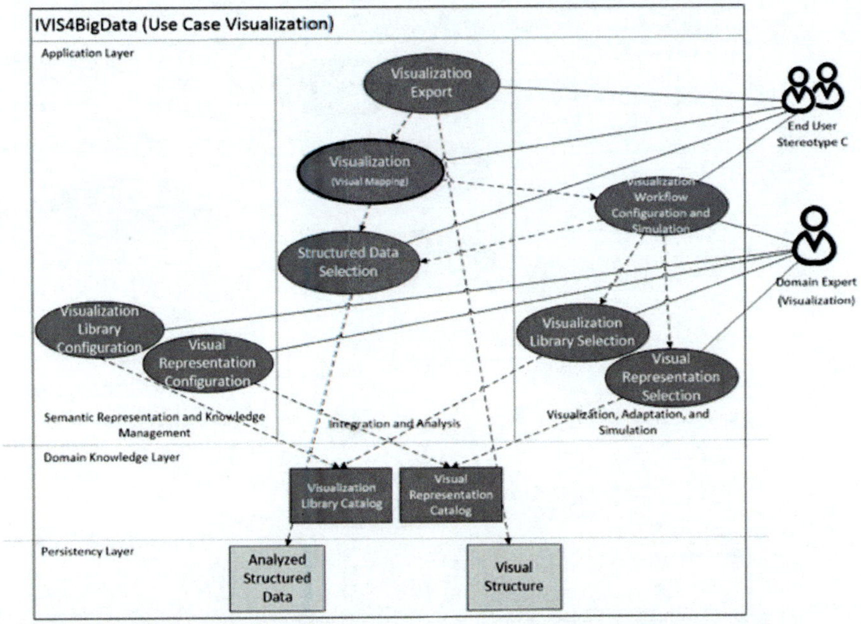

Fig. 11.8 Configuration support in IVIS4BigData use case visualization [10]

facilitate the main functionality of this informal use case and the third IVIS4BigData transformation visualization [9].

Starting from the structured and analyzed data of the heterogeneous and distributed data sources, the functions of the application layer empower end users to select structured data sets, configure and simulate Big Data visualization workflows, execute the configured workflows, and export the resulting visual structure for the consecutive use case [10]. Similar to the previous configuration use case supporting the analytics phase in IVIS4BigData, several central configuration functions for domain experts are considered within the application layer [10]. Starting with the visual representation configuration function in the semantic representation and knowledge management area, that resort to the visual representation catalog within the domain knowledge layer, this function enables domain experts to configure suitable visual representations (like, e.g., linear, tabular, hierarchical, spatial, or textual) and depending on the respective data structure within the analyzed and structured data process stage [10]. Moreover, the visualization library configuration function based on using the visualization library catalog, enables domain experts to configure visualization libraries (like, e.g., D3.js, Charts.js, dygraphs, or Google Charts) for utilization in IVIS4BigData [10].

Afterwards, visual representations as well as the visualization libraries can be selected by the visual representation selection and visualization library selection functions in the visualization, adaptation, and simulation area, by making use of configured visual representations and visualization libraries within the catalogs [10]. The last function in the visualization, adaptation, and simulation area is visualization workflow configuration and simulation [10]. This function enables domain experts as well as end users to configure and simulate individual Big Data visualization workflows as well as to store the workflows for the essential Big Data visualization depending on analyzed and structured source data as well as on the visualization and analysis purpose [10].

After the configuration and simulation of the visualization methods, the end users are empowered to perform their Big Data visualization with the aid of three functions in the integration and analysis area [10]. First, the function structured data selection enables the selection of integrated and structured raw data in the heterogeneous and distributed data sources [10]. Afterwards, the main function visualization, that represents the third transformation of the IVIS4BigData pipeline, utilizes configured and stored visualization workflows to transform analyzed and structured data to visual structures [10]. Finally, the data export function provides resulting visual structures to the visual structure process phase in the persistency layer for the consecutive use case [10].

Configuration of the Perception and Effectuation Phase. The final configuration use case perception and effectuation, which is illustrated in Fig. 11.9, is located at the fourth position in the sequence of the IVIS4BigData process stages [10]. This informal use case describes functions for end users as well as for domain experts to facilitate the generation of suitable views [10]. The main functionality view transformation, that represents the fourth IVIS4BigData transformation, transforms the visual structure into interactive views, whereby end users are empowered to interact with analyzed and visualized data of the heterogeneous and distributed raw data sources for perceiving, managing, and interpreting Big Data analysis results to support insight [10].

Starting from integrated, analyzed, and visualized raw data of the heterogeneous and distributed data sources, the configuration functions of the application layer empower end users to select visualized data sets, generate suitable views, and interact with visualized data of the heterogeneous and distributed raw data sources [10]. As well as the previous configuration use cases and before focusing on the functions for the end users, that are responsible for the execution of this IVIS4BigData workflow, one central configuration function for domain experts is considered within the application layer at first. This IVIS technique configuration function in the semantic representation and knowledge management area, that uses the IVIS technique catalog within the domain knowledge layer, enables domain experts to configure visualization techniques (like, e.g., word cloud, tree map, sunburst chart, choropleth map, or small multiples) and for the utilization in IVIS4BigData [10].

The view configuration and simulation function within the visualization, adaptation, and simulation area enables domain experts as well as end users to configure and simulate individual and suitable views as well as to store the views for the essential interaction and perception of the visualized data depending on the analysis purpose [10]. Thus, after the configuration and simulation of the visualization technology,

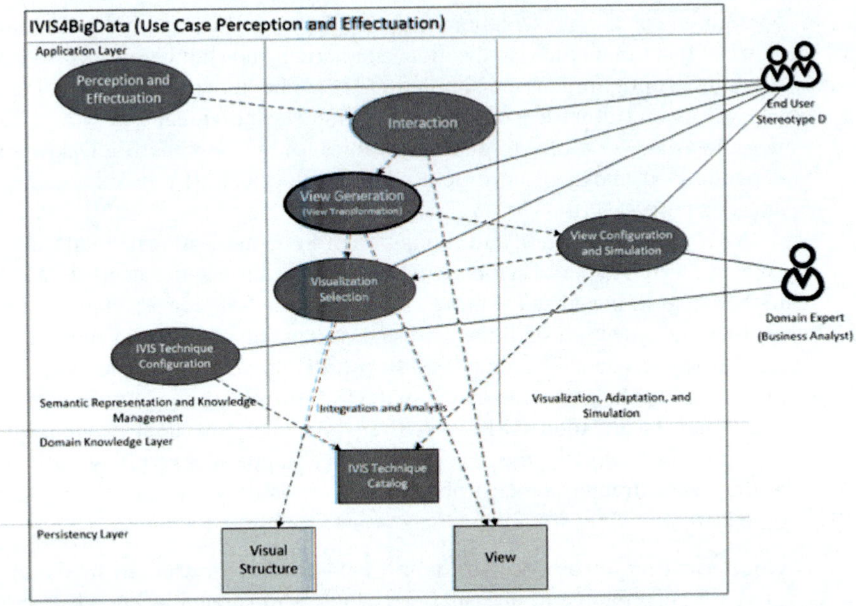

Fig. 11.9 Configuration support in IVIS4BigData use case perception and effectuation [10]

end users are empowered to perform their interaction and perception with the aid of three functions within the integration and analysis area. First, the function visualization selection enables the selection of the visual representation of the integrated and analyzed heterogeneous and distributed data sources [10]. Second, the main function view generation, that represents the fourth transformation of the IVIS4BigData pipeline, utilizes the configured and stored views to transform the visual structure to an interactive view [10]. Third, the interaction function enables end users to perceive, manage, and interpret Big Data analysis results to support insight [10].

Finally, with the aid of the perception and effectuation function within the semantic representation and knowledge management area, the emergent knowledge process, which is symbolized by the outer loop of IVIS4BigData, can be achieved by actively managing the insights created by effectuating data and integrating these effects into the knowledge base of the analysis process [10].

11.2.3.2 Conceptual IVIS4BigData Service-Oriented Architecture

For achieving a usable and sustainable reference implementation of the defined conceptual IVIS4BigData Reference Model and its conceptual reference application design, a conceptual IVIS4BigData SOA has been designed [8]. This IVIS4BigData SOA has to flexibly support the tailoring of IVIS4BigData application solutions to the requirements of its different end user stereotypes [8]. In addition, due to limited resources of Small and Medium-Sized Enterprises, the operating costs of the resulting IVIS4BigData infrastructure reference implementation has been considered as well [8]. Thus, the conceptual IVIS4BigData SOA has been technically specified and implemented based on open-source base technologies [8].

Thus, the IVIS4BigData SOA has been implemented based on open-source base technologies. Whereas existing open source Big Data technologies and frameworks in all layers of the IVIS4BigData SOA have already found their way into mainstream application and have seen wide-spread deployment in scientific communities as well as in organizations across different industry fields [60], they differ with regard to their application scenarios [8]. Therefore, the SOA approach ensures easy interoperability by adopting common existing open standards for access, analysis, and visualization for realizing a ubiquitous collaborative workspace for researchers, Data Scientists as well as business experts and decision makers which is able to facilitate the research process and its Big Data analysis applications [8].

In this way and from a global perspective, the conceptual IVIS4BigData SOA design approach is based on the design of the VERTEX Service-Oriented Architecture [8]. Based on providing and managing access to Big Data resources through open standards, the VERTEX reference architecture is materialized through existing open components gathered from successful research and development projects (such as, e.g., Smart Vortex [24], SenseCare [26] and MetaPlat [25]) dealing with resources at scale, and supported by their owners as project partners [27]. To implement an IVIS4BigData infrastructure along a conceptual SOA as outlined in Fig. 11.10, the initial VERTEX SOA architecture is refined by adding relevant CRISP- [6] and IVIS4BigData components in combination with existing **Knowledge Management Ecosystem Portal (KM-EP)** [61] services [72].

From a vertical perspective, the conceptual IVIS4BigData SOA framework defines a four-layer architecture starting from the upper application layer across the middle-ware service layer and resource layer down to the infrastructure layer [12]. Whereas both lower layers do not differ from the original VERTEX architecture, both upper layers differ from a horizontal perspective and contain the VERTEX elements only in the left area, whereas the right area is represented by the KM-EP **Content and Explicit Knowledge Management (CEKM)** extensions and the middle area is represented by the extensions related to the CRISP- and IVIS4BigData components and corresponding services [12].

Illustrated by the connection from the domain specific application within a VRE portal on the left side to the IVIS4BigData application, the alignment of the elements within the application layer emphasizes that any IVIS4BigData infrastructure represents a specific VRE research application [12] that can be managed as well as collaboratively be executed by the utilization of the built-in VRE portal functionalities. Furthermore, each IVIS4BigData infrastructure is supported by the KM-EP User Interfaces on the right side that enable end users and domain experts to configure the underlying KM-EP CEKM System, which hosts the CEKM resources for the central IVIS4BigData application [12]. The User Interfaces within the IVIS4BigData application in the central area of this layer illustrate the four IVIS4BigData process stages with their end user empowering integration and analysis as well as their spec-

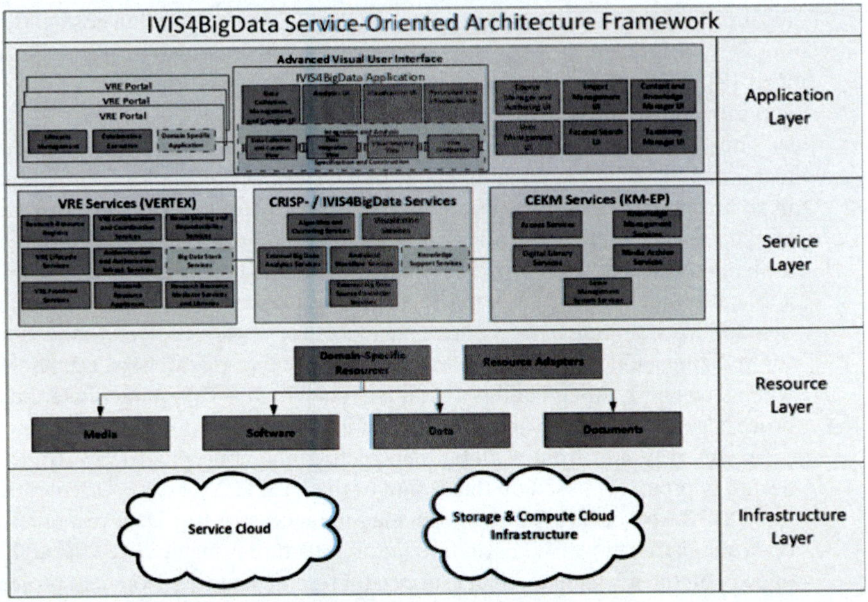

Fig. 11.10 IVIS4BigData service-oriented architecture

ification and construction functionalities over the whole Big Data analysis process [12].

The service layer contains all categories of VRE, CRISP-/ IVIS4BigData, and CEKM services that are required to access, integrate, and analyze domain specific resources as Big Data sources (e.g. documents, media objects, software, explicitly encoded knowledge resources as well as sensor data from, e.g., scientific experiments or industrial machinery settings [27]). Illustrated by the connection from the Big Data stack services from the VRE services area on the left side to the CRISP- and IVIS4BigData services area in the center as well as by the connection from its knowledge support services to the CEKM services in the right area, both connections emphasize the modular refinement and the cooperation between the three service categories at this layer [12]. Based on the KM-EP's CEKM services area that contains all services to configure and operate the basic KM-EP CEKM system, the central CRISP- and IVIS4BigData services area contains all services to configure and run Big Data analysis workflows for CRISP4BigData and IVIS4BigData [12]. Whereas these KM-EP CEKM services are consolidated as knowledge support services, the lowest external Big Data source connector services provide functions to connect distributed and heterogeneous raw data sources, and the external Big Data analytics services can be utilized to connect external Big Data algorithms or analysis workflows. Finally, the algorithm and clustering services, analysis workflow services as well as the visualization services are utilized to configure, manage, execute, and visualize the essential Big Data analysis.

Finally, the VRE services area includes the VRE related services to configure and operate a VRE environment that hosts the resulting CRISP- and IVIS4BigData research application based on the resources that are gathered, integrated, and managed as Big Data sources in the KM-EP CEKM System [12]. To be more precise, supported by the VRE life-cycle support services that are responsible to monitor and execute the actions of the VERTEX life-cycle model [27], the VRE collaboration and coordination services are utilized *"to implement the management of the VREs and the collaborative execution of the research experiments"* [27] and the VRE frontend services are responsible to execute the User Interface of the resulting VRE application. Moreover, whereas the essential Big Data stack services have been refined by the CRISP- and IVIS4BigData services, the result sharing and reproducibility services are responsible that results of the Big Data analysis may be shared and reproduced over long term and by different communities. Additionally, the **Authentication and Authorization Infrastructure (AAI)** services, research resource appliances services, and VERTEX access mediator framework services are utilized to manage the physical and logical access to the connected distributed, cross-domain, cross-organizational research resources.

Supported by the resource layer that specifies all IVIS4BigData raw data sources and the adapters for their Semantic Integration into the global Big Data source schema of the conceptual IVIS4BigData SOA environment, the lowest infrastructure layer contains the external cloud infrastructure that hosts domain specific Big Data resources as well as the deployed IVIS4BigData storage and computing services, specified at the service layer, to guarantee elastic resource consumption and deployment [27].

11.3 Modeling Anomaly Detection on Car-to-Cloud and Robotic Sensor Data

In order to develop a generic anomaly detection application for car-to-cloud and robotic sensor data that prototypically instantiated the IVIS4BigData Reference Model, specific requirements had to be considered [15]. As car-to-cloud and robotic sensor data can both be referred as being heterogeneous regarding its content, yet there are uniform characteristics applying to it (data instance type, timing, frequency, value ranges or parameter presence) [15]. In addition, this heterogeneity also results in different anomaly detection algorithms regarding accuracy, timing, and prerequisites depending on the suggested outcome [15].

Therefore and based on the IVIS4BigData Reference Model's guidelines (c.f. Sect. 11.2.3) for designing systems that utilize end users' cognitive input for Big Data analysis, only a generic reference implementation where end users as well as domain experts are empowered to gain insight, to configure involved workflows, and to provide domain knowledge will satisfy the demands in anomaly detection on car-to-cloud and robotic sensor data [15]. In a nutshell and as illustrated in Fig. 11.11,

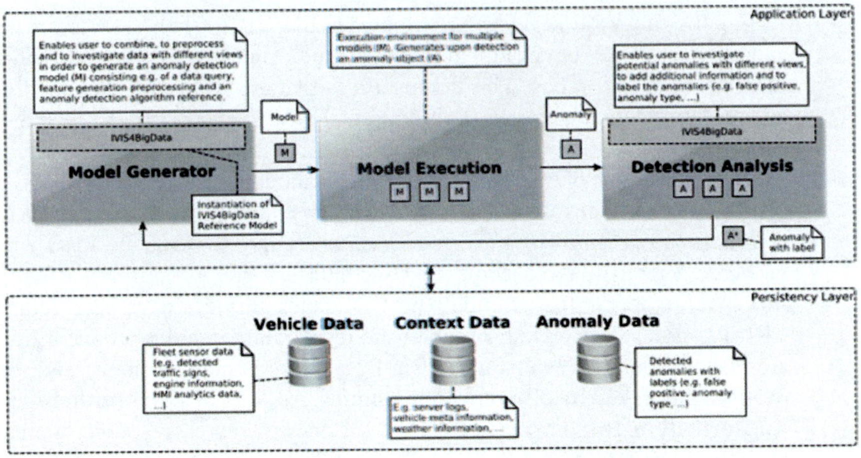

Fig. 11.11 Conceptual anomaly detection on car-to-cloud and robotic sensor data model [15]

the utilization of anomaly detection on car-to-cloud and robotic sensor data had been subdivided into three successive components [15].

The anomaly detection problem itself and the approach of how to solve it are defined in the model generator [15]. Within this component, users configure relevant input data, perform comprehensive preprocessing, select suiting algorithms, and tune their parameters [15]. Since this model stores all information that is relevant in context of the problem, a model execution component applies the same analysis on other car-to-cloud data by executing the model [15]. Once a potential anomaly is detected, it is forwarded to the third process step. The anomaly candidate will be subject to further investigations by users in the detection analysis component [15]. Through comprehensive visualization of the data and its context by application of IVIS4BigData, the users will be empowered to decide whether the detected anomaly is a true positive detection and derive necessary steps to deal with the outcome [15].

Within a closer examination and as outlined in Fig. 11.12, the model generator component consists of several logically distinct entities [15]. Some of these components instantiate the IVIS4BigData reference model and hence have User Interfaces. These are labeled with "(IVIS4BigData)" [15]. Other components only react upon artifacts generated within the former components. These components include the word "engine" within their name [15].

Following the logical flow of information from raw data to analyzed anomalies, the first component to be considered is the data integration workflow designer [15]. Within this component, raw car-to-cloud and robotic sensor data can be selected and integrated by the end users. Output of this first component is a data integration schedule as well as a data integration instruction set [15]. Both artifacts are forwarded to the second data integration engine component, which will perform the actual data integration by execution of the data integration instruction set whenever the data integration schedule is triggered [15].

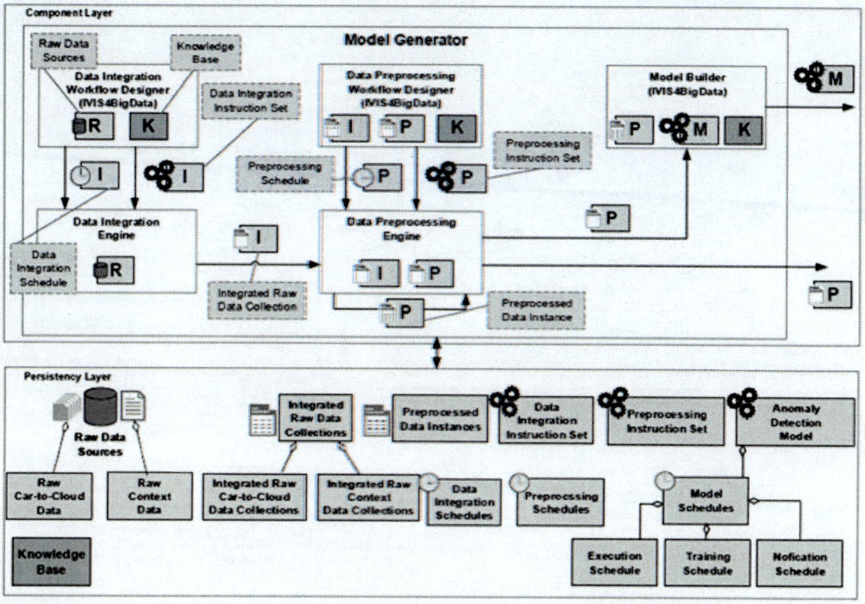

Fig. 11.12 Anomaly detection—model generator components model [15]

Within the data preprocessing workflow designer the user will be empowered to configure the data preprocessing [15]. For this objective, this component utilizes integrated raw data collections, preprocessed data instances, as well as the KB within the persistency layer and generates a preprocessing schedule and a preprocessing instruction set and forwards both artifacts to the data preprocessing engine component [15]. Output of this component are preprocessed data instances that either can be consumed by itself (chaining of preprocessing workflows) or can be forwarded to the model builder [15]. The model builder component empowers end users to construct an analysis model [15]. Therefore, it utilizes preprocessed data instances from the data preprocessing engine, already generated anomaly detection models (from previous component executions) as well as the KB [15]. The model execution core component consists of six components as visualized within Fig. 11.13 [15]. It consumes as input besides the user interaction (through which the user configures the system) the anomaly detection model as well as preprocessed data instances and generates label candidate data instances as output [15].

Following again the logical information flow for anomaly detection, the first model training workflow designer component utilizes preprocessed data instances, the anomaly detection model as well as the Knowledge Base in order to enable the training of anomaly detection models that use a semi-supervised or supervised anomaly detection algorithm [15]. This component provides an User Interface to the users in order to perform the model training use cases [15].

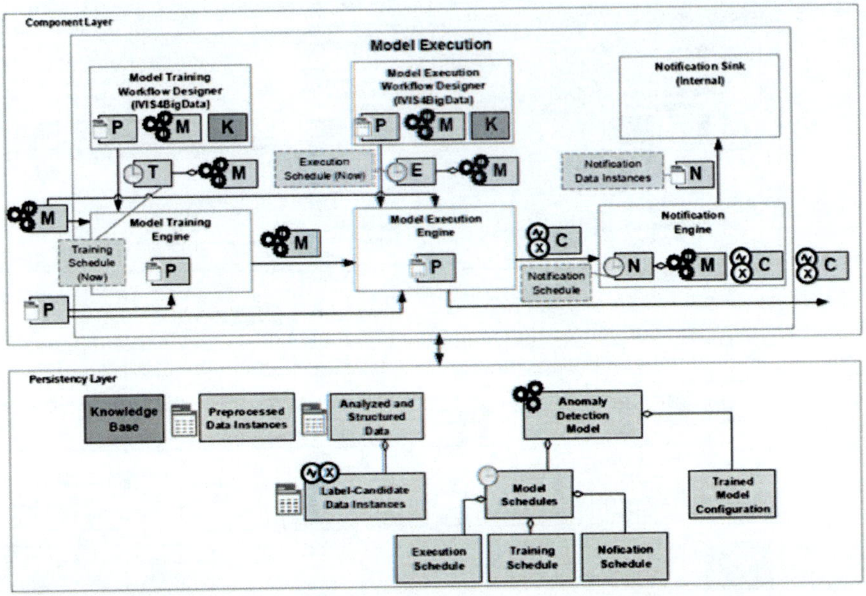

Fig. 11.13 Anomaly detection—model execution components model [15]

As output, this component forwards a special training schedule that triggers an immediate training execution as part of an anomaly detection model to the model training engine that utilizes preprocessed data instances from the first core component as well as from the persistency layer and trains the model whenever it is triggered by a training schedule [15]. Once training is executed, the trained model is accompanied with a trained model configuration and combined forwarded to the model execution engine [15].

Within the model execution workflow designer, end users are empowered to execute and configure the model execution [15]. Since this component is similar to the model training workflow designer component, it utilizes the same artifacts from the persistency layer [15]. Once the user interaction successfully concludes, the component forwards a special execution schedule (triggers immediate execution) as part of the anomaly detection model to the model execution engine that utilizes either an anomaly detection model from the first core component, from the model training engine, or from the model execution workflow designer component and applies the model execution on preprocessed data instances in order to generate label-candidate data instances [15]. Afterwards, these results will be forwarded to a notification engine as well as to the third and last core component anomaly analysis [15].

The notification engine utilizes existing label candidate data instances from the persistency layer together with the new ones generated by the model execution engine in order to generate notifications [15]. Once notifications are generated, they are transferred as notification data instances to notification sinks that can be either internal (part of this system) or external (e.g. external API Calls or mail notifications) [15].

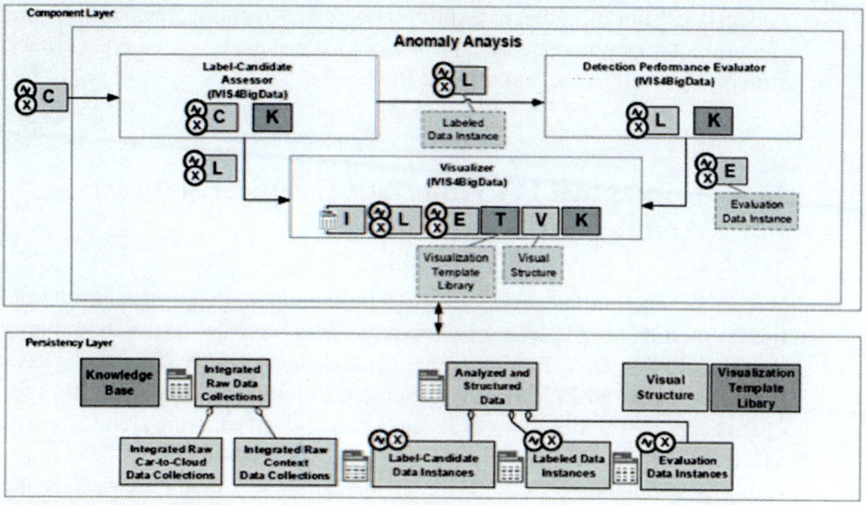

Fig. 11.14 Anomaly detection—anomaly analysis components model [15]

As part of this system and if configured within the model accordingly, the notification data instances are transmitted to a notification sink (internal) component, where they will be stored for further inspection by the end users [15].

The anomaly analysis core component is the last core component of this system and consists of three components that are visualized within Fig. 11.14 [15]. The core components receive label candidate data instances from the model execution core component and generates beside persisting its results no external output [15].

The first label candidate assessor component utilizes the existing Knowledge Base from the persistency layer, existing label-candidate data instances as well as new label-candidate data instances from the model execution core component [15]. The objective of this component is to empower end users to assess the proposed labels [15]. Once the users conclude the assessment, labeled data instances are generated and transmitted to both remaining components of this core component [15].

Within the detection performance evaluator, the new labeled data instances from the label candidate assessor component as well as the existing labeled data instances from the persistency layer are utilized in order to provide the user the possibility to perform a detection performance evaluation [15]. As in all IVIS4BigData instantiating components, the user is supported by and contributes to the knowledge base located in the persistency layer [15]. Output of this component are evaluation data instances, highly aggregated information on the detection performance (e.g. a confusion matrix), that are transmitted to a visualizer [15].

The last visualizer component focuses on the visualization of true and false positives and negatives (within labeled data instances from the persistency layer and label-candidate assessor component) as well as on the visualization of evaluation data instances (from the persistency layer and detection performance evaluator component) [15]. For this objective, the component utilizes integrated raw data collections (for application of reverse transformations), a visualization template library, a visual structure as well as the Knowledge Base [15].

11.4 Conceptual IVIS4BigData Technical Software Architecture

Within this section, the architecture of the exemplary proof-of-concept implementation as well as an exemplary prototypical reference application based on the introduced IVIS4BigData Reference Model will be outlined. Therefore, based on the defined conceptual IVIS4BigData Service-Oriented Architecture, the specification of the general exemplary proof-of-concept technical software architecture that serves as a basis for the resulting exemplary prototypical reference application for demonstrations and hands-on exercises will be presented. Finally, based on the defined use cases as well as on the design of the conceptual architecture, the design of an exemplary prototypical reference application will be outlined to demonstrate the general feasibility and applicability as well as to evaluate the resulting IVIS4BigData infrastructure in practice.

In this way, for implementing the interaction as well as for implementing the Big Data analysis and Information Visualization functionalities of the different IVIS4BigData **Human-Computer Interaction (HCI)** process stages supporting a variety of end user stereotypes spanning from those that are not trained in developing their own Big Data analysis and Information Visualization application solution to support the generated appropriate visualizations of the analysis results for their data to those which have the necessary technical competences and skills for programming virtually any type of special Big Data analysis or Information Visualization applications that they consider the best for supporting their intended Visual Analysis [20], a generic software architecture model has been defined. Figure 11.15 outlines this generic software architecture model including the specific software components.

11.4.1 Technical Specification of the Client-Side Software Architecture

With focus on the upper client-side, the IVIS4BigData software architecture is specified to be divided in two functional areas [20]. Whereas the left-sided GUI components are focusing on providing functions for interacting with the different IVIS4BigData HCI process stages (c.f. Fig. 11.5) of the resulting IVIS4BigData application solution [20] (e.g. menus, configuration dialogs, views, or template panels), both components of the Information Visualization area are focusing on providing functions to configure, to simulate, and to interact with the different multiple

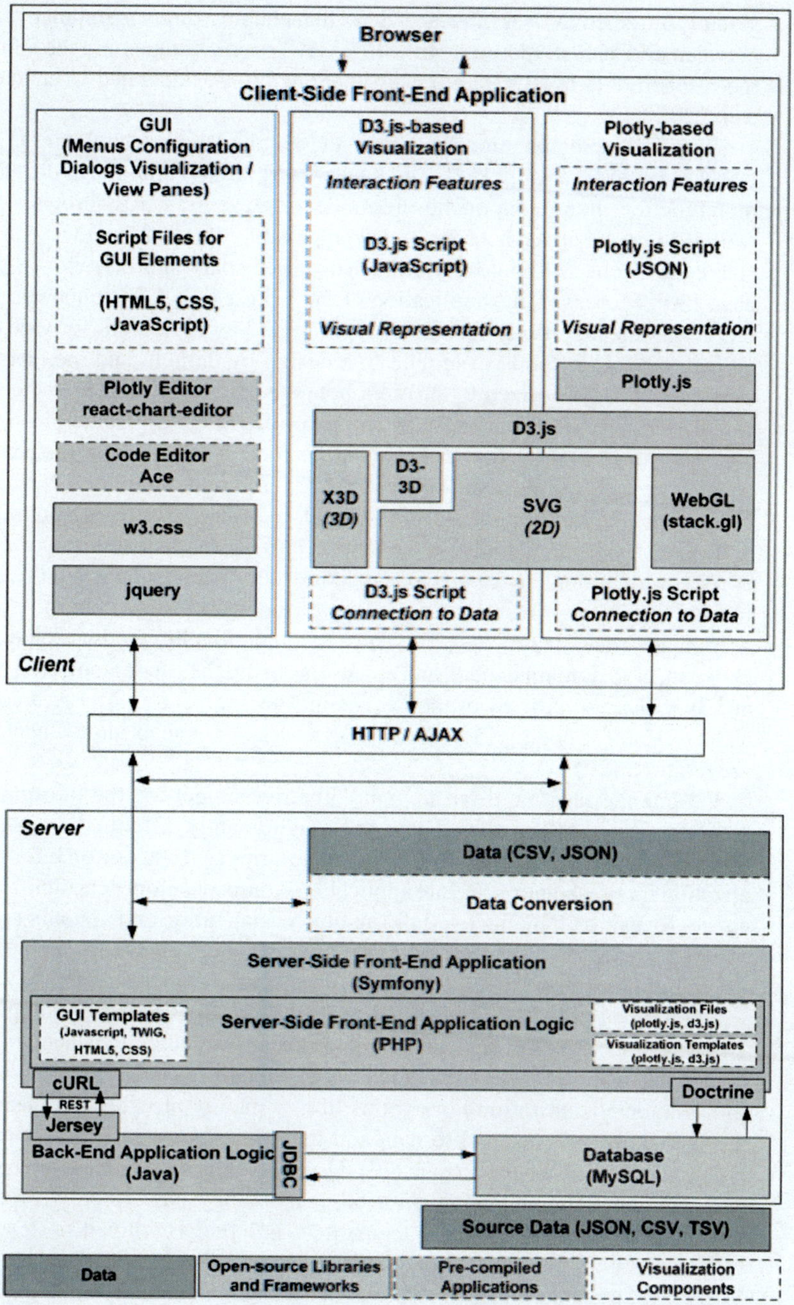

Fig. 11.15 Exemplary technical IVIS4BigData software architecture [20]

visually-interactive User Interface views that enable a direct manipulative interaction between end user stereotypes with single HCI process stages and the adjustments of the respective IVIS4BigData transformations by user-operated User Interface controls [20].

The GUI implementation is based on the standard web technologies HTML5 [78], CSS [73], and JavaScript [52]. The jQuery [65] library supports in developing common functions like AJAX on the client side for preventing cross-browser problems as well as to create an asynchronous web application that can interact with the server-side components without interfering with the display and behavior of the existing page [20]. Moreover, the application of the w3.css [73] CSS framework which differs from other solutions like, e.g., bootstrap [71] by its straight-forward concept, its built-in modern responsive mobile first design by default, and the fact that it only uses CSS [73], is utilized to ensure a homogeneous appearance of the GUI. Thus, it makes it very easy, for example, to adapt the IVIS4BigData color scheme and, e.g., add new colors and to use it with the form component of the Symfony PHP framework [53] used on the server side [20]. To create individual views within the D3.js visualization library, the IVIS4BigData front-end software architecture integrates the Ace code editor [1] and connects it to a view pane that empowers, e.g., end users as well as domain experts to, e.g., construct, specify, and simulate certain views for discussion, improvement, as well as for the utilization within the essential integration and analysis process [20]. Thus, this functionality facilitates the cooperation between, e.g., domain experts and end users during the user empowered construction and specification process of the IVIS4BigData application. The Ace code editor is also suitable for coding Plotly.js-based views in case such more advanced features are required [20].

Despite the fact that there are many libraries supporting the Information Visualization process of the gathered data, these significantly differ in the way they support utilization and licensing [20]. While some libraries are focusing on Information Visualization for presentation and are applicable without developer capabilities, some others are focusing on interactive Information Visualization and are only applicable for expert usage with software development competences and skills [21]. Thus, the initial exemplar prototypical Information Visualization functionality of IVIS4BigData will be implemented using the D3.js as well as Plotly.js base technologies [20].

D3.js is a JavaScript based drawing library for visualizing data and manipulating documents using HTML, SVG, and CSS. Although D3.js is commonly known as an Information Visualization library, this library mainly provides the means to enable the HTML DOM[1] respond to data and thus can also be utilized for manipulating HTML documents based on data [20]. Usually, D3.js based charts are utilizing SVG [79] as an Information Visualization base technology. Nevertheless, even if the SVG Information Visualization base technology is largely confined to 2D graphics, as D3.js mainly takes on DOM manipulation, the utilization of other 3D Information Visualization technologies is just as concise and conceptually simple as using SVG for supporting 2D Information Visualization [20]. Therefore, in this exemplar pro-

[1] Document Object Model.

totypical proof-of-concept implementation approach D3.js will be combined with X3DOM [31] that enables the integration of 3D content into the webpage's HTML code and *"allows you to manipulate the 3D content by only adding, removing, or changing DOM elements"* [31] in a similar way as SVG does it for 2D content [20]. Moreover, and as an alternative, e.g., for user stereotypes who are SVG literate but have no experience with X3DOM, the d3-3d.js [51] library is additionally utilized, which *"adds 3d transformations to SVG"* [51].

Built on top of D3.js, the high-level and declarative open source Plotly.js charting library, *"that ships with 20 chart types, including 3D charts, statistical graphs, and SVG maps"* [39] is utilized as alternative Information Visualization technology [20]. In this library, the charts are described declaratively as JSON objects where each aspect of the chart, such as, e.g., colors, grid lines, and the legend, has a corresponding set of JSON attributes [39]. Plotly.js uses D3.js (SVG) as well as WebGL for 3D graphics rendering [20]. While D3.js is more practical for up to tens of thousands of points and vector-quality image export [49], WebGL allows interactive rendering of hundreds of thousands to millions of x-y points [43].

11.4.2 Technical Specification of the Server-Side Software Architecture

In order to ensure a smooth integration into the KM-EP, the prototypical proof-of-concept specification and corresponding implementation of the IVIS4BigData technical software architecture and all associated lower server-side components are based on the Symfony PHP framework [20]. Thus, the integration into the KM-EP enables the utilization of its VRE features as well as the utilization of its solutions in its underlying CEKM system components and services like its digital library, media archive, user and rights management, as well as its learning management [20].

Starting from a data source perspective at the bottom of the server-side components and in addition to the visualization ability of analyzed and structured raw data from the built-in data collections of the current IVIS4BigData analysis project by utilizing the JSON[2] [23] data exchange format, this architecture also supports the visualization of external analyzed and structured raw data (like, e.g., external Big Data analysis results or exported Big Data analysis results of other IVIS4BigData projects) by utilizing common CSV[3] [38] or TSV[4] [48] standards [20].

With focus on the central Symfony-based core of the server-side front-end software architecture and in particular to generate the Graphical User Interface of the resulting web-based IVIS4BigData application, the Twig [54] PHP template engine in addition with Symfony's built-in form component is utilized as well as the traditional HTML5 and CSS markup languages and JavaScript for creating static and

[2] JavaScript Object Notation.

[3] Comma Separated Values.

[4] Tabulator Separated Values.

dynamic websites [20]. This open source template engine, which has been developed by Fabien Potencier (creator of the Symfony framework) extends the traditional PHP framework with useful functionalities for templating environments [54]. Twig can be easily included in Symfony and is already utilized within the latest KM-EP software architecture. Within the IVIS4BigData front-end software architecture, *"TWIG templates will be utilized to define the overall structure of the Graphical User Interfaces for the main window and the tabs for the individual functions"* [20].

To implement the Information Visualization application logic as well as for persisting general Information Visualization knowledge and information on user generated views into the underlying information model within a MySQL database, the open source Doctrine [22] PHP libraries are utilized *"that are primarily focused on providing persistence services and related functionality"* [22]. With its main components object relational mapper and database abstraction layer, Doctrine provides functionalities for database storage and object mapping and can be easily included in Symfony [22].

Additionally, visualization files and templates can also be stored in the form of Plotly.js [39] and D3.js [49] scripts as well as JSON configuration files, which associate visualization templates to data sets and store the visual mapping, i.e., the assignment of the data attributes to the visual properties represented in the resulting view. Both Plotly.js and D3.js come with built-in functions for reading data in JSON and CSV format [20]. In order to visualize structured, integrated, and analyzed raw data sources as well as corresponding IVIS4BigData analysis results stored in the common XLS format within an IVIS4BigData information visualization web application, the specification of the prototypical proof-of-concept IVIS4BigData technical software architecture also provides functions for converting XLS files to the JSON format [20].

11.5 IVIS4BigData Supporting Advanced Visual Big Data Analytics

After deriving and qualitatively evaluating the conceptual IVIS4BigData Reference Model, its Service-Oriented Architecture, and its conceptual application design, two prototypical reference applications for demonstrations and hands-on exercises for previous identified e-Science user stereotypes with special attention to the overall user experience to meet the users' expectation and way-of-working will be outlined within this section. In this way and based on the requirements as well as data know-how and other expert know-how of an international leading automotive original equipment manufacturer and a leading international player in industrial automation, two specific industrial Big Data analysis application scenarios (**anomaly detection on car-to-cloud data** and **predictive maintenance analysis on robotic sensor data**) will be utilized to demonstrate the practical applicability of the IVIS4BigData Reference Model and proof this applicability through a comprehensive evaluation.

11.5.1 *Application Scenario: Anomaly Detection on Car-to-Cloud Data*

Based on an international leading automotive original equipment manufacturer's requirements as well as data know-how and other expert know-how and by instantiation of a prototypical IVIS4BigData infrastructure, the outlined reference application was designed to perform anomaly detection on car-to-cloud data that empowers different end user stereotypes in the automotive application domain to gain insight from detected anomalies, anomaly candidates, and car-to-cloud data overall over the entire processing chain of anomaly detection (c.f. Fig. 11.16) [15].

In order to evaluate the prototype, a relevant use case with corresponding test data has to be selected. In this way, the first exposing faults in test drives use case scenario that enables vehicle manufacturer & supplier user stereotypes to detect defect-caused anomalies based on in-vehicle recordings transmission in series-production car-to-cloud data will been utilized for the quantitative evaluation.

As this use case tried to find uncommon resource usage behavior over a broad car fleet, there exist two corner cases regarding the vehicle **Electronic Control Unit (ECU)** resource capabilities [15]. Whereas ECUs with resource capabilities that are comparable to a PC (required for rendering images and calculating navigation routes) are able to execute a huge number of tasks concurrently where each of them only consuming a minor share of the overall available resources, other ECUs are comparable to an old-fashioned pocket calculator that are executing only a few tasks where all of them are consuming all of its resources [15]. In order to cover both bookends, suitable synthetic in-vehicle recording test data for both ECU types have been generated by an international leading automotive original equipment manufacturer [15]. Table 11.1 illustrates the configuration parameter of the synthetic test data generation based on the international leading automotive original equipment manufacturer's knowledge.

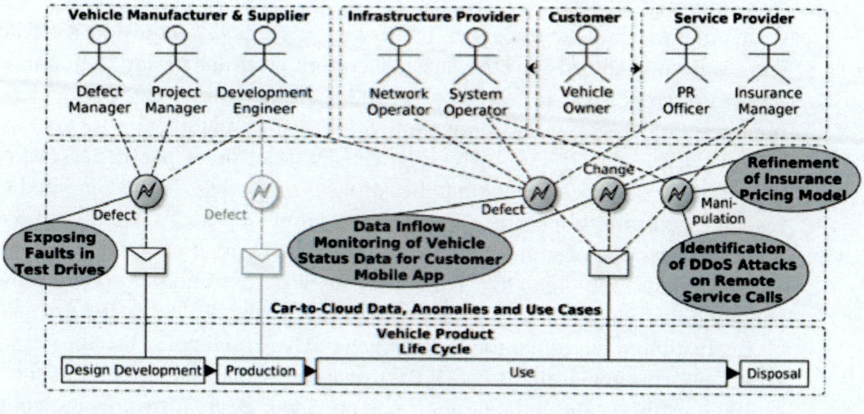

Fig. 11.16 Exemplary Anomaly Detection Use Cases along the Vehicle Product Life Cycle [15]

Table 11.1 Use case exposing faults in test drives—synthetic test data generation parameter [15]

Parameter name	Evaluation manifestation
Mean number of tasks (ECU 1)	5
Mean number of tasks (ECU 2)	80
Deviation number of tasks	5%
CPU resource usage (ECU 1)	80%
CPU resource usage (ECU 2)	10%
Deviation number of CPU	6%
RAM resource usage (ECU 1)	85%
RAM resource usage (ECU 2)	25%
Deviation number of RAM	9%
Anomaly deviation number of tasks	80%
Anomaly deviation RAM	40%
Anomaly deviation CPU	60%
Number of anomalies	5 per ECU%
Number of vehicles	10
Number of drives per vehicle	10
Mean drive duration	20 min
Deviation drive duration	95%
Sampling interval	60 s

Once single tasks consume uncommon shares of resources (very few or very much), these situations are of interest for the vehicle development engineers and can be caused by all reasons for anomaly occurrence: Change (the software of the ECU was updated), defect (a situation occurred for which the ECU's software was not prepared for), or manipulation (a tester introduced without alignment additional tasks into a ECU) [15].

In order to synthetically generate these anomalies within the data set, four parameters were introduced into the data generating program beside the common mean number of tasks, CPU resource usage, and RAM resource usage parameters in combination with their random derivation configuration parameters (deviation number of tasks, CPU, and RAM) of both ECUs. Three of these parameters were utilized to determine the effect on the number of tasks as well as on the CPU and the RAM resource consumption (anomaly derivation number of tasks, CPU, and RAM) and one parameter has been utilized to determine the number of anomalies that shall be introduced into the data [15]. All anomalies were assumed to endure only for one sampling interval and the anomalies are even distributed within the drives [15].

Evaluation of classification tasks in general require labeled instances and concise **Key Performance Indicators (KPIs)** or figures that enable comparison between multiple methods and indicate their performance. Basic measures are counting the number of **True Positive (TP)** and **True Negative (TN)** as well as **False Positive**

Table 11.2 Quantitative performance evaluation—detection performance [15]

Algorithm type	Confusion matrix		TPR	FPR
Semi supervised	TP: 5	FN: 5	0.5	0.000234
	FP: 1	TN: 4 267		
Supervised	TP: 5	FN: 5	0.5	0.0
	FP: 0	TN: 4 268		
Unsupervised	TP: 10	FN: 0	1.0	0.0
	FP: 0	TN: 4 268		

(FP) and **False Negative (FN)** instances including their resulting true positive rate ($TPR = \frac{TP}{TP+FN}$) an false positive rate ($FPR = \frac{FP}{FP+TN}$). In the context of relevant anomaly algorithms of the model execution, three common classification algorithms (unsupervised $\hat{=}$ k-nearest neighbors, semi-supervised $\hat{=}$ artificial neural network, supervised $\hat{=}$ one-class support vector machine) have been compared. Although the computing performance can be considered as an important anomaly detection metric and has also been evaluated, Table 11.2 illustrates the results of the most important detection performance metric.

Concluding the results of both performance metrics under consideration of the respective parametrization, regarding detection performance, the unsupervised k-nearest neighbors algorithm with its ability to perfectly separate anomalous data instances from normal ones can be identified as the most effective anomaly detection algorithm [15]. With focus on computing performance, the training of the semi-supervised one-class SVM[5] algorithm has been identified as the most time-efficient process step whereas the most time-consuming processing step is represented by the execution of the unsupervised k-nearest neighbors algorithm [57]. Nevertheless, due to the fact *"that it is important to distinguish between training of Machine Learning models and deploying such models for prediction"* [45], the higher time consumption of the k-nearest neighbors algorithm in context to both other algorithms can be explained by the fact that unsupervised algorithms combine training and execution within one single step [15]. In this way, although the final decision for the right algorithm depends on the source data as well as on the analysis scenario [16], the prototypical proof-of-concept reference implementation has successfully proven that it is able to reach the use case's objective with reliably identifying anomalous data instances, even if not all parameters found by all of them.

[5] nu = 0.5; kernel function = Radial Basis Function (RBF).

11.5.2 Application Scenario: Predictive Maintenance Analysis on Robotic Sensor Ata

Additionally, based on the requirements as well as data know-how and other expert know-how of a leading international player in industrial automation and by instantiation of a prototypical IVIS4BigData infrastructure, the second reference application is designed to perform predictive maintenance analysis on robotic sensor data that empowers different end user stereotypes in the robotics application domain to gain insight from robotic sensor data [15].

To accomplish this quantitative evaluation on real-world data, a controlled defect-oriented experiment [2] has been executed where an 6-axis industrial robot (c.f. Fig. 11.17) was operated beyond its regular operation configuration until one of its components develops a fault and the entire system breaks down [58]. Afterwards, and to identify the existing but unknown anomalies, the data generated during this controlled defect-oriented experiment has been analyzed by the aid of the IVIS4BigData proof-of-concept reference implementation.

To be more precise, the controlled defect-oriented experiment has been conducted under pre-defined scope conditions and aimed at identifying relevant parameters that support the predictive maintenance of the robot wrist. The pre-defined scope conditions are the utilization of the TX2-40 6-axis robot, the exclusive consideration of the axis five (wrist), as well as the radius of movement inside a fixed reference path.

With focus on the potential anomalies, the global mechatronics solution provider's domain experts expected noticeable drifts of the analyzed and visualized robot sensor data when the wrist would be operated beyond its regular operation configuration over

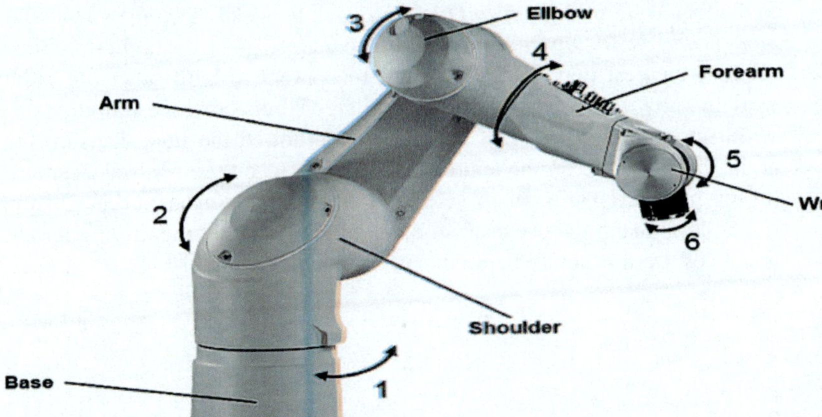

Fig. 11.17 Exemplary 6-Axis industrial robot [64] [56]

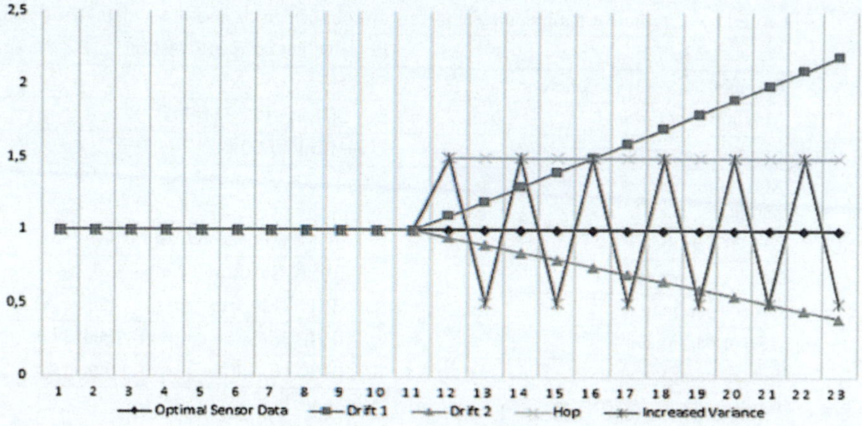

Fig. 11.18 Exemplary drift possibilities of analyzed and visualized robotic sensor data [56]

a considerable time period [56]. Based on the domain expert's existing knowledge on robotics sensor data, on the one hand drifts can appear as small differences between consecutive sensor data measurements that are resulting in a big deviation of the future measurements in relation to the optimal value. On the other hand, drifts can also appear as a spontaneous hop as well as a change of the measurement's variance at a certain measurement. Figure 11.18 illustrates an overview of exemplary drift possibilities of analyzed and visualized robotic sensor data.

In contrary to the first anomaly detection on car-to-cloud data application scenario on synthetic test data where hidden but well-known anomalies have to be detected, the expected anomalies (drifts of robot wrist sensors) of this additional predictive maintenance analysis on robotic sensor data application scenario are unknown (unknown wrist sensor parameter, unknown drift appearance) at the beginning of this controlled experiment. Therefore, the domain experts are utilizing the prototypical IVIS4BigData reference application to analyze different wrist sensor signals and compare the analysis results with the aid of different Information Visualization configurations of the IVIS4BigData reference application.

Regarding to the wrist sensor test data, five relevant sensor parameters were recorded over a total time period of 25 days. Table 11.3 illustrates the parameter of the recorded 6-axis robot wrist sensor data.

Table 11.3 Predictive maintenance analysis on robotic sensor data—test data description [56]

Parameter name	Evaluation manifestation
Number of sensor parameters	5
Maximum duration	30 Days (2 592 000 s)
Sampling interval	10 min (600 s)
Maximum number of measurements per sensor parameter	4 320
Sensor parameter 1	PCMD (Position Command)
Sensor parameter 2	PFBK (Position Feedback)
Sensor parameter 3	IPHA (Electric Current—Phase A)
Sensor parameter 4	IPHB (Electric Current—Phase B)
Sensor parameter 5	IPHC (Electric Current—Phase C)

Whereas the first PCMD[6] parameter is utilized to control the movement of the wrist actor, the corresponding PFBK[7] parameter identifies the actual wrist actor position. Additionally, the three IPHA, B, and C[8] parameter are utilized for the measurement of the actual electric wrist actor current. Nevertheless, and based on the scope conditions of the controlled experiment to move the axis five (wrist) within a fixed reference path in this specific example, only the parameters of the phases A and C (IPHA and IPHC) of the 6-axis robot wrist sensor data are relevant in addition to both position command and feedback (PCMD and PFBK) sensor parameter.

After operating the 6-axis robot beyond its regular operation configuration until one of its components develops a fault and the entire system breaks down [58], the raw data of the identified sensor parameters have been integrated, analyzed, as well as visualized by the aid of the IVIS4BigData proof-of-concept reference implementation. Nevertheless, whereas the first application scenario on anomaly detection on car-to-cloud data already evaluated the general applicability of the IVIS4BigData proof-of-concept reference implementation with focus on the first data collection, management, and curation and the second analytics IVIS4BigData HCI process stage, this evaluation focuses on comparing the analysis results by the aid of different Information Visualization configurations within the third visualization and fourth perception and effectuation IVIS4BigData HCI process stage.

Nevertheless, the choice of the right chart that fits to the inherent structure of the data which suggests the resulting shape is sophisticated due to the high number of available representations. Therefore, based on Tidwell's schema for visual representations [69] according to the organizational model of the source data, the line graph visualization has been selected and utilized to visualize the linear integrated and analyzed wrist sensor data parameters. In this way, after integrating and analyzing the recorded sensor data by the aid of domain specific analysis algorithms (fast

[6]Position Command.

[7]Position Feedback.

[8]Electric Current—Phase A, B, and C.

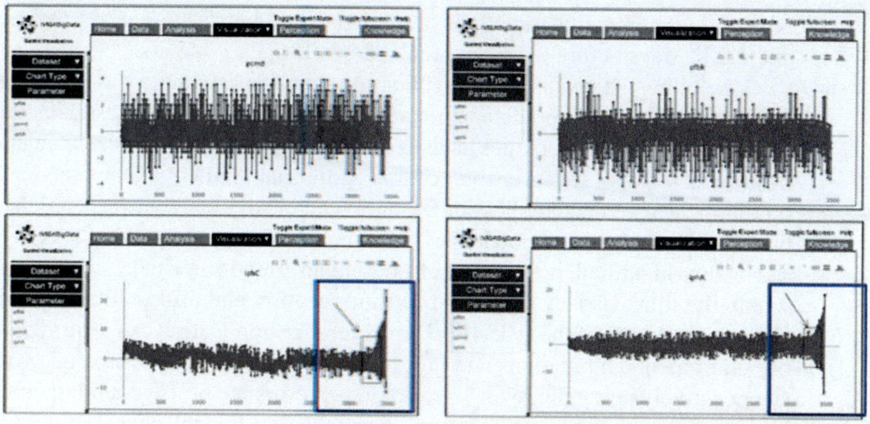

Fig. 11.19 Robotic sensor data analysis result—parameter PCMD, PFBK, IPHC, and IPHA [56]

fourier transformation, dimension reduction, and multi-dimensional reduction) that are applied in a row to all sensor parameters, Fig. 11.19 illustrates the results of the third visualization HCI process stage.

In addition to Tidwell's schema for visual representations, the visualization approach is strongly influenced by the research results of Ben Shneiderman [8]. As computer speed and display resolution increases, Shneiderman [63] denotes that *"Information Visualization and graphical interfaces are likely to have an expanding role"* because the bandwidth of information presentation is potentially higher in the visual domain than for media other senses [8]. Users can scan, recognize, and recall images rapidly and can detect changes in size, color, shape, movement, or texture and they can point to a single pixel, even in a megapixel display, and can drag one object to another to perform an action [63]. As a result of his research he summarizes the basic visual design guideline principles as the Visual Information Seeking Mantra—***"Overview first, zoom and filter then details-on-demand"*** [63]. Thus, by default the visualizations of the integrated and analyzed sensor parameters are visualizing an overview about the entire data starting from the first sample until the last sample before the entire system breaks down. Nevertheless, the integrated zooming features of the utilized Plotly.js visualization library enables a custom zoom capability [8].

Whereas the visualization of both position related parameters (PCMD and PFBK) do not allow conclusions about potential anomalies, both parameters that are measuring the electric current (IPHA and IPHC) are showing significant deviations before the wrist develops a fault and the entire system breaks down [8]. Whereas no anomalies can be identified within the analyzed and visualized sensor data at the beginning of this experiment, the values of both parameters start to drift between sample 3 200 and 3 300. Moreover, based on the knowledge that the system breaks down after sample 3 475 as well as on the sample interval of 10 min (600 s), the start of the potential equipment degradations and failures can be circumscribed between 1.91[9] and 1.22[10] days before the wrist breaks down [56].

Concluding the results of this application scenario, in contrary to the first anomaly detection on car-to-cloud data application scenario on synthetic test data where hidden but well-known anomalies have to be detected, the unknown but expected anomalies (drifts of robot wrist sensors) of this additional predictive maintenance analysis on robotic sensor data application scenario can clearly be identified [56]. Moreover, based on their existing knowledge the domain experts agreed that the results of the sensor data identified by the aid by of suitable Information Visualization opportunities within the third visualization and fourth perception and effectuation IVIS4BigData HCI process stage of the IVIS4BigData reference application can be utilized to identify equipment degradations and failures early in their aging or erosion process that can negatively affect the robot's precision performance or its general operation ability [56]. Nevertheless, and to enable a reliable equipment degradation and failure identification threshold, the identified anomalies have to be confirmed by further wrist sensor experiments [56].

11.6 Conclusion and Discussion

After deriving and qualitatively evaluating the conceptual IVIS4BigData Reference Model, its Service-Oriented Architecture, and its conceptual application design, two prototypical reference applications for demonstrations and hands-on exercises for previous identified e-Science user stereotypes with special attention to the overall user experience to meet the users' expectation and way-of-working will be outlined within this chapter.

With focus on the evaluation of the resulting prototypical proof-of-concept reference implementation for demonstrations and hands-on exercises for the identified e-Science user stereotypes with special attention to the overall user experience to meet the users' expectation and way-of-working, supported by the requirements as well as data know-how and other expert know-how of a leading international player in industrial automation, the specific industrial Big Data analysis application scenario anomaly detection on car-to-cloud data [57] has been utilized to demonstrate the practical applicability of the IVIS4BigData Reference Model and proof this applicability through a comprehensive evaluation [8]. Although the final decision for the right analysis algorithm depends on the source data as well as on the analysis scenario [16], based on the results of both quantitative performance metrics under consideration of the respective parametrization the prototypical proof-of-concept reference implementation has successfully proven that it is able to reach the use case's objective with respect to reliably identifying anomalous data instances [8]. Moreover, even future improvements of the prototype implementation's User Interface are

[9] sample 3 475—sample 3 200 = 275 samples $\widehat{=}$ 2 750 min $\widehat{=}$ 45.83 h $\widehat{=}$ 1.91 days.

[10] sample 3 475—sample 3 300 = 175 samples $\widehat{=}$ 1 750 min $\widehat{=}$ 29.17 h $\widehat{=}$ 1.22 days.

identified in order to address the discovered issues [57], the implemented prototypical proof-of-concept IVIS4BigData reference application empowers the evaluator to perform the specific industrial Big Data analysis application scenario in the subject area of anomaly detection on car-to-cloud data [57]. In this way, the results of the qualitative usability evaluation assess the usability of the implemented prototypical proof-of-concept IVIS4BigData reference application [8].

Additionally, and in contrary to the first evaluation of the IVIS4BigData proof-of-concept reference implementation in a precise anomaly detection on car-to-cloud data application scenario on synthetic test data where hidden but well-known anomalies had to be detected to evaluate the general applicability of the IVIS4BigData proof-of-concept reference implementation, and additional predictive maintenance analysis on robotic sensor data application scenario has been utilized to identify existing but unknown anomalies in real-world data [8]. Therefore, and supported by the requirements as well as data know-how and other expert know-how of a leading international player in industrial automation, the results of this evaluation assess the usability of the implemented prototypical proof-of-concept IVIS4BigData reference application which combine data analysis as well as information visualization approaches that are utilized to find previously unrecognized patterns in data (ill-defined Information Need [59]) in combination within Knowledge Management approaches to utilize the recognized patterns (well-defined Information Need [59]) as an iterative configuration and specification process in a specific area of interest (predictive maintenance) which supports an organization to gain insight [8]. On the other hand, this additional evaluation within a further specific industrial Big Data analysis application scenario predictive maintenance analysis on robotic sensor data also assess the practical applicability of the IVIS4BigData Reference Model within an additional application domain [8]. In this way and as outlined in Sect. 11.2.3.2, also the application design illustrated by the alignment of the elements within the application layer of the IVIS4BigData Service-Oriented Architecture that any IVIS4BigData infrastructure represents a specific VRE research application can successfully be evaluated [12].

In this way, after deriving the theoretical reference model which covers the new conditions of the present situation by identifying advanced visual User Interface opportunities for perceiving, managing, and interpreting distributed Big Data analysis results, as well as specifying and developing its corresponding prototypical proof-of-concept reference implementation, the evaluation based on two precise industrial application scenarios documents its applicability in context to the identified e-Science use cases and end user stereotypes with special attention to the overall user experience to meet the users' (students, graduates, as well as scholars, and practitioners) expectation and way-of-working [8].

References

1. Ajax.org.: Ace (Ajax.org Cloud9 Editor) (Version 1.2.6) (2010). Last accessed 27 Feb 2018
2. Albert, W., Tullis, T.: Measuring the User Experience: Collecting, Analyzing, and Presenting Usability Metrics. Newnes (2013)
3. Apache Software Foundation.: Apache Hadoop (Version: 2.6.3) (2014). Last accessed 10 Jan 2016
4. Apache Software Foundation.: Apache Spark (Version: 1.6.1) (2016). Last accessed 18 April 2016
5. Ardito, C., Buono, P., Costabile, M.F., Lanzilotti, R., Piccinno, A.: End users as co-designers of their own tools and products. J. Vis. Lang. Comput. **23**(2), 78–90 (2012). Special issue dedicated to Prof. Piero Mussio
6. Berwind, K.: A Cross Industry Standard Process to support Big Data Applications in Virtual Research Environments (forthcoming). Ph.D. thesis, University of Hagen, Faculty of Mathematics and Computer Science, Chair of Multimedia and Internet Applications, Hagen, Germany (2019)
7. Bornschlegl, M.X.: Ivis4bigdata: Qualitative evaluation of an information visualization reference model supporting big data analysis in virtual research environments. In: Advanced Visual Interfaces: Supporting Big Data Applications, vol. 10084 of Lecture Notes in Computer Science. Springer International Publishing, pp. 127–142 (2016)
8. Bornschlegl, M.X.: A Cross Industry Standard Process to support Big Data Applications in Virtual Research Environments (forthcoming). Ph.D. thesis, Advanced Visual Interfaces Supporting Distributed Cloud-Based Big Data Analysis, Hagen, Germany (2019)
9. Bornschlegl, M.X., Berwind, K., Hemmje, M.L.: Modeling end user empowerment in big data applications. In: 26th International Conference on Software Engineering and Data Engineering (SEDE: San Diego, CA, USA, 2–4 Oct 2017 (Winona, MN, USA, 2017), pp. 47–54. International Society for Computers and Their Applications, International Society for Computers and Their Applications (2017)
10. Bornschlegl, M.X., Berwind, K., Hemmje, M.L.: Modeling end user empowerment in big data analysis and information visualization applications. In: International Journal of Computers and Their Applications (Winona, MN, USA, 2018), International Society for Computers and Their Applications, International Society for Computers and Their Applications, pp. 30–42
11. Bornschlegl, M.X., Berwind, K., Kaufmann, M., Engel, F.C., Walsh, P., Hemmje, M.L., Riestra, R., Werkmann, B.: Ivis4bigdata: a reference model for advanced visual interfaces supporting big data analysis in virtual research environments. In: Advanced Visual Interfaces. Supporting Big Data Applications. Lecture Notes in Computer Science, vol. 10084, pp. 1–18. Springer International Publishing (2016)
12. Bornschlegl, M.X., Dammer, D., Lejon, E., Hemmje, M.L.: Ivis4bigdata infrastructures supporting virtual research environments in industrial quality assurance. In: Proceedings of the Joint Conference on Data Science, JCDS 2018, 22–23 May 2018. Edinburgh, UK (2018)
13. Bornschlegl, M.X., Engel, F.C., Bond, R., Hemmje, M.L.: Advanced Visual Interfaces. Supporting Big Data Applications (2016)
14. Bornschlegl, M.X., Manieri, A., Walsh, P., Catarci, T., Hemmje, M.L.: Road mapping infrastructures for advanced visual interfaces supporting big data applications in virtual research environments. In: Proceedings of the International Working Conference on Advanced Visual Interfaces, AVI 2016, Bari, Italy, 7–10 June 2016. pp. 363–367 (2016)
15. Bornschlegl, M.X., Reis, T., Hemmje, M.L.: A prototypical reference application of an ivis4bigdata infrastructure supporting anomaly detection on car-to-cloud data. In: 27th International Conference on Software Engineering and Data Engineering (SEDE: New Orleans, LA, USA, 8–10 Oct 2017 (Winona, MN, USA, 2018), pp. 108–115. International Society for Computers and Their Applications, International Society for Computers and Their Applications (2018)
16. Brownlee, J.: Supervised and unsupervised machine learning algorithms (2016). Last accessed 23 Aug 2018

17. Card, S.K., Mackinlay, J.D., Shneiderman, B.: Information visualization. In: Card, S.K., Mackinlay, J.D., Shneiderman, B. (eds.) Readings in Information Visualization, pp. 1–34. Morgan Kaufmann Publishers Inc., San Francisco, CA, USA (1999)
18. Chang, R., Ziemkiewicz, C., Green, T., Ribarsky, W.: Defining insight for visual analytics. IEEE Comput. Graph. Appl. **29**(2), 14–17 (2009)
19. Costabile, M.F., Mussio, P., Parasiliti Provenza, L., Piccinno, A.: Supporting end users to be co-designers of their tools. In: End-User Development. Lecture Notes in Computer Science, vol. 5435, pp. 70–85. Springer Berlin Heidelberg (2009)
20. Dammer, D.: Big data visualization framework for cloud-based big data analysis to support business intelligence. Master's thesis, University of Hagen, Faculty of Mathematics and Computer Science, Chair of Multimedia and Internet Applications, Hagen, Germany (2018)
21. Dendelion Blu Ltd.: Big data visualization: review of the 20 best tools (2015). Last accessed 13 Sept 2016
22. Doctrine Team.: Doctrine (Version 2.5.4) (2016). Last accessed 07 Feb 2018
23. ECMA International. Standard ECMA-404, the JSON data interchange format
24. European Commission.: Scalable semantic product data stream management for collaboration and decision making in engineering. FP7-ICT-2009-5, Proposal Number: 257899, Proposal Acronym: SMART VORTEX (2009)
25. European Commission.: Development of an easy-to-use metagenomics platform for agricultural science. H2020-MSCA-RISE-2015, Proposal Number: 690998, Proposal Acronym: MetaPlat (2015)
26. European Commission.: Sensor enabled affective computing for enhancing medical care. H2020-MSCA-RISE-2015, Proposal Number: 690862, Proposal Acronym: SenseCare (2015)
27. European Commission.: Virtual environment for research interdisciplinary exchange. EINFRA-9-2015, Proposal Acronym: VERTEX (2015)
28. Fischer, G.: In defense of demassification: empowering individuals. Hum.-Comput. Interact. **9**(1), 66–70 (1994)
29. Fischer, G.: Meta-design: empowering all stakeholder as codesigners. In: Handbook on Design in Educational Computing. pp. 135–145. Routledge, London (2013)
30. Fischer, G., Nakakoji, K.: Beyond the macho approach of artificial intelligence: empower human designers - do not replace them. Knowl.-Based Syst. **5**(1), 15–30 (1992)
31. Fraunhofer Institute for Computer Graphics Research IGD.: X3DOM (Version: 1.2) (2009). Last accessed 11 Aug 2017
32. Fraunhofer Institute for Computer Graphics Research IGD.: Visual business analytics (2015). Last accessed 02 Dec 2015
33. Freiknecht, J.: Big Data in der Praxis. Carl Hanser Verlag GmbH & Co. KG, München, Deutschland (2014)
34. Friendly, M.: Milestones in the history of data visualization: a case study in statistical historiography. In: Weihs, C., Gaul, W. (eds.) Classification: The Ubiquitous Challenge, pp. 34–52. Springer, New York (2005)
35. Harris, H., Murphy, S., Vaisman, M.: Analyzing the Analyzers: An Introspective Survey of Data Scientists and Their Work. O'Reilly Media, Inc. (2013)
36. Hutchins, E.: Cognition in the Wild. MIT Press (1995)
37. Illich, I.: Tools for Conviviality. World Perspectives. Harper & Row (1973)
38. Internet Engineering Task Force.: Common Format and MIME Type for Comma-Separated Values (CSV) Files (2005). Last accessed 07 Feb 2018
39. Johnson, A., Parmer, J., Parmer, C., Sundquist, M.: Plotly.js (Version: 1.31.2) (2012). Last accessed 29 Oct 2017
40. Keim, D., Andrienko, G., Fekete, J.-D., Görg, C., Kohlhammer, J., Melançon, G.: Visual analytics: definition, process, and challenges. In: Kerren, A., Stasko, J., Fekete, J.-D., North, C. (eds.) Information Visualization. Lecture Notes in Computer Science, vol. 4950, pp. 154–175. Springer Berlin Heidelberg (2008)
41. Keim, D., Mansmann, F., Schneidewind, J., Ziegler, H.: Challenges in visual data analysis. In: Information Visualization, 2006. IV 2006. Tenth International Conference on Information Visualisation (IV'06), pp. 9–16 (2006)

42. Keim, D.A., Mansmann, F., Thomas, J.: Visual analytics: how much visualization and how much analytics? SIGKDD Explor. Newsl. **11**(2), 5–8 (2010). May
43. Khronos Group Inc.: WebGL (Version: 2.0) (2011). Last accessed 08 Feb 2018
44. Kuhlen, R.: Informationsethik: umgang mit Wissen und Information in elektronischen Räumen. UTB / UTB. UVK-Verlag-Ges. (2004)
45. Machine Learning Group at the University of Waikato.: Weka (Version (3.7) (1992). Last accessed 01 Aug 2018
46. Manieri, A., Demchenko, Y., Wiktorski, T., Brewer, S., Hemmje, M., Ferrari, T., Riestra, R., Frey, J.: Data science professional uncovered: how the EDISON project will contribute to a widely accepted profile for data scientists
47. Manyika, J., Chui, M., Brown, B., Bughin, J., Dobbs, R., Roxburgh, C., Hung Byers, A.: The next frontier for innovation, competition, and productivity. McKinsey Global Institute, Big Data (2011)
48. Microsoft Corporation.: Microsoft office excel (version 2016) (1985). Last accessed 07 Feb 2018
49. Mike Bostock.: d3.js (Version: 4.2.3) (2011). Last accessed 16 Sept 2016
50. Ng, A.: User friendliness? user empowerment? how to make a choice? Technical report, Graduate School of Library and Information Science, University of Illinois at Urbana-Champaign (2004)
51. Nieke, S.: d3-3d (Version 0.0.7) (2017). Last accessed 27 Feb 2018
52. Oracle Corporation.: JavaScript (Version 1.8.5) (1995)
53. Potencier, F.: Symfony (Version: 4.0.1) (2005). Last accessed 09 Dec 2017
54. Potencier, F.: Twig (Version: 2.4.4) (2009). Last accessed 09 Dec 2017
55. Prajapati, V.: Big Data Analytics with R and Hadoop. Packt Publishing (2013)
56. Puchtler, P.: Predictive-maintenance-analysis of robotic sensor data based on a prototype reference application of the ivis4bigdata infrastructure. Master's thesis, University of Hagen, Faculty of Mathematics and Computer Science, Chair of Multimedia and Internet Applications, Hagen, Germany (2018)
57. Reis, T.: Anomaly detection in car-to-cloud data based on a prototype reference application of the ivis4bigdata infrastructure. Master's thesis, University of Hagen, Faculty of Mathematics and Computer Science, Chair of Multimedia and Internet Applications, Hagen, Germany (2018)
58. Robert Bosch GmbH.: Stress test for robots (2014). Last accessed 03 Dec 2018
59. Robertson, S.E.: Information Retrieval Experiment. In: The Methodology of Information Retrieval Experiment, pp. 9–31. Butterworth-Heinemann, Newton, MA, USA (1981)
60. Ryza, S., Laserson, U., Owen, S., Wills, J.: Advanced Analytics with Spark, vol. 1. O'Reilly Media, Inc., Sebastopol, CA, USA, 3 (2015)
61. Salman, M., Star, K., Nussbaumer, A., Fuchs, M., Brocks, H., Vu, B., Heutelbeck, D., Hemmje, M.: Towards social media platform integration with an applied gaming ecosystem. In: SOTICS 2015 : The Fifth International Conference on Social Media Technologies, Communication, and Informatics, pp. 14–21. IARIA (2015)
62. SAS Institute Inc.: Data visualization: what it is and why it is important (2012). Last accessed 21 Dec 2015
63. Shneiderman, B.: The eyes have it: a task by data type taxonomy for information visualizations. In: Proceedings, IEEE Symposium on Visual Languages, pp. 336–343 (1996)
64. Staubli International AG.: Staubli tx2-40 6-axis industrial robot (2018)
65. The jQuery Foundation.: jQuery (Version 3.2.1) (2006). Last accessed 27 Feb 2018
66. The R Foundation.: The R Project for Statistical Computing (Version 3.2.5) (1993). Last accessed 28 April 2016
67. Thomas, J.J., Cook, K., et al.: A visual analytics agenda. IEEE Comput. Graph. Appl. **26**(1), 10–13 (2006). Jan
68. Thomas, J.J., Cook, K.A.: Illuminating the Path: The Research and Development Agenda for Visual Analytics. National Visualization and Analytics Ctr (2005)
69. Tidwell, J.: Designing Interfaces. O'Reilly Media, Inc. (2005)

70. Tufte, E.: Visual Explanations: Images and Quantities, Evidence, and Narrative. Graphics Press (1997)
71. Twitter, I.: Bootstrap (Version 4.0.0) (2011). Last accessed 27 Feb 2018
72. Vu, D.B.: Realizing an applied gaming ecosystem: extending an education portal suite towards an ecosystem portal. Master's thesis, Technische Universität Darmstadt (2016)
73. W3Schools.: W3.CSS (Version 4) (2015). Last accessed 27 Feb 2018
74. Wang, W.: Big data, big challenges. In: Semantic Computing (ICSC), 2014 IEEE International Conference on Semantic Computing, p. 6 (2014)
75. Wiederhold, G.: Mediators in the architecture of future information systems. Computer **25**(3), 38–49 (1992). March
76. Wong, P.C., Thomas, J.: Visual analytics. IEEE Comput. Graph. Appl. **5**, 20–21 (2004)
77. Wood, J., Andersson, T., Bachem, A., Best, C., Genova, F., Lopez, D.R., Los, W., Marinucci, M., Romary, L., Van de Sompel, H., Vigen, J., Wittenburg, P., Giaretta, D., Hudson, R.L.: Riding the wave: how Europe can gain from the rising tide of scientific data. Final report of the high level expert group on scientific data; a submission to the European commission
78. World Wide Web Consortium (W3C).: HTML (Version 5) (2014). Last accessed 12 Sept 2016
79. World Wide Web Consortium (W3C).: SVG (Version 2) (2015). Last accessed 16 Sept 2016

Chapter 12
Classification of Pilot Attentional Behavior Using Ocular Measures

Kavyaganga Kilingaru, Zorica Nedic, Lakhmi C. Jain, Jeffrey Tweedale, and Steve Thatcher

Abstract Revolutionary growth in technology has changed the way humans interact with machines. This can be seen in every area, including air transport. For example, countries such as United States are planning to deploy NextGen technology in all fields of air transport. The main goals of NextGen are to enhance safety, performance and to reduce impacts on environment by combining new and existing technologies. Loss of Situation Awareness (SA) in pilots is one of the human factors that affects aviation safety. There has been a significant research on SA indicating that pilot's perception error leading to loss of SA is a one of the major causes of accidents in aviation. However, there is no system in place to detect these errors. Monitoring visual attention is one of the best mechanisms to determine a pilot's attention and hence perception of a situation. Therefore, this research implements computational models to detect pilot's attentional behavior using ocular data during instrument flight scenario and to classify overall attention behavior during instrument flight scenarios.

Keywords Attention classification · Pilot situation awareness classification · Scan path analysis · Knowledge discovery in data · Attention focusing · Attention blurring

---

K. Kilingaru · Z. Nedic
University of South Australia, Adelaide, Australia

L. C. Jain (✉)
University of Technology Sydney, Ultimo, Australia
e-mail: jainlakhmi@gmail.com; jainlc2002@yahoo.co.uk

Liverpool Hope University, Liverpool, UK

KES International, Shoreham-by-Sea, UK

J. Tweedale
Defence Science and Technology Group, Adelaide, Australia

S. Thatcher
Central Queensland University, Rockhampton, Australia

G. Phillips-Wren et al. (eds.), *Advances in Data Science: Methodologies and Applications*, Intelligent Systems Reference Library 189,
https://doi.org/10.1007/978-3-030-51870-7_12

12.1 Introduction

Air travel is a common mode of transport in the modern era and considered one of the safest. Even though aviation accidents are not as common as road accidents, associated losses have a greater impact. One civil aircraft accident can claim the lives of hundreds of people and cause millions of dollars of economic loss. Therefore, airlines are bound to abide by strict safety policies and guidelines. Safety breaches by airlines are just one of the causes of aviation accidents.

Other causes include technical faults, human error and environmental conditions [1]. Past investigations have shown that more than 70% of accidents are caused by human error [2]. Given their devastating effects, research into improving safety is a priority in aviation. In order to enhance safety, performance, and to reduce impacts on environment, countries like the United States are planning to deploy Next Generation (NextG) technologies in all fields of air transport. This research investigates the feasibility of improving aviation safety by designing a novel system to monitor pilot visual behaviour and detect possible errors in instrument scan pattern that could potentially lead to loss of pilot Situation Awareness (SA).

From the previous researches, it is evident that ocular measures are effective measures in determining attentional behavior [3]. Identified attentional behaviours can further be used as to detect potential pilot errors. With the ongoing research in embedded eye trackers and technology growth, it can be foreseen that aircraft will include such advanced recording devices in the near future [4].

In this research study, knowledge discovery in data process was used to collect ocular data and extract attention patterns. Flight simulator experiments were conducted with trainee pilots and ocular data were collected using eye trackers. In the absence of readily available classifications of existing data, we developed a feature extraction and decision model based on the observed data and inputs from the subject matter experts. Different attributes from the instrument scan sequence are also used to aggregate and devise models for scoring attention behaviors.

This is a significant step towards detection of perceptual errors in aviation human factors. Based on this model, further applications can be developed to assess the performance of trainee pilots by flight instructors during simulator training. Also, the model can be further developed into a SA monitoring and alerting system in future aircrafts and in such way reducing the risk of accidents due to loss of SA.

12.2 Situation Awareness and Attention in Aviation

Situation Awareness (SA) is defined as awareness of all the factors that help in flying an aircraft safely under normal and non-normal conditions [5]. In aviation, minor deviations and trivial failures may cause major threats over time if not attended to in a timely manner [6]. Therefore, it is important that a pilot should perceive, comprehend, and project correctly what he or she has perceived to assess the situation correctly.

Attention is a very important human cognitive function. It enables the human brain to control thoughts and actions at any given time [7]. Attending to something is considered the most significant task and has a major impact on performance of other tasks. During any task, when humans attend, they perceive. Perception is saved in memory and translated into understanding, which is used for planning actions. In aviation, a pilot's level of attentiveness contributes to the overall SA.

Humans use various senses to perceive; however, visual attention is considered the predominant source of perception [8, 9]. Vision system data are information rich and hence useful in a number of areas. In particular, these data can be highly beneficial in monitoring drivers' attention [10–12] and pilots' attention [13–15]. Vision system data can to some extent be detected and related to attention via physiological factors of human eyes.

12.2.1 Physiological Factors

Human errors during driving or flying may occur because of multiple causes, including spatial disorientation, workload, fatigue, or inattention. Although there is no system in place to correctly identify the causes before these result in mishaps, there have been research studies focusing on related areas. Monitoring physiological factors has proved an effective way of measuring possible causes of human error during driving or flying. In the early 1980s, an experiment was conducted to relate differences in heart rate to different levels of workload [16]; however, no exact relationship between heart rate and workload was established because of the difficulty in defining workload. Nevertheless, the author concluded that pilot activity, task demand, and effort did result in varying heart rates. In another experiment conducted to diagnose mental workload of pilots, researchers collected cardiac, eye, brain, and subjective data during an actual flight scenario [17]. The researchers found eye movements to be a more reliable diagnostic method than heart rate, indicating high visual demand on pilots during flight operations. The results from Electroencephalogram (EEG) did not provide statistically significant results.

Another study investigated brain wave activity associated with a simulated driving task [18]. The study found that the brain loses capacity and slows as a person fatigue. Eye movements and pupil dilation are other popular measures used when monitoring workload, fatigue, and attention [19–21]; for example, differences in pupil diameter and fixation time, eye movement distance and speed under different levels of mental workload were analysed in [22]. The research review shows that as in other operator-driven environments, many behavioural changes in pilots during flight operations can be observed by measuring various physiological parameters, such as heart rates, brain waves, eye movements, and facial expression. However, monitoring heart rates and brain waves are intrusive methods and are generally regarded as not feasible to use in real-time situations inside the cockpit, when pilots are operating the aircraft. Although many methods are intrusive, most attentional characteristics can be observed by monitoring pilot eye movements in a non-intrusive way.

12.2.2 Eye Tracking

It is evident that pilots are more prone to misperceptions during poor visual conditions. Although pilots are aware of this, researchers have found that "*pilots continue to confidently control their aircraft on the basis of visual information and fail to utilize the instruments right under their noses*" [23]. It is not only visual misperceptions that play a major role in aviation mishaps but the over confidence of pilots, as well. Simulators are already in place to help pilots practise instrument scanning. However, training alone has not been able to significantly change pilots' vulnerability to such mishaps [23]. Consequently, there is a need to evaluate trainee pilots' instrument scanning skills on simulators and also monitor scan patterns during flights. The evaluation of pilots' scan patterns should help identify mistakes during the training stage and may improve the training. Monitoring pilots' instrument scans is also important to help reduce in-flight human error considerably.

Capturing a pilot's eye movements through non-intrusive eye tracking methods is the best way to identify pilot SA behavioural characteristics. Under normal conditions, a person looking at an object for a length of time classified as a gaze will perceive information from that object or area of interest [24]. Specific behaviour and possible causes can be identified by observing where, when, and what a person is seeing (where seeing is interpreted to mean looking at an object long enough to be defined as a gaze). Therefore, during flight operations, the position and duration of a pilot's gaze can indicate the pilot's behaviour at that time. The major task pilots perform during flight is perceiving information from different instruments. It is necessary to maintain the correct timing and proper sequence of instrument scanning throughout the flight. If the correct scan sequence is not followed, pilots may not perceive the required information, or may fail to detect incorrect information, which may lead to loss of SA. Mapping eye movements—glance, gaze and stare to cognitive behaviours is discussed in detail in a previous article [4].

From the flying manuals [25] and inputs from the SMEs, the key instruments that must be scanned during flight are Artificial Horizon (AH), Airspeed Indicator (ASI), Turn Coordinator (TC), Vertical Speed Indicator (VSI), Altimeter (ALT) and Navigator (NAV). Distributed attention and perception during an instrument scan are essential for pilots to master. The required instrument scan varies depending on the flight phase, as different instruments play critical roles during each phase of the flight. An anomalous instrument scan pattern can be mapped with erroneous behaviours such as attention focusing, attention blurring and misplaced attention, which are attentional indicators that a pilot could lose SA [3]. These indicators are defined as:

Attention focusing: A sequence of fixations with few or no transitions is considered fixation on a single instrument and hence indicates attention focusing. Continuous fixations on a particular instrument in a limited time period are clustered to identify the instrument being interrogated. Figure 12.1 shows a sample fixation pattern on a particular instrument during attention focusing.

Fig. 12.1 Attention focus

Attention blurring: This behaviour is characterised by a small number of fixations and increased number of transitions between instruments. The fixation spans are very short and not sufficient to actually perceive the information. The pilot is simply glancing at instruments or observing them via peripheral vision. Figure 12.2 illustrates a sample instrument scan pattern during attention blurring.

Misplaced attention: This behaviour is characterised by very short fixation spans inside the instrument panel. More time is spent fixating outside the instrument regions in the instrument panel rather than fixating on the relevant instruments. Figure 12.3 shows a sample scan pattern during the event of misplaced attention.

To translate fixation data into behaviour patterns, it is necessary to continuously monitor fixations and represent them in digital form. This research study shows that implicit knowledge can be derived by periodically monitoring the position and sequence of fixation data. This time-stamped data stream was analysed to digitally classify pilot behaviour.

12.3 Knowledge Discovery in Data

Data can be conceived of as a set of symbols, but data alone do not convey meaning. To produce useful insights, data need to pass through a series of steps to extract the relevant information and convert it into wisdom. This process is called 'Knowledge Discovery in Data (KDD)' [26] and it involves the development of methodologies and

Fig. 12.2 Attention blurring

Fig. 12.3 Misplaced attention

tools to help extract wisdom from data. Fundamental purpose of KDD is to reduce a large volume of raw data into a form that is easily understood. The end results are produced in the form of a visualisation or descriptive report or a combination of both. This process is aided by data-mining techniques to discover patterns [26].

Terminologies of the knowledge discovery process were introduced before 1990. Popular definitions from early studies include [27–30]:

- Data: Data correspond to symbols, such as text or numbers, and are always raw. They have no meaning unless they are associated with a domain or situation in the real world.
- Information: Data that are processed and have relational connections, so they are more meaningful and useful. Information, as a result of data processing, can provide facts such as 'who' did 'what', 'where' and 'when'.
- Knowledge: Extracted useful patterns are called knowledge. This is used to derive further understanding. New knowledge can be influenced by old knowledge.
- Wisdom: This is an evaluated understanding of knowledge. Wisdom comes from analytical processes based on human understanding. Wisdom is nondeterministic and can be used in prediction processes.

The steps used in the KDD process are described below:

- Data acquisition: In general, this step involves collection of raw data for processing.
- Data pre-processing: Incomplete and inconsistent data are removed from the data set as preparation for further processing. This step can involve removal of outliers and extraneous parameters to clean and reduce the size of the target data set.
- Feature extraction and data transformation: Useful features are extracted from the data during feature extraction. Huge set of data is reduced and converted to meaningful information appropriate for recognition algorithms to process as part of data transformation.
- Data Mining/Pattern recognition: Information is processed by algorithms to discover new knowledge. The knowledge can be in the form of patterns, or rules, or predictive models.
- Evaluation: The knowledge, or pattern, is evaluated to derive useful reports or other outcomes such as predictions or ratings.

12.3.1 Knowledge Discovery Process for Instrument Scan Data

KDD has evolved over time, and in recent years, with a huge amount of data becoming available in every field, there has been much appreciation for KDD. Therefore, research in KDD usually involves overlapping of two or more fields such as artificial intelligence, machine learning, pattern recognition, databases, statistics, and data visualisation [26]. This study applied KDD principles on pilot's instrument

Fig. 12.4 KDD process for instrument scan

scan data. The study established methodology to convert instrument scan data into a sequence of behaviours to identify flight operator attentiveness during instrument flight. Figure 12.4 shows how the methodology in the study applied steps of the KDD process. The main steps involved in the process are vision data acquisition, cleansing, ocular gesture extraction, cognitive behaviour recognition through temporal analysis, and behaviour evaluation. The results provide an insight into the attention levels of the operator.

Instrument Scan Data Acquisition

Instrument scan data were collected using the EyeTribe Tracker [31] while participants performed instrument flying scenarios on Prepar3d [32] flight simulator.

The steps followed included:

1. **Participant Briefing**: Each participant was briefed about the scenario before each simulator session; for example, details on the departure and landing airports and weather condition settings. For the practise sessions, the participant was asked to perform some of the chosen instrument scans in a known order for the purpose of verifying eye tracker output. After the practise session participants were asked to perform preconfigured scenarios.
2. **Gaze Calibration**: Calibration is an important step prior to conducting any eye tracking experiment. Calibration involves software set up based on the participant's eye characteristics and lighting conditions in the area, for improved gaze estimate accuracy. Therefore, in the experiment, the student operator's eye movements were calibrated with the simulator screen coordinates prior to the first simulator operation. The calibration and verification step involved:
   a. Asking the participant to sit in a comfortable position in front of the simulator.
   b. Adjusting the eye tracker so that the eyes of the participant were detected and well captured, with both eyes almost at the centre of the green area, as shown in Fig. 12.5.
   c. Calibrating eye movements of the participant using the on-screen calibration points on the simulator monitor. On successful calibration, the EyeTribe tracker shows the calibration rating in stars, as shown in Fig. 12.6. Verifying the calibration was done by asking the participant to look at the points and confirm the tracker was detecting the gaze correctly.
3. **Simulator Configuration**: Prepar3D simulator is configured to launch aircraft in Instrument Flight Rules (IFR) mode, with different departure and destination airports. The participant was asked to perform the instrument flying using just the instrument panel. Weather conditions and failures were preconfigured for different scenarios without the knowledge of participant.

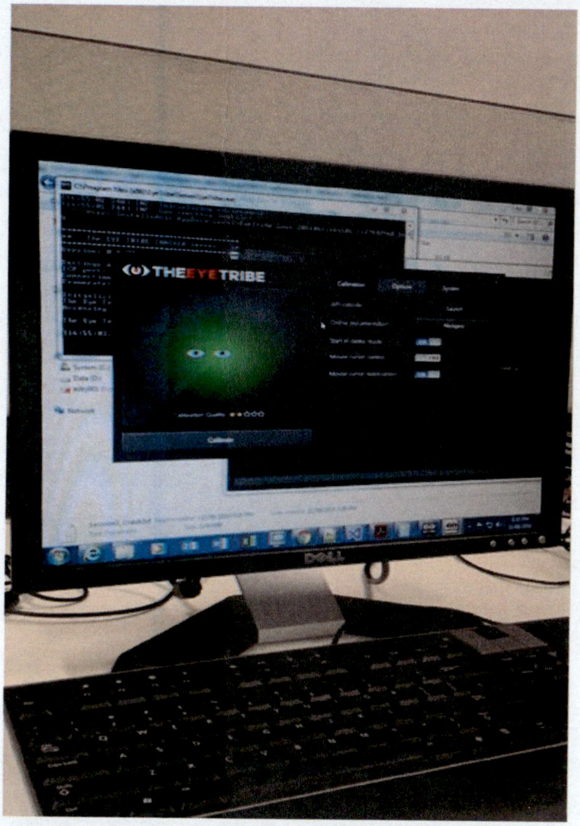

Fig. 12.5 EyeTribe tracker calibration [31]

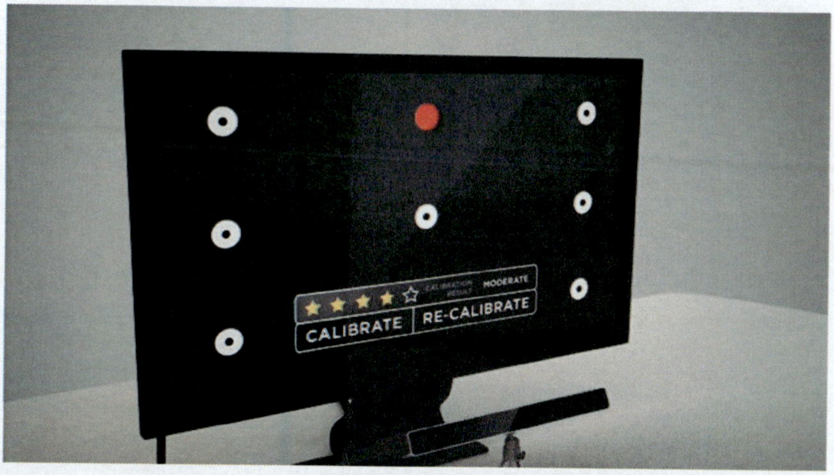

Fig. 12.6 Calibration results screen [31]

4. **Gaze Tracking**: Gaze tracking was commenced from the EyeTribe tracker console immediately after the scenario started. Gaze records were saved into a file named after the time stamp. The end result (crash or successful landing) and the simulator configurations for each scenario were also recorded.

The eye tracker provides data on the gaze coordinates for each frame, the time stamp and the pupil diameter in Java Script Object Notation (JSON) format, as shown below in Fig. 12.7.

In the sample, 'category' and 'request' indicate the type of request sent to EyeTribe tracker. A successful request receives a response message with status code 200. 'Values' contains the main data with gaze coordinates for each eye, averaged coordinates and time stamp of the current frame and the state indicating which state the tracker is in.


```
{"category":"tracker","request":"get","statuscode
   ":200,
 "values":
      {"frame":
           {"avg":{"x":872.6505,"y":193.9748},"fix":
              false,
           "lefteye":{"avg":{"x":810.9225,"y
              ":157.5539},
             "pcenter":{"x":0.3660,"y":0.5334},
             "psize":17.5429,"raw":{"x":966.7793,"y
                ":336.9954}},
           "righteye":{"avg":{"x":993.5639,"y
              ":300.1396},
         "pcenter":{"x":0.4917,"y":0.5179},
         "psize":16.7462,"raw":{"x":905.120,"y
            ":257.6098}},
         "state":7,
         "timestamp":"2016-09-16 13:18:13.064"}}}
```


Fig. 12.7 Sample readings in JSON format from EyeTribe tracker

Data Preparation and Cleansing
With eye tracking, there is a possibility of missing frames when eye movement is not captured, or corrupted data are captured. To process the data further, the raw Java Script Object Notation (JSON) data were first converted into comma separated values using an online 'JSON to CSV' conversion tool [33]. The raw data were filtered to eliminate corrupted data and interpolate the missing data with approximate values based on the values of the previous and next frames. Invalid frames were eliminated via Structured Query Language (SQL) transformation scripts and missing values were cleaned by applying multiple imputation by chained equation based on average gaze coordinates from the left and right eyes and pupil size.

Mapping Area of Interest (AOI) and Sequential Representation
As the initial step for transforming the raw visual data into information, each frame from the continuous stream of vision data was mapped onto the Area of Interest (AOI) on the flight simulator screen. The following regions were marked as important AOIs for the purpose of this experiment: instruments—AH, ASI, TC, VSI, ALT and NAV, any other points on the instrument panel and the horizon.

From the gaze data, it is evident that instrument scan path is comprised of a series of gaze data frames. Gaze data in order of time can be represented as a temporal sequence based on AOI transitions. A Finite State Machine (FSM) recogniser is implemented to represent and track transitions. A state transition model is defined as a directed graph represented by:

$$G = (S, Z, T, S0, F) \tag{12.1}$$

where:

- S represents a finite set of states. For the regular instrument scan, S = {S0, S1, S2,...,Sn}.
- Z represents a set of output symbols. For the current model, these are instruments such as AH and they trigger transitions from Si to Sj.
- T represents a set of transitions {T00, T01,...,Tm}, where Tij is the transition from Si to Sj.
- S0 is the initial state.
- F is the final state.

Each transition from one state to another was triggered by an event, principally the defined gaze changing from instrument to instrument or another gaze point. Figure 12.8 shows various states and the change in instrument fixation as the events that trigger transition from one state to another. Further, the instrument scan for the whole scenario is transformed into a set of state transitions triggered by change of the AOI.

Attention Behaviour Identification
The next step in the process is to analyse the gaze data and identify the attentional indicators. In the past, studies have focused on representation sequences of gaze

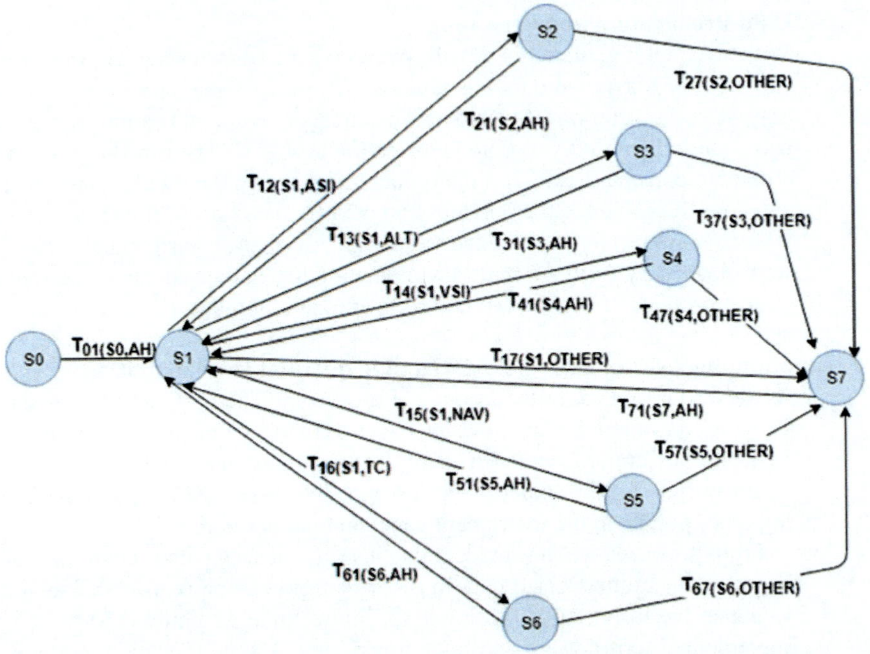

Fig. 12.8 Instrument scan state transition model

in different ways; for example, visual representation as a sequence of transitions represented as AOI rivers in [34, 35]. In these approaches, analysis must still be done manually. Since data were collected from the eye tracker at a rate of 30 frames per second, even small scenarios of 20 min result in large data sets, making decisions based on visualisation is challenging. Therefore, gaze data was translated into sequences of AOI transitions, further this study investigated the use of methods of sequential pattern mining to analyse gaze sequences [36, 37].

Although common mistakes made by pilots in different situations are listed in flight manuals [25], there has been no mention of defined wrong or bad instrument cross check. Therefore, the analysis focused on detecting attentional indicators of Misplaced Attention (MA), Attention Focusing (AF), Attention Blurring (AB) and the Distributed Attention (DA) from the gaze transition sequence.

Behavior Evaluation

The final step in this experiment process is to evaluate the recognised attention indicators as behaviours. To achieve this, the repeated attention patterns are awarded scores, and the scores are aggregated to relatively rank each pattern as poor, average or good. However, the study refrains from classifying scan patterns as good or bad in a general context because of the lack of decisive measures in aviation human factors.

12.4 Simulator Experiment Scenarios and Results

This section will further describe the experiment with flight simulator scenarios and the relevant results. During the experiments, trainee pilots were briefed on the simulator, but not directed to perform a particular instrument scan. Each participant was asked to use only the six-instrument display to perform multiple scenarios. The experiment set-up procedure and calibration process were described in instrument data acquisition section. Some of the scenarios also had failures injected into instruments such as ALT or ASI. The operators were not informed of the failures. Table 12.1 shows the different scenarios performed by each student.

12.4.1 Fixation Distribution Results

To extract the fixation distribution values, averaged left and right eye coordinates and state of the frame were used from the eye tracker output. These values, along with the pre-configured AOIs, were passed as input to a mapping program developed in Java. The program maps the pilot's gaze to respective instruments. The six instruments in the experiment had AOIs marked as AH, ASI, ALT, NAV, TC, VSI and OTHER, indicating all other areas on the screen. For each scenario, the instrument mapping record was used to create a fixation distribution chart showing percentage fixation

Table 12.1 Flight simulator scenarios performed by trainee pilots

Student	Trial	Scenario	Sample name
Student A	1	Clear skies	Student A Trial 1
Student A	2	Clear skies, instrument failures	Student A Trial 2
Student A	3	Storm dusk, instrument failures	Student A Trial 3
Student B	1	Clear skies	Student B Trial 1
Student B	2	Clear skies, instrument failures	Student B Trial 2
Student B	3	Storm dusk, instrument failures	Student B Trial 3
Student C	1	Clear skies	Student C Trial 1
Student C	2	Clear skies, instrument failure	Student C Trial 2
Student C	3	Clear Skies, instrument failures	Student C Trial 3
Student D	1	Clear skies	Student D Trial 1
Student D	2	Clear skies, instrument failures	Student D Trial 2
Student D	3	Storm dusk, instrument failures	Student D Trial 3
Student E	1	Clear skies	Student E Trial 1
Student E	2	Storm dusk, instrument failures	Student E Trial 2
Student F	1	Clear skies	Student F Trial 1
Student F	2	Clear skies, instrument failures	Student F Trial 2

on each AOI. The charts were created using Power BI which is Microsoft Business Intelligence service [38]. The percentage fixation distribution charts are shown in Figs. 12.9 and 12.10.

From the charts in Fig. 12.9, Student A showed totally different fixation distributions during the three different scenarios. Student A spent 71–92% of the time gazing at areas other than the six primary instruments during all three scenarios. Student C had similar fixation distribution in both Trial 2 and Trial 3. Both the trials had the same simulator scenarios with clear skies and instrument failures. Also, Student C spent more time gazing AH. All the participating students had approximately 30–40 number of hours of flight operating experience. On observing the fixation distribution from Student E and Student F in Fig. 12.10, it is clear that Student E exhibited better fixation distribution on chosen AOIs. Student E spent 83–88% of the time gazing at areas other than the six primary instruments. On the other hand, Student E spent less

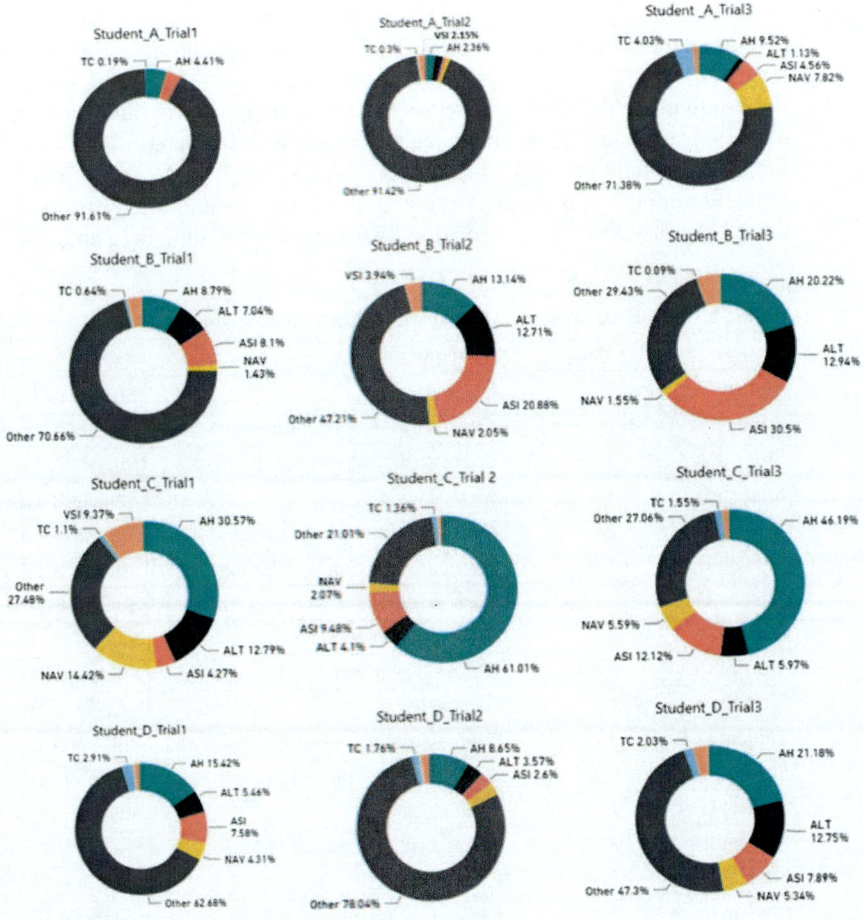

Fig. 12.9 Fixation distribution students A to D

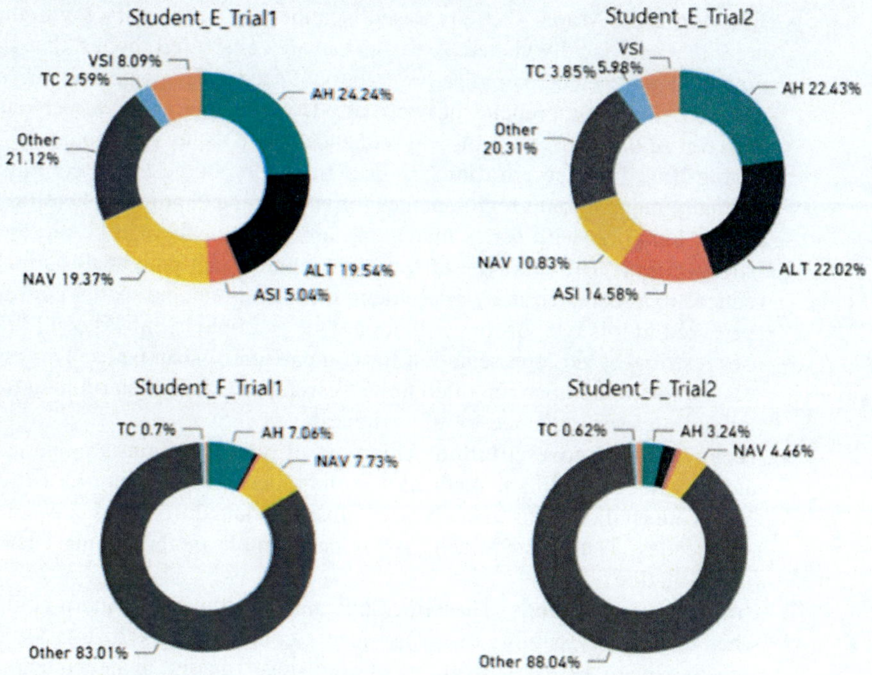

Fig. 12.10 Fixation distribution student E and F

than 22% of the time gazing at areas other than the six primary instruments. Further, Student E spent more time scanning AH and ALT.

It is also observed that Student participants spent more time scanning chosen AOI instruments and less time on 'other' during scenarios with failures. It appears that most of the student participants maintained similar fixation distributions for the different scenarios. However, fixation distributions extensively vary between different students. In other words, each student tends to follow his/her own individual fixation pattern regardless of the scenario. It was found that using the fixation distribution method to represent instrument scan is not sufficient to identify attentional behaviours. Therefore, the study further investigated the possibility of sequential representation and sequential analysis of instrument scan.

12.4.2 Instrument Scan Path Representation

There are different ways to represent a scan path or gaze trajectories, including but not limited to:

- **Fixation Heat Maps**: These represent spatial gaze behaviours, highlighting the areas that are visually visited. Areas visited are considered 'hotter' than the other areas and represented by indicative colours. If used in scan path comparisons, it is easy to visually comprehend the heat maps. However, there are no clear boundaries between AOIs. Also, the temporal sequences of AOIs are not captured.
- **String-Based Representations**: In gaze trajectory studies, gaze coordinates are normally mapped onto region names for each frame captured. Therefore, a scan path is temporal with series of region names and hence can be represented in 'string' form. With this type of representation, a scan path analysis problem is reduced to a sequence analysis problem. Both temporal and spatial information is preserved in this type of representation. One example is SubsMatch [39], which uses a string-based representation for comparison of scan paths. This algorithm was applied in comparison of complex search patterns by determining transition probabilities for sequences of transitions.
- **Vector-Based Representation**: This type of representation is numerically fast and easy to process mathematically. Normal measures in vector-based representations are euclidean distances between fixations and differences between lengths of saccades. The Multi Match [40] is an example method using vector-based representation.
- **Probabilistic Methods**: These methods are used for scan pattern comparisons when there is a possibility of each sequence containing repetitive tasks. They are also used when there is a possibility of a high level of noise in the sequence. One of the examples of probablistic representation is Hidden Markov Model (HMM) used for represent learning behaviours while comparing high versus low performers [37].

In this research, a combination of string-based and state transition model was used for the representation of instrument scan sequence. Figure 12.8 provided an overview of the chosen representation. The scan sequences are then classified into attentional behaviours and rated as poor, average, or good.

12.5 Attentional Behaviour Classification and Rating

Attention level of pilots during instrument flying is useful information that can be derived from the instrument scan sequence. Different attentional error indicators have been identified as Misplaced Attention (MA), Attention Blurring (AB) and Attention Focusing (AF). A number of classification methods have been used in classifying similar data, including known classifications and qualitative methods relying on human judgement [41]. However, no previous data with classifications were available that could be matched with the instrument scan data from this research experiment to meet the required research objectives. Therefore, a supervised classification model could not be used for this study. In the absence of readily available classifications of existing data, this research study developed a feature extraction and decision model based on the observed data and inputs from the Subject Matter Expert (SME).

Further, the study used different attributes from the instrument scan sequence to aggregate and devise models for scoring attention indicators. Figures 12.11, 12.12, and 12.13 show how an instrument scan sequence segment contributes to the attention score.

Finally, a rating model is used to classify pilot attention based on scan sequence. Of the available set of records, different scan sequences are rated as poor, average and good, by aggregating individual attention errors and attention distribution scores to compute an overall attention score. The attention ratings are defined by rules derived relative to the mean value of the attention scores. The attention scoring model is based on two components: the attention error indicator scores and the attention distribution score. The two measures are aggregated to derive the overall attentional score.

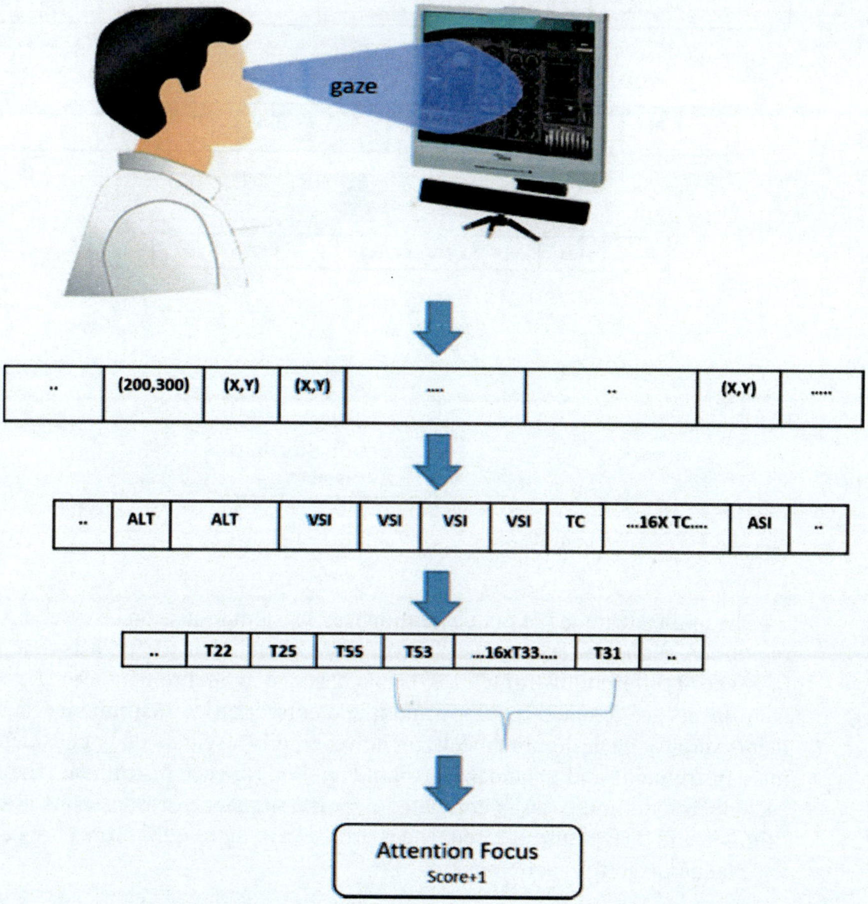

Fig. 12.11 Attention focus score

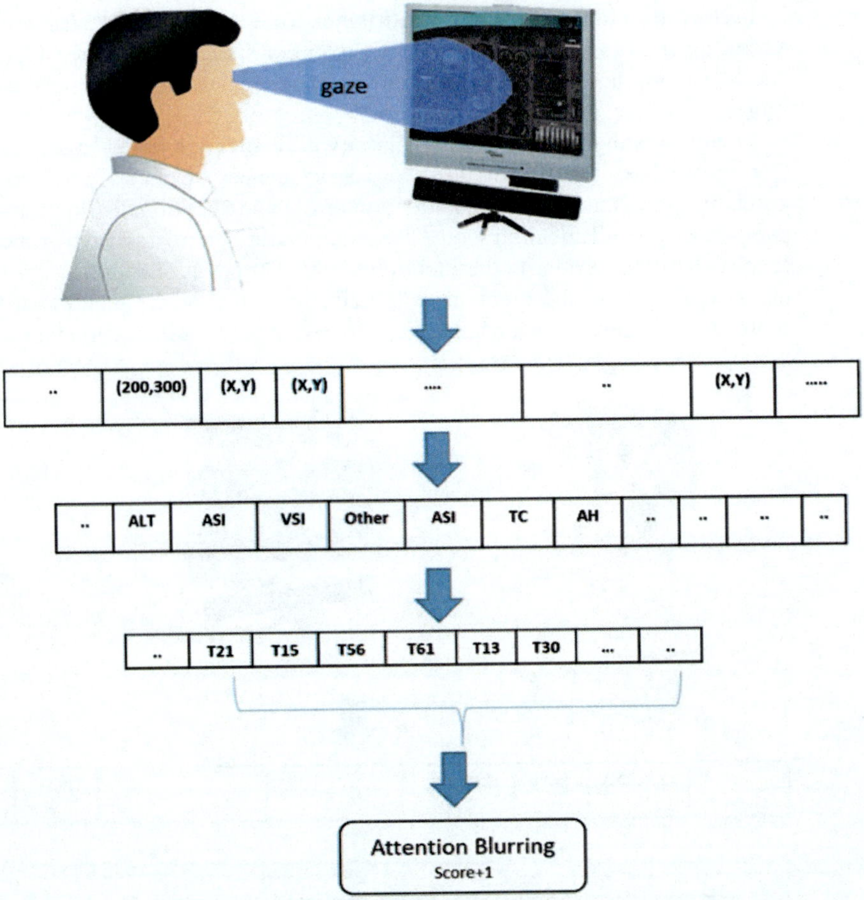

Fig. 12.12 Attention blurring score

One of the attributes of good attention is consistent transition between instrument regions. A higher attention distribution score means that the pilot is able to regularly check different instrument AOIs and has a good attention pattern. This ensures instruments are scanned regularly and in the correct order. Instrument scan requirements vary for each flight manoeuvre; however, this research study considers the six main instruments and a standard threshold interval for each instrument. The score on each attentional indicator is computed over the sequence of transitions as in Algorithms 1 and 2. Because the sequences are of varying length, scores are calculated and standardised for each transition.

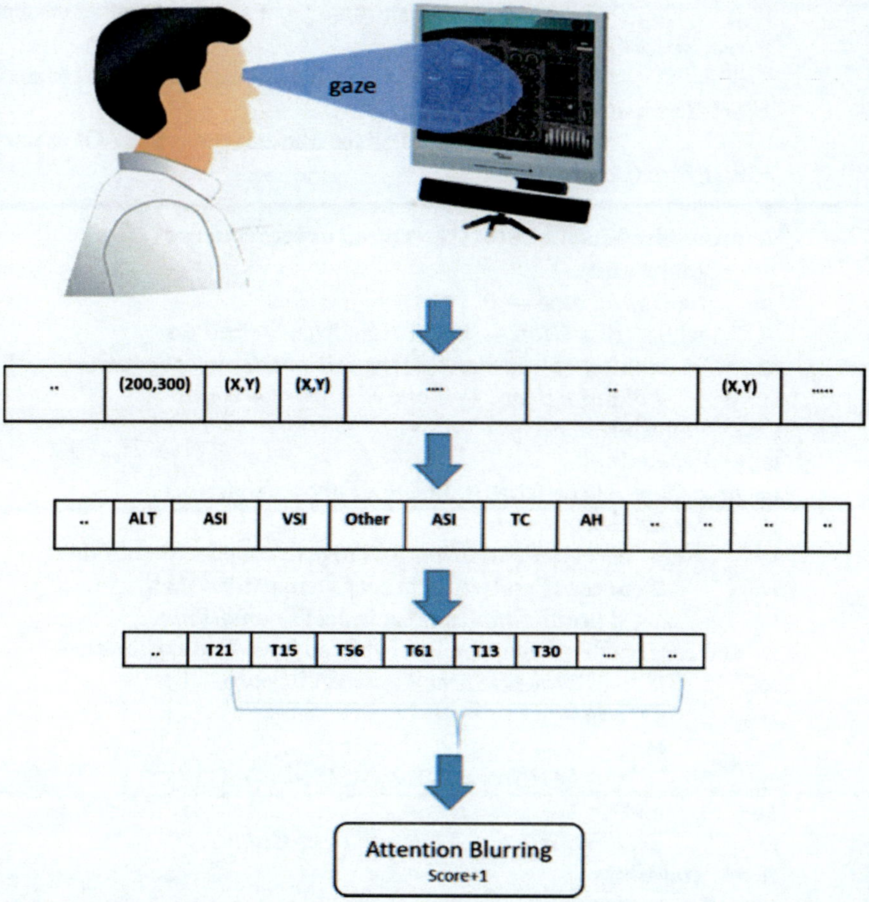

Fig. 12.13 Misplaced attention score

```
                              ▷ Set of transitions from and to AOI other example:T06
transitionOthers : list
                                     ▷ Set of transitions to same AOI example:T11
selfTransitions : list
                              ▷ Set of all the transitions to same AOI example:T56
allTransitions : list

procedure MisplacedAttention(TargetSequence)
    count : int
    totalMAScore ← 0
    while transition ← transitionOther ≠ null do
        count ← OccurrenceoftransitioninTargetSequence
        TotalMAScore ← TotalMAScore + count
    end while
end procedure
procedure AttentionBlurring(TargetSequence)
    totalABScore ← 0
    while currentTransition ← TargetSequence ≠ null do
        if currentTransition in selfTransitions then
            if prevTransition not in selfTransitions
and currentTransitionincludesAOIsofprevTransition then
                totalABScore ← totalABScore − 1
            end if
        else
            totalABScore ← totalABScore + 1
        end if
        prevTransition ← currentTransition
    end while
end procedure
```

Algorithm 1: Misplaced Attention and Attention Blurring Score

```
procedure ATTENTIONFOCUSING(TargetSequence)
    intAFRepeatThreshold ← num
    AddinstrumentAFScore : objectlist(instrument, score)  ▷ Attention focus score is computed for each AOI instrument and then aggregated to find the overall score
    totalAFScore ← 0
    while currentTransition ← selfTransitions ≠ null do
        currentTransitionRepeat ← 0
        currentAFScore ← 0
        repeat
            targetTransition ← next_transition_in_TargetSequence
            if currentTransition = targetTransitions then
                currentTransitionRepeat ← currentTransitionRepeat + 1
                if currentTransitionRepeat ≥ intAFRepeatThreshold then
                    currentAFScore ← currentAFScore + 1
                end if
            else
                instrumentAFScore ← (instrument, currentAFScore)
                currentAFScore ← 0
            end if
        until End_of_TargetSequences
    end while
    totalAFScore ← mean_of_scores_in_instrumentAFScore
end procedure
procedure ATTENTIONDISTRIBUTION(TargetSequence)
    totalADScore ← 0
    minADThreshold ← num1
    maxADThreshold ← num2                  ▷ num2 is greater than num1
    while currentTransition ← AllTransitions ≠ null do
        currentIndex ← 0
        previousIndex ← 0
        if currentTransition not in selfTransitions then
            currentIndex =
index_of_next_occurence_of_currentTransition_in_TargetSequence

            distance ← currentIndex − previousIndex
            if distance ≤ maxADThresholdanddistance ≥ minADThreshold then
                totalADScore ← totalADScore + 1
            end if
            previousIndex ← currentIndex
        end if
    end while
end procedure
```

Algorithm 2: Attention focusing and Attention Distribution Score

Attention errors indicate lower attentiveness. Therefore, AB, AF, and MA scores are inversely proportional to the overall attention rating. However, attention levels should increase with higher values of Attention Distribution (AD) score. Based on the above interpretation, the attention rating is modelled as a function of AD scores and the aggregation of attention error indicator scores.

The formula below is applied to generate the attention rating:

$$S = AD/(AF + AB + MA) \tag{12.2}$$

where,

- S is the overall attention score,
- AD is the attention distribution score,
- AF is the attention focusing score,
- MA is the misplaced attention score,
- AB is the attention blurring score.

The purpose of the attention score is to provide metrics for attention classification. Because there are no predefined metrics and labels in classifying attention during pilot instrument scans, a rule-based engine is defined on the basis of sample observations. Mean of the computed attention scores is calculated and a threshold constant is defined around the mean. The sample with an attention score in the range of the mean threshold is classified as average attention, above the threshold as good attention and below the threshold as poor attention. This method provides the flexibility to rate attention based on the sample data instead of a predefined value.

12.5.1 Results

This section covers attention scores and rating of instrument scan sequences recorded from trainee pilots. The attention rating model developed in Java traversed individual scan sequences and computed the attention error indicator scores and the attention distribution scores for each sequence. The score for each indicator was computed over the sequence of transitions as specified earlier in the section. Because the sequences are of varying lengths, scores were calculated and standardised for each transition sequence. Table 12.2 shows the computed attention error indicators and attention distribution.

Observed levels of attention during instrument scan are considered good indicators of Situation Awareness (SA). Therefore, it is implied that attention errors indicate potential loss of SA and lower attention ratings. In contrast, shared attention between AOI indicates a good level of attention. The overall attention score was derived as the attention distribution score over aggregated attention error scores. Each sample was checked to see if the overall attention score stood above or below the decision range computed over the sample arithmetic mean. Then instrument scan patterns were classified as having good, average or poor attention depending on the overall attention score being greater than the decision range, within the decision range or below the decision range respectively. Table 12.3 provides the attention score and classification as ratings of instrument scan sequences.

It can be observed that samples from Student B (Student B Trial 2 and Student B Trial 3) have the lowest attention scores, and hence, are categorised as having poor attention. These scenarios also have higher attention focusing scores and lower attention distribution scores compared with other scenarios. Student A Trial 3, Student D

Table 12.2 Attention indicator scores as certainty factor

Sample name	Misplaced attention	Attention blurring	Attention focusing	Attention distribution
Student_A Trial 1	0.035294118	0.226190476	0.064705882	0.035714286
Student_A Trial 2	0.023839398	0.626884422	0.026139691	0.148241206
Student_A Trial 3	0.067286652	0.476190476	0.056254559	0.119868637
Student_B Trial 1	0.023897059	0.307550645	0.069852941	0.060773481
Student_B Trial 2	0.006261181	0.076991943	0.1326774	0.00179051
Student_B Trial 3	0.008616047	0.098060345	0.132920481	0.012931034
Student_C Trial 1	0.05046805	0.401058632	0.074684575	0.102605863
Student_C Trial 2	0.033639144	0.281776417	0.087410805	0.058192956
Student_C Trial 3	0.068697868	0.366353841	0.084124961	0.083651952
Student_D Trial 1	0.035443038	0.557667934	0.037974684	0.086185044
Student_D Trial 2	0.03113325	0.66084788	0.02117061	0.134663342
Student_D Trial 3	0.0368	0.461538462	0.047466667	0.120192308
Student_E Trial 1	0.044352044	0.278419593	0.094479094	0.067236599
Student_E Trial 2	0.012714207	0.139933628	0.122719735	0.016039823
Student_F Trial 1	0.062305296	0.40625	0.052959502	0.115625
Student_F Trial 2	0.03878976	0.404503106	0.072019653	0.083850932

Table 12.3 Overall attention score and rating

Sample name	Attention score	Attention rating
Student_A Trial 1	0.035545024	Poor
Student_A Trial 2	0.057962946	Good
Student_A Trial 3	0.05909799	Good
Student_B Trial 1	0.045903065	Average
Student_B Trial 2	0.004006455	Poor
Student_B Trial 3	0.024225496	Poor
Student_C Trial 1	0.059330765	Good
Student_C Trial 2	0.046623157	Average
Student_C Trial 3	0.051693226	Good
Student_D Trial 1	0.037405251	Poor
Student_D Trial 2	0.049954955	Good
Student_D Trial 3	0.062262241	Good
Student_E Trial 1	0.053681508	Good
Student_E Trial 2	0.02307329	Poor
Student_F Trial 1	0.066441038	Good
Student_F Trial 2	0.048501777	Good

Trial 3 and Student F Trial 1 have the top three attention scores. Though Student A Trial 3 and Student F Trial 1 did not have the top fixation distribution percentage, the instrument scan sequences showed consistent scanning of instruments of interest. Student E Trial 1 has the good overall attention rating where as the second trial from the same Student E Trial 2 resulted in poor attention rating.

This shows that attention behaviour varies in different scenarios. Although, Student E Trial 2 was the had good fixation density distribution as shown Fig. 12.9, the attention rating is not consistent with the fixation density distribution results. This further strengthens the hypothesis that attention is dependent on the duration and the order of the scan and not only the aggregated fixation duration over a time period.

12.6 Conclusions

The motivation for the experiments discussed in this chapter was to arrive at a reliable measure and method that provide a better mechanism to identify pilot's attention distribution, and attention error indicators such as Attention Blurring (AB), Attention Focusing (AF) and Misplaced Attention (MA). During the course of the research, it was proved that ocular measures are effective measures in determining attentional behaviour.

The study also highlighted the importance of sequential representation of gaze data and not only the aggregated fixation distribution on AOIs. Attention indicator score models were designed and applied to the sequences to identify various attentional behaviours. It has been observed from the results that attention indicators can overlap during instrument scan. However, using the scoring model helps to determine the frequently exhibited attention indicators. The computation of attention provides a comparative rating of attention within the data set. The attention scores from the data set were categorised as good, average or poor relative to other participants in the group. However, the study refrains from labelling the behaviour as good or poor in general scenarios because, so far in aviation, there has been no clear distinction between expected good attention behaviour and poor attentional behaviour.

There were a few challenges that arose during this study. Currently, there is no standard definition of expected patterns during instrument scan. In addition, there are no real-time data or known classifications available in the aviation literature. Therefore, the study was based on the recommended instrument scans in instrument flying manuals and input from aviation Subject Matter Experts (SMEs). The six primary instrument scans during instrument flying was used as the case for this thesis. However, the system could be easily extended to include other instruments and additional AOIs. One future extension could involve the development of an expert system that includes other scenarios during instrument scan and integrates the attention scoring and rating algorithms for the purpose of analysis of pilot behaviour. The scope of this study included only ocular measures, as eye tracking is a proven method of detecting visual attention. Along with ocular measures, integration of speech processing or other physiological measures such as facial expressions recognition systems may help in developing a robust futuristic SA monitoring system.

This research investigated the possibility of identifying attention errors but did not attempt to provide feedback to the pilot. However, in the future, a system based on this research could be developed that could monitor pilots' behaviour in real time, and provide timely feedback and alerts to the pilots, which could prove to be lifesaving.

References

1. Ancel, E., Shih, A.T., Jones, S.M., Reveley, M.S., Luxhøj, J.T., Evans, J.K.: Predictive safety analytics: inferring aviation accident shaping factors and causation. J. Risk Res. **18**(4), 428–451 (2015)
2. Shappell, S.A., Wiegmann, D.A.: Human factors analysis of aviation accident data: developing a needs-based, data-driven, safety program. In: 3rd Workshop on Human Error, Safety, and System Development (HESSD'99) (1999)
3. Thatcher, S., Kilingaru, K.: Intelligent monitoring of flight crew situation awareness. Adv. Mater. Res. **433**(1), 6693–6701 (2012). Trans Tech Publications
4. Kilingaru, K., Tweedale, J.W., Thatcher, S., Jain, L.C.: Monitoring pilot "situation awareness". J. Intell. Fuzzy Syst. **24**(3), 457–466 (2013)
5. Regal, D.M., Rogers, W.H., Boucek. G.P.: Situational awareness in the commercial flight deck: definition, measurement, and enhancement. SAE Technical Paper (1988)
6. Sarter, N.B., Woods, D.D.: Situation awareness: a critical but ill-defined phenomenon. Int. J. Aviat. Psychol. **1**(1), 45–57 (1991)
7. Oakley, T.: Attention and cognition. J. Appl. Attention **17**(1), 65–78 (2004)
8. Mack, A., Rock, I.: In Attentional Blindness. MIT press (1998)
9. Lamme, V.A.: Why visual attention and awareness are different. Trends Cognitive Sci. **7**(1), 12–18 (2003)
10. Underwood, G., Chapman, P., Brocklehurst, N., Underwood, J., Crundall, D.: Visual attention while driving: sequences of eye fixations made by experienced and novice drivers. Ergonomics **46**(6), 629–646 (2003)
11. Smith, P., Shah, M., da Vitoria, Lobo N.: Determining driver visual attention with one camera. IEEE Trans. Intell. Transp. Syst. **4**(4), 205–218 (2003)
12. Ji, Q., Yang, X.: Real-time eye, gaze, and face pose tracking for monitoring driver vigilance. Real-time imaging. **8**(5), 357–377 (2002)
13. Yu, C.S., Wang, E.M., Li, W.C., Braithwaite, G.: Pilots' visual scan patterns and situation awareness in flight operations. Aviat. Space Environ. Med. **85**(7), 708–714 (2014)
14. Haslbeck, A., Bengler, K.: Pilots' gaze strategies and manual control performance using occlusion as a measurement technique during a simulated manual flight task. Cogn. Technol. Work **18**(3), 529–540 (2016)
15. Ho, H.F., Su, H.S., Li, W.C., Yu, C.S., Braithwaite, G.: Pilots' latency of first fixation and dwell among regions of interest on the flight deck. In: International Conference on Engineering Psychology and Cognitive Ergonomics. Springer, Cham (2016)
16. Roscoe, A.H.: Heart rate as an in-flight measure of pilot workload. Royal Aircraft Establishment Farnborough (United Kingdom) (1982)
17. Hankins, T.C., Wilson, G.F.: A comparison of heart rate, eye activity, EEG and subjective measures of pilot mental workload during flight. Aviat. Space Environ. Med. **69**(4), 360–367 (1998)
18. Craig, A., Tran, Y., Wijesuriya, N., Nguyen, H.: Regional brain wave activity changes associated with fatigue. Psychophysiology **49**(44), 574–582 (2012)

19. Diez, M., Boehm-Davis, D.A., Holt, R.W., Pinney, M.E., Hansberger, J.T., Schoppek, W.: Tracking pilot interactions with flight management systems through eye movements. In: Proceedings of the 11th International Symposium on Aviation Psychology, vol. 6, issue 1. The Ohio State University, Columbus (2001)
20. Van De Merwe, K., Van Dijk, H., Zon, R.: Eye movements as an indicator of situation awareness in a flight simulator experiment. Int. J. Aviat. Psychol. **22**(1), 78–95 (2012)
21. Fitts, P.M., Jones, R.E., Milton, J.L.: Eye movements of aircraft pilots during instrument-landing approaches. Ergon. Psychol. Mech. Models Ergon. **3**(1), 56 (2005)
22. de Greef, T., Lafeber, H., van Oostendorp, H., Lindenberg, J.: Eye movement as indicators of mental workload to trigger adaptive automation. In: International Conference on Foundations of Augmented Cognition, pp. 219–228. Springer, Berlin, Heidelberg (2009)
23. Gibb, R., Gray, R., Scharff, L.: Aviation Visual Perception: Research, Misperception and Mishaps. Routledge (2016)
24. Rayner, K., Pollatsek, A.: Eye movements and scene perception. Can. J. Psychol. **46**(3), 342 (1992)
25. Instrument flying handbook: faa-h-8083-15a, United States Department of Transport Federal Aviation Administration (2012)
26. Fayyad, U., Piatetsky-Shapiro, G., Smyth, P.: From data mining to knowledge discovery in databases. AI Mag. **17**(3), 37 (1996)
27. Ackoff, R.L.: From data to wisdom. J. Appl. Syst. Anal. **16**(1), 3–9 (1989)
28. Bellinger, G., Castro, D., Mills, A.: Data, information, knowledge, and wisdom (2004)
29. Cleveland, H.: Information as a resource. Futurist **16**(6), 34–39 (1982)
30. Zeleny, M.: Management support systems: towards integrated knowledge management. Hum. Syst. Manage. **7**(1), 59–70 (1987)
31. Eyetribe: Eyetribe tracker, Available Online: https://s3.eu-central-1.amazonaws.com/theeyetribe.com/theeyetribe.com/dev/csharp/index.html. Last accessed on 27 July 2019
32. Lockheed-Martin: Prepar3d, Available Online: http://www.prepar3d.com. Last accessed on 27 July 2019
33. Mill, E.: Json to CSV tool. Online: https://konklone.io/json/. Last accessed on 02 April 2018
34. Burch, M., Kull, A., Weiskopf, D.: AOI rivers for visualizing dynamic eye gaze frequencies. Comput. Graph. Forum **32**(3), 281–290 (2013)
35. Kurzhals, K., Weiskopf, D.: Aoi transition trees. In: Proceedings of the 41st Graphics Interface Conference, pp. 41–48. Canadian Information Processing Society (2015)
36. Abbott, A., Hrycak, A.: Measuring resemblance in sequence data: An optimal matching analysis of musicians' careers. Am. J. Sociol. **96**(1), 144–185 (1990)
37. Kinnebrew, J.S., Biswas, G.: Comparative action sequence analysis with hidden markov models and sequence mining. In: Proceedings of the Knowledge Discovery in Educational Data Workshop at the 17th ACM SIGKDD Conference on Knowledge Discovery and Data Mining (KDD 2011). San Diego, CA (2011)
38. Power BI: [Available online], https://powerbi.microsoft.com/en-us/. Last accessed 26 August 2019
39. Kübler, T., Eivazi, S., Kasneci, E.: Automated visual scanpath analysis reveals the expertise level of micro-neurosurgeons. In: MICCAI Workshop on Interventional Microscopy, pp. 1–8 (2015)
40. Dewhurst, R., Nyström, M., Jarodzka, H., Foulsham, T., Johansson, R., Holmqvist, K.: It depends on how you look at it: Scanpath comparison in multiple dimensions with MultiMatch, a vector-based approach. Behav. Res. Methods **44**(4), 1079–1100 (2012)
41. Li, H.: A short introduction to learning to rank. IEICE Trans. Inform. Syst. **94**(10), 1854–1862 (2011)

Chapter 13
Audio Content-Based Framework for Emotional Music Recognition

Angelo Ciaramella, Davide Nardone, Antonino Staiano, and Giuseppe Vettigli

Abstract Music is a language of emotions and music emotional recognition has been addressed by different disciplines (e.g., psychology, cognitive science and musicology). Nowadays, the music fruition mechanism is evolving, focusing on the music content. In this work, a framework for processing, classification and clustering of songs on the basis of their emotional contents, is explained. On one hand, the main emotional features are extracted after a pre-processing phase where both Sparse Modeling and Independent Component Analysis based methodologies are applied. The approach makes it possible to summarize the main sub-tracks of an acoustic music song (e.g., information compression and filtering) and to extract the main features from these parts (e.g., music instrumental features). On the other hand, a system for music emotion recognition based on Machine Learning and Soft Computing techniques is introduced. One user can submit a target song, representing his conceptual emotion, and obtain a playlist of audio songs with similar emotional content. In the case of classification, a playlist is retrieved from songs belonging to the same class. In the other case, the playlist is suggested by the system exploiting the content of the audio songs and it could also contains songs of different classes. Experimental results are proposed to show the performance of the developed framework.

A. Ciaramella (✉) · A. Staiano
Department of Science and Technology, University of Naples "Parthenope",
Centro Direzionale, Isola C4, 80143 Naples, Italy
e-mail: angelo.ciaramella@uniparthenope.it

A. Staiano
e-mail: antonino.staiano@uniparthenope.it

D. Nardone
Blue Reply, Cognitive & Data Warehouse, Turin, Italy

G. Vettigli
Centrica Hive, 50/60 Station Rd, Cambridge CB1 2JH, UK

G. Phillips-Wren et al. (eds.), *Advances in Data Science: Methodologies and Applications*, Intelligent Systems Reference Library 189,
https://doi.org/10.1007/978-3-030-51870-7_13

13.1 Introduction

One of the main channels for accessing reality and information about people and their social interaction is the Multimedia content [1]. One special medium is music that is essential for an independent child and adult life [2] for its extraordinary ability to evoke powerful emotions [3]. Recently, music emotion recognition have been studied in different disciplines such as psychology, physiology, cognitive science and musicology [4] where emotion usually has a short duration (seconds to minutes) while mood has a longer duration (hours or days). Several studies in neuroscience, by exploiting the current technique of neuroimaging, found interesting biological property triggered by specific areas of the brain when listening at emotional music. While authors in [5] demonstrated that the amygdala plays an important role in the recognition of fear, when scary music is played, authors in [3] found that music creating high pleasurable feeling emotion, stimulates dopaminergic pathways in the human brain such the *mesolimbic*, which is involved in reward and motivation.

Another study [6], for example, exploits the Electroencephalography (EEG) data, for the emotional response of terminally ill cancer patients to a music therapy intervention and a recent study confirms an anti-epileptic effect of Mozart music on the EEG in children, suggesting the "Mozart therapy" as a treatment for drug-resistant epilepsy warrants [7]. In these last years, several websites have tried to combine social interaction with music and entertainment. For example *Stereomood* [8] is a free emotional internet radio. Moreover, in [9] the authors introduced a framework for mood detection of acoustic music data, based on a music psychological theory in western cultures. In [10] the authors proposed and compared two fuzzy classifiers determining emotion classes by using a *Arousal and Valence* (AV) scheme. While, in [4] the authors focus on a music emotion recognition system based on fuzzy inferences. Recently, a system for music emotion recognition based on *machine learning* and *computational intelligence* techniques has been introduced in [11]. In that system, one user formulates a query providing a target audio song with similar emotions to the ones he wishes to retrieve while the authors use supervised techniques on labeled data or unsupervised techniques on unlabeled data. The emotional classes are a subset of the model, proposed by Russell, showed in Fig. 13.1. According to it, emotions are explained as combinations of arousal and valence, where arousal measures the level of activation and valence measures pleasure/displeasure.

Moreover, in [12] a robust approach of features extraction from music recordings has been introduced. The approach permits to extract the representative sub-tracks by compression and filtering.

Aim of this chapter is to explain a robust framework for processing, classification and clustering of musical audio songs by their emotional contents. The main emotional features are obtained after a pre-processing on the sub-tracks of an acoustic music song using both Sparse Modeling [13] and Independent Component Analysis [14]. This mechanism permits to compress and filter the main music features corresponding to their content (i.e., music instruments). The framework take in input a target song, representing a conceptual emotion, and permits to obtain a playlist of

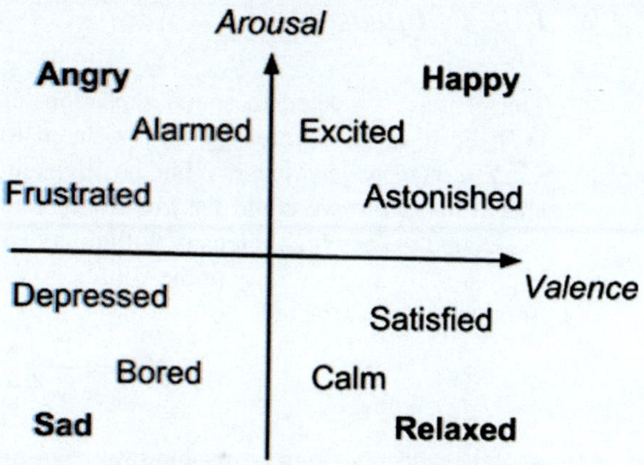

Fig. 13.1 Russell's model representing two-dimensional emotion classes

audio songs with similar emotional content. In the case of classification, a playlist is obtained from the songs belonging to the same class. In the other case, the playlist is suggested by the system exploiting the content of the audio songs and it could also contain songs of different classes.

The paper is organized as follows. In Sect. 13.2 the music emotional features are described. In Sects. 13.3 and 13.4 the overall system and the used techniques are described. Finally, in Sects. 13.5 and 13.6 several experimental results and considerations are proposed, respectively.

13.2 Emotional Features

The music emotion recognition systems are based on the extraction of emotional features from acoustic music recordings. In particular, an Arousal-Valence plane is considered for describing the emotional classes and some features are extracted for further analyisis (*intensity*, *rhythm*, *key*, *harmony* and *spectral centroid* [4, 9, 10, 15]).

13.2.1 Emotional Model

The Arousal-Valence plane is composed by a two-dimensional emotion space of 4 quadrants and emotions are classified on the plane (see Fig. 13.1). On the adopted Russell model [16] the right (left) side of the plane refers to the positive (negative) emotion, whereas the upper (lower) side of the plane refers to the energetic (silent) emotion.

13.2.2 Intensity

This feature is related to sound sensation and the amplitude of the audio waves [17]. Formally, low intensity is associated to sensations of *sadness*, *melancholy*, *tenderness* or *peacefulness*. While positive emotions are correlated to high intensity, and in particular we could list *joy*, *excitement* or *triumph*, while *anger* or *fear* are associated to very high intensity with many variations. The intensity of the sound is expressed by the regularity of the volume in the song. In particular, the mean energy of the wavel is extracted

$$AE(x) = \frac{1}{N}\sum_{t=0}^{N} |x(t)^2|, \tag{13.1}$$

where $x(t)$ is the value of the amplitude at time t and N is the length of signal.

The standard deviation of AE is calculated

$$\sigma(AE(x)) = \sqrt{\frac{1}{N}\sum_{t=0}^{N}(AE(x) - x(t))^2}. \tag{13.2}$$

This value expresses the regularity of the volume in the song: high volume, regularity of loudness, and loudness frequency.

13.2.3 Rhythm

The rhythm of a song is described by *beat* and *tempo*. The beat is the regularly occurring pattern of rhythmic stresses in music and tempo is the speed of the beat, expressed in *Beats Per Minute* (BPM).

Regular beats makes listeners *peaceful* or even *melancholic*, but irregular beats could make some listeners feel *aggressive* or *unsteady*. The approach used in our framework pemits to track beats estimating the beat locations [18, 19].

13.2.4 Key

In a song, a group of pitches in ascending order form a scale, spanning an octave. In our framework we adopt a key detection system for estimating key associated with the maximum duration in the song for each key change [20].

13.2.5 Harmony and Spectral Centroid

Harmony refers to the way chords are constructed and how they follow each other in a song. The harmony can be estimated analysing the overtones and evaluating the following function

$$HS(f) = \sum_{k=1}^{M} \min(||X(f)||^2, ||X(kf)||^2), \tag{13.3}$$

where, f is the frequency, X is the Short Time Fourier Transform of the source signal, and M denotes the maximum number of frequencies for which the mean of $X(f)$ is higher than a value θ [1] (only those frequencies are used in the computation). At the end, the standard deviation of $HS(f)$ is obtained.

For estimating the fundamental pitch of the signal the *spectral centroid* is considered [21].

13.3 Pre-processing System Architecture

The emotional features are applied after a pre-processing on the audio tracks. The main objective is to extract robust features representing the music content of the audio songs.

13.3.1 Representative Sub-tracks

In the proposed system a Sparse Modeling (SM) has been considered for extracting information from music audio tracks [13, 22]. In a SM schema a data matrix $\mathbf{Y} = [\mathbf{y}_1, \ldots, \mathbf{y}_N]$, where $\mathbf{y}_i \in R^m$, $i = 1, \ldots, N$, is considered. The aim is to evaluate a compact dictionary $\mathbf{D} = [\mathbf{d}_1, \ldots, \mathbf{d}_N] \in R^{m \times N}$ and coefficients $\mathbf{X} = [\mathbf{x}_1, \ldots, \mathbf{x}_N] \in R^{N \times N}$, for representing the collection of data $\mathbf{Y}$ and in particular, minimizing the following objective function

$$\sum_{i=1}^{N} \|\mathbf{y}_i - \mathbf{D}\mathbf{x}_i\|_2^2 = \|\mathbf{Y} - \mathbf{D}\mathbf{X}\|_F^2, \tag{13.4}$$

so that, the best representation of the data can be obtained. In the sparse dictionary learning framework, one requires the coefficient matrix $\mathbf{X}$ to be sparse by solving

[1] In the experiment we used $\theta = 10 \times 10^{-3}$.

$$\begin{aligned} &\min_{\mathbf{D},X} \|\mathbf{Y} - \mathbf{D}\mathbf{X}\|_F^2 \\ &\text{s.t. } \|\mathbf{x}_i\|_0 \leq s,\ \|\mathbf{d}_j\|_2 \leq 1 \forall i, j, \end{aligned} \tag{13.5}$$

where $\|\mathbf{x}_i\|_0$ indicates the number of nonzero elements of $\mathbf{x}_i$. In particular, dictionary and coefficients are learned simultaneously such that each data point $\mathbf{y}_i$ is written as a linear combination of at most s atoms of the dictionary [13]. Now we stress that from the following reconstruction error

$$\sum_{i=1}^{N} \|\mathbf{y}_i - \mathbf{Y}\mathbf{c}_i\|_2^2 = \|\mathbf{Y} - \mathbf{Y}C\|_F^2, \tag{13.6}$$

with respect to the coefficient matrix $\mathbf{C} \triangleq [\mathbf{c}_1, \ldots, \mathbf{c}_N] \in R^{N \times N}$, each data could be expressed as a linear combination of all the data. To find $k \ll N$ representatives we use the following optimization problem

$$\begin{aligned} &\min \|\mathbf{Y} - \mathbf{Y}C\|_F^2 \\ &\text{s.t. } \|\mathbf{C}\|_{0,q} \leq k,\ \mathbf{1^T C} = \mathbf{1^T}, \end{aligned} \tag{13.7}$$

where $\|\mathbf{C}_{0,q}\| \triangleq \sum_{i=1}^{N} I(\|\mathbf{C}\|_q) > 0$, $\mathbf{c}^i$ denotes the i-th row of $\mathbf{C}$ and $I(\dot{})$ denotes the indicator function. In particular, $\|\mathbf{C}_{0,q}\|$ counts the number of nonzero rows of $\mathbf{C}$. Since this is an NP-hard problem, a standard l_1 relaxation of this optimization is adopted

$$\begin{aligned} &\min \|\mathbf{Y} - \mathbf{Y}C\|_F^2 \\ &\text{s.t. } \|\mathbf{C}\|_{1,q} \leq \tau,\ \mathbf{1^T C} = \mathbf{1^T}, \end{aligned} \tag{13.8}$$

where $\|\mathbf{C}_{1,q}\| \triangleq \sum_{i=1}^{N} \|\mathbf{c}^i\|_q$ is the sum of the l_q norms of the rows of $\mathbf{C}$, and $\tau > 0$ is an appropriately chosen parameter. The solution of the optimization problem 13.8, not only indicates the representatives as the nonzero rows of $\mathbf{C}$, but also provides information about the ranking, i.e., relative importance of the representatives for describing the dataset. We can rank k representatives $\mathbf{y}_{i_1}, \ldots, \mathbf{y}_{i_k}$ as $i_1 \geq i_2 \geq \cdots \geq i_k$, i.e., $\mathbf{y}_{i_1}$ has the highest rank and $\mathbf{y}_{i_k}$ has the lowest rank. In this work, by using the Lagrange multipliers, the optimization problem is defined as

$$\begin{aligned} &\min \tfrac{1}{2}\|\mathbf{Y} - \mathbf{Y}C\|_F^2 + \lambda\|\mathbf{C}\|_{1,q} \\ &\text{s.t. } \mathbf{1^T C} = \mathbf{1^T}. \end{aligned} \tag{13.9}$$

implemented in an Alternating Direction Method of Multipliers (ADMM) optimization framework (see [22] for further details).

13.3.2 Independent Component Analysis

Blind Source Separation of instantaneous mixtures has been well addressed by Independent Component Analysis (ICA) [14, 23]. ICA is a computational method for separating a multivariate signal into additive components [23]. In general for various real-world applications, convolved and time-delayed versions of the same sources can be observed instead of instantaneous ones [24–26] as in a room where the multipath propagation of a signal causes reverberations. This scenario is described by a convolutive mixture model where each element of a mixing matrix $\mathbf{A}$ in the model $\mathbf{x}(t) = \mathbf{A}\mathbf{s}(t)$, is a filter rather than a scalar

$$x_i(t) = \sum_{j=1}^{n} \sum_{k} a_{ikj} s_j(t-k), \tag{13.10}$$

for $i = 1, \ldots, n$.

We note that for inverting the convolutive mixtures $x_i(t)$ a set of similar FIR filters should be used

$$y_i(t) = \sum_{j=1}^{n} \sum_{k} w_{ikj} x_j(t-k). \tag{13.11}$$

The output signals $y_1(t), \ldots, y_n(t)$ of the separating system are the estimates of the source signals $s_1(t), \ldots, s_n(t)$ at discrete time t, and w_{ikj} are the coefficients of the FIR filters of the separating system. In the proposed framework, for estimating the w_{ikj} coefficients, the approach introduced in [26] (named Convolved ICA, CICA) is adopted. In particular, the approach represents the convolved mixtures in the frequency domain ($X_i(\omega, t)$) by a Short Time Fourier Transform (STFT). STFT permits to observe the mixtures both in time (frame) and frequency (bin). For each frequency bin, the observations are separated by ICA model in the complex domain. One problem to solve is related with the permutation indeterminacy [23], that is solved in this approach by an Assignment Problem (e.g., Hungarian algorithm) with a Kullback-Leibler divergence [24, 25].

13.3.3 Pre-processing Schema

In Fig. 13.2 a schema of the proposed pre-processing system is shown. First of all, each music track is segmented into several frames and the latters are allocated in a matrix of observations $\mathbf{Y}$. The matrix is processed by a SM approach (see Sect. 13.3.1), for extracting the representative frames (sub-tracks) of the music audio songs. This step is fundamental for improving information storage (e.g., for mobile devices) and to avoid unnecessary information.

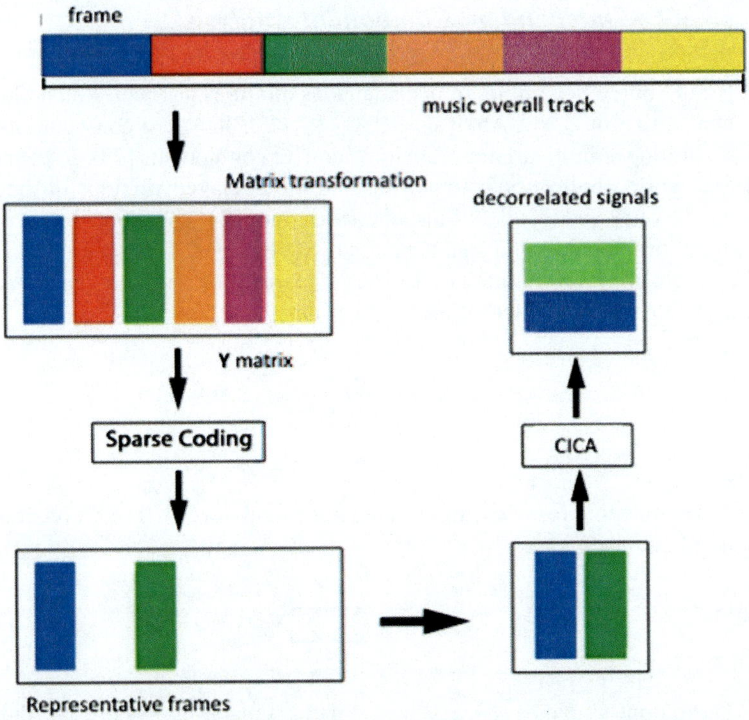

Fig. 13.2 Pre-processing procedure of the proposed system

Successively, for separating the components from the extracted sub-tracks, the CICA approach described in Sect. 13.3.2 is applied. Aim is to extract the fundamental information of the audio songs (e.g., those related to singer voice and music instrumentals). Moreover, the emotional features (see Sect. 13.2) of each extracted component are evaluated before agglomerating or classification [27].

13.4 Emotion Recognition System Architecture

In Fig. 13.3 a schema of the emotion recognition system is summarized. It has been designed for the Web, aiming for social interactions.
The aim is to provide a framework for retrieving audio songs from a database by using emotional information in two different scenarios:

- supervised—songs are emotional labeled by the users
- unsupervised—no information about the emotion information is given.

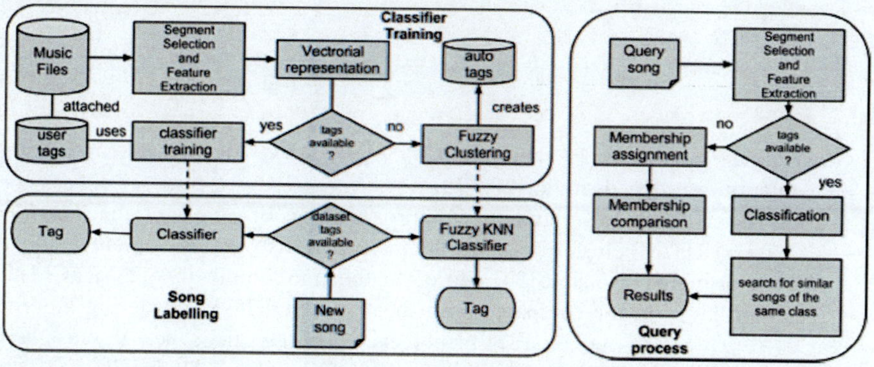

Fig. 13.3 System archiecture

The query engine allows to submit a target audio song and suggests a playlist of emotional similar songs.

On one hand, the classifier is used to identify the class of the target song and the results are shown as the most similar songs in the same class. Hence, the most similar songs are ranked by a fuzzy similarity measure based on the Łukasiewicz product [28–30].

On the other hand, a clustering algorithm computes the memberships of each song that finally are compared to select the results [31]. We considered three techniques to classify the song in the supervised case: Multi-Layer Perceptron (MLP), Support Vector Machine (SVM) and Bayesian Network (BN) [27, 32], while we considered Fuzzy C-Means (FCM) and Rough Fuzzy C-Means (RFCM) for the clustering task [33, 34].

13.4.1 *Fuzzy and Rough Fuzzy C-Means*

The *Fuzzy C-Means* (FCM) is a fuzzification of the C-Means algorithm [33]. Aim is partitioning a set of N patterns $\{\mathbf{x}_k\}$ into c clusters by minimizing the objective function

$$J_{\mathbf{FCM}} = \sum_{k=1}^{N} \sum_{i=1}^{c} (\mu_{ik})^m \|\mathbf{x}_k - \mathbf{v}_i\|^2 \tag{13.12}$$

where $1 \leq m < \infty$ is the fuzzifier, $\mathbf{v}_i$ is the i-th cluster center, $\mu_{ik} \in [0, 1]$ is the membership of the k-th pattern to it, and $\|\cdot\|$ is a distance between the patterns, such that

$$\mathbf{v}_i = \frac{\sum_{k=1}^{N} (\mu_{ik})^m \mathbf{x}_k}{\sum_{k=1}^{N} (\mu_{ik})^m} \tag{13.13}$$

and

$$\mu_{ik} = \frac{1}{\sum_{j=1}^{c} (\frac{d_{ik}}{d_{jk}})^{\frac{2}{m-1}}} \tag{13.14}$$

with $d_{ik} = \|\mathbf{x}_k - \mathbf{v}_i\|^2$, subject to $\sum_{i=1}^{c} \mu_{ik} = 1, \forall k$. The algorithm to calculate these quantities proceeds iteratively [33].

Based on the lower and upper approximations of *rough set*, the *Rough Fuzzy C-Means* (RFCM) clustering algorithm makes the distribution of membership function become more reasonable [34]. Moreover, the time complexity of the RFCM clustering algorithm is lower compared with the traditional FCM clustering algorithm. Let $X = \{x_1, x_2, \ldots, x_n\}$ be a set of objects to be classified, the i-th class be denoted by w_i, its centroid be v_i, and the number of class be k. Define $\underline{R}w_i = \{x_j | x_j \in w_i\}$, $\overline{R}w_i = \{x_j | \|x_j - v_i\| \leq A_i, A_i > 0\}$, we have

1. if $x_j \in \underline{R}w_i$, then $\forall l \in \{1, \ldots, k\}, l \neq j, x_j \in \overline{R}w_i, x_j \in' \overline{R}w_l$
2. if $x_j \in \overline{\underline{R}}w_i$, then at least exist $l \in \{1, \ldots, k\}$, make $x_j \in \underline{R}w_i$.

Provided that A_i is called the upper approximate limit, which characterizes the border of all possible objects possibly belonging to the i-th class. If some objects do not belong to the range which is defined by the upper approximate limit, then they belong to the negative domain of this class, namely, they do not belong to this class. The objective function of RFCM clustering algorithm is:

$$J_{\mathbf{RFCM}} = \sum_{k=1}^{N} \sum_{i=1, \mathbf{x}_k \in \overline{R}w_i}^{c} (\mu_{ik})^m \|\mathbf{x}_k - \mathbf{v}_i\|^2 \tag{13.15}$$

where the constraints are $0 \leq \sum_{j=1}^{n} \mu_{ij} \leq N$, $\sum_{i=1, \mathbf{x}_k \in \overline{R}w_i}^{c} \mu_{ik} = 1$. We can also get the membership formula of RFCM algorithm as follows

$$\mathbf{v}_i = \frac{\sum_{k=1}^{N} (\mu_{ik})^m \mathbf{x}_k}{\sum_{k=1}^{N} (\mu_{ik})^m} \tag{13.16}$$

and

$$\mu_{ik} = \frac{1}{\sum_{l=1, \mathbf{x}_k \in \overline{R}w_i}^{c} (\frac{d_{ik}}{d_{lk}})^{\frac{2}{m-1}}} \tag{13.17}$$

Also in this case the algorithm proceeds iteratively.

13.4.2 Fuzzy Memberships

After the FCM (or FRCM) process is completed, the i-th object in the c class has a membership μ_{ic}. In fuzzy classification, we assign a fuzzy membership μ_{uc} for a target input x_u to each class c (on C total classes) as a linear combination of the fuzzy vectors of k-nearest training samples:

$$\mu_{uc} = \frac{\sum_{i=1}^{k} w_i \mu_{ic}}{\sum_{i=1}^{k} w_i} \tag{13.18}$$

where μ_{ic} is the fuzzy membership of a training sample x_i in class c, x_i is one of the k-nearest samples, and w_i is the weight inversely proportional to the distance d_{iu} between x_i and x_u, $w_i = d_{iu}^{-2}$. With Eq. 13.18 we get the $C \times 1$ fuzzy vector μ_u indicating music emotion strength of the input sample: $\mu_u = \{\mu_{u1}, \ldots, \mu_{uC}\}$ such that $\sum_{c=1}^{C} \mu_{uc} = 1$. The corresponding class is obtained considering the maximum of μ_u.

13.5 Experimental Results

In this Section we report some experimental results obtained by using the music emotion recognition framework. At first, we highlight the performance of the pre-processing step considering the first 120 seconds of the songs with a sampling frequency of 44100 Hz and 16 bit of quantization. The aim is to agglomerate music audio songs by adopting three criteria

1. without pre-processing;
2. applying SM;
3. applying SM and CICA.

In a first experiment, 9 popular songs, as listed in Table 13.1, are considered.

In Fig. 13.4 we report the agglomerations obtained by three criteria. From a simple analysis we deduced that, in all cases, songs with labels 1, 9 and 6 get agglomerated together for their well defined musical content (e.g., rhythm).

Table 13.1 Songs used for the first experiment

Author	Title	Label
AC/DC	Back in Black	1
Nek	Almeno stavolta	2
Led Zeppelin	Stairway to Heaven	3
Louis Armstrong	What a wonderful world	4
Madonna	Like a Virgin	5
Michael Jackson	Billie Jean	6
Queen	The Show Must Go On	7
The Animals	The House of the Rising Sun	8
Sum 41	*Still Waiting*	9

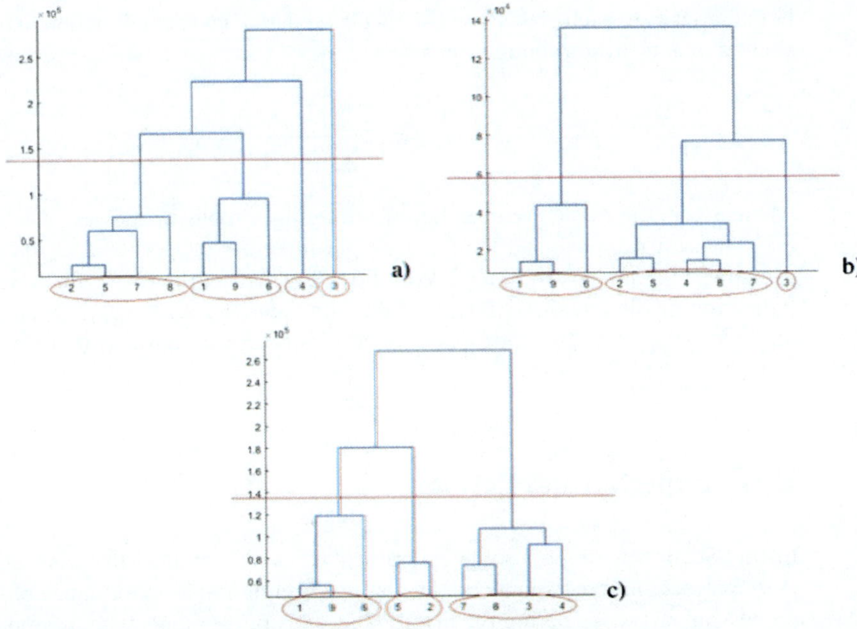

Fig. 13.4 Hierarchical clustering on the dataset of 9 songs applying three criteria: **a** overall song elaboration; **b** sparse modeling; **c** sparse modeling and CICA

Later on, we explored the agglomeration differences considering the musical instruments content. Thus, we inferred the similarity among the musical tracks 3 (without its last part) and 4 (i.e., by SM and CICA) (Fig. 13.4c), due particularly to the rhythmic content and the presence in 3 of a predominant synthesized wind musical instrument, also present as wind musical instruments in 4, both belonging to the same cluster. Moreover, this cluster is closed to another cluster composed by traks 7 and 8, sharing a musical keyboard content.

In the second experiment, we considered 28 musical audio songs of different genres

- 10 *children songs*,
- 10 *classic music*,
- 8 *easy listening* (multi-genre class).

The results are shown in Fig. 13.5. First of all we observed the waveform of song 4 (see Fig. 13.6), showing two different loudnesses. In this case, the SM approach allows to have a more robust estimation. In particular, from Figs. 13.5a (overall song elaboration) and 13.5b (sparse modeling)) we noticed that song number 4 is in a different agglomerated cluster. Moreover, by applying CICA we also obtained the agglomeration of the children and classic songs in two main classes (Fig. 13.5c). The first cluster gets separated in two subclasses, namely classic music and easy

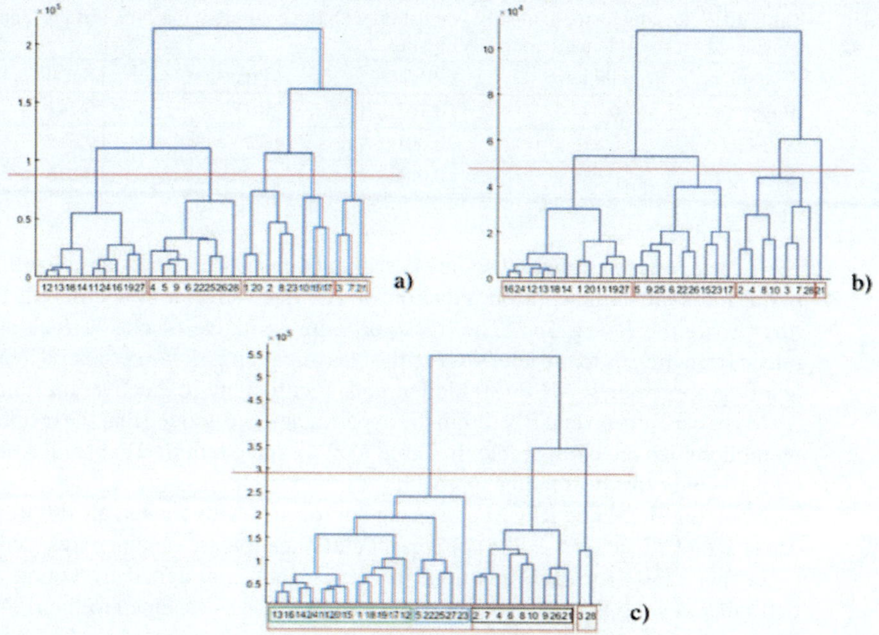

Fig. 13.5 Hierarchical clustering on the dataset of 28 songs applying three criteria: **a** overall song elaboration; **b** sparse modeling; **c** sparse modeling and CICA

Fig. 13.6 Waveform of song 4

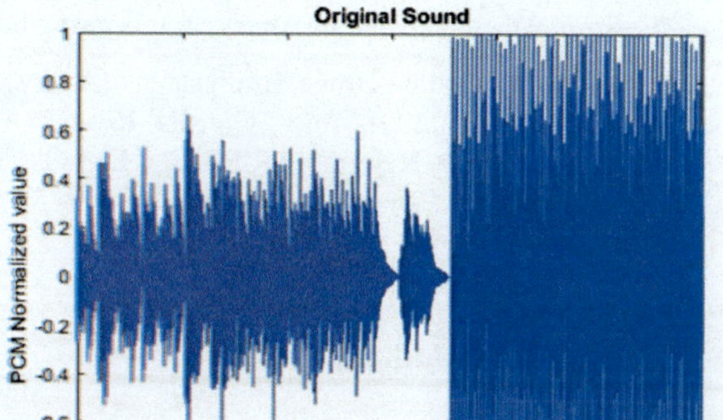

listening. In the second cluster, we find all children songs except songs 1 and 5. The mis-classification of song 1 is due to the instrumental feature of the song (without a singer voice), like a classic song, while song 5, instead, is a children song with an adult man singer voice thus it is classified as easy listening.

Table 13.2 Results for 10-fold cross-validation with three different machine learning approaches considered for the automatic song labeling task

Classifier	TP rate	FP rate	Precision	Recall
Bayes	0.747	0.103	0.77	0.747
SVM	0.815	0.091	0.73	0.815
MLP	0.838	0.089	0.705	0.838

Successively, we reported some experimental results obtained applying the emotional retrieval framework on a dataset of 100 audio tracks of 4 different classes: *Angry, Happy, Relax, Sad.* The tracks are representative of classic rock and pop music from the 70s to the late 90s. For the classification task we compared 3 machine learning approaches: MLP (30 hidden nodes with sigmoidal activation functions), SVM (linear Kernel) and BN. From the experiments, we noticed that the results of the methodologies are comparable. In Table 13.2 we have reported the results obtained by a 10-fold cross-validation approach [32].

Applying FCM and RFCM clustering approaches with a mean on 100 iterations, 71.84% and 77.08% ($A = 0.5$) of perfect classification are obtained, respectively.

In this cases, for each iteration, the class label is assigned by voting and, in particular, a song is considered *perfect classified* if it is assigned to the right class. We stress that in this case the emotional information is suggested by the system and that it may also suggests songs belonging to different classes. In the experiments for one querying song we considered at most one ranked song for the same author. For example we could consider a querying song as "Born in the USA" of "Bruce Springsteen" labeled as Angry. In this case, the first 4 similar songs retrived are:

- "Born to Run—Bruce Springsteen" (Angry)
- "Sweet Child O' Mine—Guns N' Roses" (Angry)
- "Losing My Religion—R.E.M." (Happy)
- "London Calling—The Clash" (Angry).

13.6 Conclusions

In this Chapter we introduced a framework for processing, classification and clustering of songs on the basis of their emotional contents. The main emotional features are extracted after a pre-processing phase where both Sparse Modeling and Independent Component Analysis based methodologies are used. The approach makes it possible to summarize the main sub-tracks of an acoustic music song and to extract the main features from these parts. The musical features took into account were intensity, rhythm, scale, harmony and spectral centroid. The core of the query engine takes in input a target audio song provided by the user and returns a playlist of the most similar songs. A classifier is used to identify the class of the target song, and then the most similar songs belonging to the same class are obtained . This is achieved by using a fuzzy similarity measure based on the Łukasiewicz product. In the case of classification, a playlist is obtained from the songs of the same class. In the other cases, the playlist is suggested by the system by exploiting the content of the audio songs, which could also contain songs of different classes. The obtained results with clustering are not comparable with those obtained with the supervised techniques. However, we stress that in the first case, the playlist is obtained by songs contained in the same class and in the second case the emotional information is suggested by the system. The approach can be considered a real alternative to human based classification systems (i.e., stereomood). In the next future the authors will focus the attention on a greater database of songs, further musical features and the use of semi-supervised approaches. Moreover they will experiment new approaches as the Fuzzy Relational Neural Network [28], that allows to extract automatically memberships and IF-THEN reasoning rules.

Acknowledgements This work was partially funded by the University of Naples Parthenope (*Sostegno alla ricerca individuale per il triennio 2017–2019* project).

References

1. Vinciarelli, A., Pantic, M., Heylen, D., Pelachaud, C., Poggi, I., D'Errico, F.: Marc schroeder. A survey of social signal processing. IEEE Trans. Affect. Comput. Bridging Gap Between Soc. Anim. Unsoc. Mach. (2011)
2. Barrow-Moore, J.L.: The Effects of Music Therapy on the Social Behavior of Children with Autism. Master of Arts in Education College of Education California State University San Marcos, November 2007
3. Blood, A.J., Zatorre, R.J.: Intensely pleasurable responses to music correlate with activity in brain regions implicated in reward and emotion. Proc. Natl. Acad. Sci. **98**(20), 11818–11823 (2001)
4. Jun, S., Rho, S., Han, B.-J., Hwang, E.: A fuzzy inference-based music emotion recognition system. In: 5th International Conference on In Visual Information Engineering—VIE (2008)
5. Koelsch, S., Fritz, T., v. Cramon, D.Y., Müller, K., Friederici, A.D.: Investigating emotion with music: an fMRI study. Hum. Brain Mapp. **27**(3), 239–250 (2006)
6. Ramirez, R., Planas, J., Escude, N., Mercade, J., Farriols, C.: EEG-based analysis of the emotional effect of music therapy on palliative care cancer patients. Front. Psychol. **9**, 254 (2018)
7. Grylls, E., Kinsky, M., Baggott, A., Wabnitz, C., McLellan, A.: Study of the Mozart effect in children with epileptic electroencephalograms. Seizure—Eur. J. Epilepsy **59**, 77–81 (2018)
8. Stereomood Website
9. Lu, L., Liu, D., Zhang, H.-J.: Automatic mood detection and tracking of music audio signals. IEEE Trans. Audiom Speech Lang. Process. **14**(1) (2006)
10. Yang, Y.-H., Liu, C.-C., Chen, H.H.: Music Emotion Classification: a fuzzy approach. Proc. ACM Multimed. **2006**, 81–84 (2006)
11. Ciaramella, A., Vettigli, G.: Machine learning and soft computing methodologies for music emotion recognition. Smart Innov. Syst. Technol. **19**, 427–436 (2013)
12. Iannicelli, M., Nardone, D., Ciaramella, A., Staiano, A.: Content-based music agglomeration by sparse modeling and convolved independent component analysis. Smart Innov. Syst. Technol. **103**, 87–96 (2019)
13. Ciaramella, A., Gianfico, M., Giunta, G.: Compressive sampling and adaptive dictionary learning for the packet loss recovery in audio multimedia streaming. Multimed. Tools Appl. **75**(24), 17375–17392 (2016)

14. Ciaramella, A., De Lauro, E., De Martino, S., Falanga, M., Tagliaferri, R.: ICA based identification of dynamical systems generating synthetic and real world time series. Soft Comput. **10**(7), 587–606 (2006)
15. Thayer, R.E.: The Biopsichology of Mood and Arousal. Oxfrod University Press, New York (1989)
16. Russell, J.A.: A circumplex model of affect. J. Pers. Soc. Psychol. (1980)
17. Revesz, G.: Introduction to the Psychology of Music. Courier Dover Publications (2001)
18. Bello, J.P., Daudet, L., Abdallah, S., Duxbury, C., Davies, M., Sandler, M.B.: Tutorial on onset detection in music signals. IEEE Trans. Speech Audio Process. (2005)
19. Davies, M.E.P., Plumbley, M.D.: Context-dependent beat tracking of musical audio. IEEE Trans. Audio, Speech Lang. Process. **15**(3), 1009–1020 (2007)
20. Noland, K., Sandler, M.: Signal processing parameters for tonality estimation. In: Proceedings of Audio Engineering Society 122nd Convention, Vienna (2007)
21. Grey, J.M., Gordon, J.W.: Perceptual effects of spectral modifications on musical timbres. J. Acoust. Soc. Am. **63**(5), 1493–1500 (1978)
22. Elhamifar, E., Sapiro, G., Vidal, R. See all by looking at a few: sparse modeling for finding representative objects. In: Proceedings of the IEEE Computer Society Conference on Computer Vision and Pattern Recognition, art. no. 6247852, pp. 1600–1607 (2012)
23. Hyvarinen, A., Karhunen, J., Oja, E.: Independent Component Analysis. Wiley, Hoboken, N. J. (2001)
24. Ciaramella, A., De Lauro, E., Falanga, M., Petrosino, S.: Automatic detection of long-period events at Campi Flegrei Caldera (Italy). Geophys. Res. Lett. **38**(18) (2013)
25. Ciaramella, A., De Lauro, E., De Martino, S., Di Lieto, B., Falanga, M., Tagliaferri, R.: Characterization of Strombolian events by using independent component analysis. Nonlinear Process. Geophys. **11**(4), 453–461 (2004)
26. Ciaramella, A., Tagliaferri, R.: Amplitude and permutation indeterminacies in frequency domain convolved ICA. Proc. Int. Joint Conf. Neural Netw. **1**, 708–713 (2003)
27. Duda, R.O., Hart, P.E., Stork, D.G.: Pattern Classification. Wiley-Interscience (2000)
28. Ciaramella, A., Tagliaferri, R., Pedrycz, W., Di Nola, A.: Fuzzy relational neural network. Int. J. Approx. Reason. **41**, 146–163 (2006)
29. Sessa, S., Tagliaferri, R., Longo, G., Ciaramella, A., Staiano, A.: Fuzzy similarities in stars/galaxies classification. In: Proceedings of IEEE International Conference on Systems, Man and Cybernetics, pp. 494–4962 (2003)
30. Turunen, E.: Mathematics behind fuzzy logic. Adv. Soft Comput. Springer (1999)
31. Ciaramella, A., Cocozza, S., Iorio, F., Miele, G., Napolitano, F., Pinelli, M., Raiconi, G., Tagliaferri, R.: Interactive data analysis and clustering of genomic data. Neural Netw. **21**(2–3), 368–378 (2008)
32. Bishop, C.M.: Pattern Recognition and Machine Learning. Springer (2006)
33. Bezdek, J.C.: Pattern Recognition with Fuzzy Objective Function Algorithms. Plenum Press, New York (1981)
34. Wang, D., Wu, M.D.: Rough fuzzy c-means clustering algorithm and its application to image. J. Natl. Univ. Def. Technol. **29**(2), 76–80 (2007)

Chapter 14
Neuro-Kernel-Machine Network Utilizing Deep Learning and Its Application in Predictive Analytics in Smart City Energy Consumption

Miltiadis Alamaniotis

Abstract In the smart cities of the future artificial intelligence (AI) will have a dominant role given that AI will accommodate the utilization of intelligent analytics for prediction of critical parameters pertaining to city operation. In this chapter, a new data analytics paradigm is presented and being applied for energy demand forecasting in smart cities. In particular, the presented paradigm integrates a group of kernel machines by utilizing a deep architecture. The goal of the deep architecture is to exploit the strong capabilities of deep learning utilizing various abstraction levels and subsequently identify patterns of interest in the data. In particular, a deep feedforward deep neural network is employed with every network node to implement a kernel machine. This deep architecture, named neuro-kernel machine network, is subsequently applied for predicting the energy consumption of groups of residents in smart cities. Obtained results exhibit the capability of the presented method to provide adequately accurate predictions despite the form of the energy consumption data.

Keywords Deep learning · Kernel machines · Neural network · Smart cities · Energy consumption · Predictive analytics

14.1 Introduction

Advancements in information and communication technologies have served as the vehicle to move forward and implement the vision of smart and interconnected societies. In the last decade, this vision has been shaped and defined as a "smart city" [28]. A smart city is a fully connected community where the exchange of information aims at improving the operation of the city and improve the daily life of the citizens [18]. In particular, exploitation of information may lead to greener, less polluted and

M. Alamaniotis (✉)
Department of Electrical and Computer Engineering, University of Texas at San Antonio, UTSA Circle, San Antonio, TX 78249, USA
e-mail: miltos.alamaniotis@utsa.edu

G. Phillips-Wren et al. (eds.), *Advances in Data Science: Methodologies and Applications*, Intelligent Systems Reference Library 189,
https://doi.org/10.1007/978-3-030-51870-7_14

more human cities [4, 16]. The latter is of high concern and importance because it is expected that the population of cities will increase in the near future [21].

In general, the notion of smart city may be considered as the assembly of a set of service groups [1]. The coupling of the city services with information technologies have also accommodated the characterization of those groups with the term "smart." In particular, a smart city is comprised of the following service groups: smart energy, smart healthcare, smart traffic, smart farming, smart transportation, smart buildings, smart waste management, and smart mobility [25].

Among those groups, smart energy is of high interest [8, 10]. Energy is the cornerstone of the modern civilization, upon which the modern way of life is built [12]. Thus, it is normal to assume that smart energy is of high priority compared to the rest of the smart city components; in a visual analogy, Fig. 14.1 denotes smart energy as the fundamental component of smart cities [6]. Therefore, the optimization of the distribution and the utilization of electrical energy within the premises of the city is essential to move toward self-sustainable cities.

Energy (load) prediction has been identified as the basis for implementing smart energy services [9]. Accurate prediction of the energy demand promotes the efficient utilization of the energy generation and distribution by making optimal decisions. Those optimal decisions are made by taking into consideration the current state of the energy grid and the anticipated demand [13]. Thus, energy demand prediction accommodates fast and smart decisions with regard the operation of the grid [5]. However,

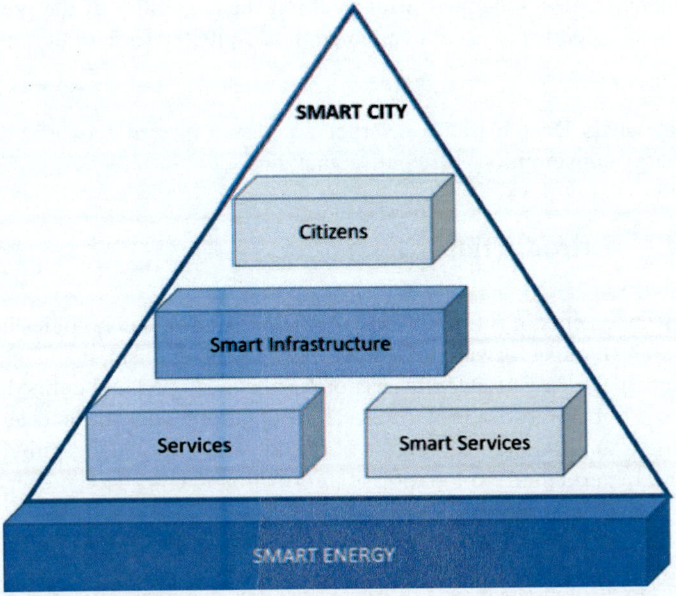

Fig. 14.1 Visualization of a smart city as pyramid with smart energy consist of the fundamental component

the integration of information technologies and the use of smart meters from each consumer has added further uncertainty and volatility in the demand pattern. Hence, intelligent tools are needed that will provide high accurate forecasts [20].

In this chapter, the goal is to introduce a new demand prediction methodology that is applicable to smart cities. The extensive use of information technologies in smart cities, as well as the heterogeneous behavior of consumers even in close geographic vicinity will further complicate the forecasting of the energy demand [27]. Furthermore, predicting the demand of a smart city partition (e.g. a neighborhood) that includes a specific number of consumers will impose high challenges in energy forecasting [5]. For that reason, the new forecasting methodology adopts a set of various kernel machines that are equipped with different kernel functions [7]. In addition, it assembles the kernel machines into a deep neural network architecture that is called *neuro-kernel-machine network* (NKMN)). The goal of the NKMN is to analyze the historical data aiming at capturing the energy consumption behavior of the citizens by using a set of kernel machines—with each machine to model a different set of data properties—[2]. Then, the kernel machines interact via a deep neural network that accommodates the interconnection of kernel machines via a set of weights. This architecture models the "interplay" of the data properties in the hope that the neural driven architecture will identify the best combination of kernel machines that captures the citizens' stochastic energy behavior [11].

The current chapter is organized as follows. In the next section, kernel machines and more specifically the kernel modeled Gaussian processes are presented, while Sect. 14.3 presents the newly developed NKMN architecture. Section 14.4 provides the test results obtained on a set of data obtained from smart meters, whereas Sect. 14.5 concludes the paper and provides its main points.

14.2 Kernel Modeled Gaussian Processes

14.2.1 Kernel Machines

Recent advancements in machine learning and in artificial intelligence in general have boosted the use of intelligent models in several real-world applications. One of the traditional learning models is the kernel machines, which is a set of parametric models that may be used in regression or classification problems [17].

In particular, kernel machines are analytical models that are expressed as a function of a kernel function (a.k.a. kernel) [17], whereas a kernel function is any valid analytical function that is cast into the so-called dual form as given below:

$$k(x_1, x_2) = f(x_1)^T \cdot f(x_2) \tag{14.1}$$

where $f(x)$ is any valid mathematical function known as the basis function, and T denotes its transpose. Therefore, the selection of the basis function determines also the form of the kernel and implicitly models the relation between the two input variables x_1 and x_2. From a data science point of view, the kernel models the similarity between the two parameters, hence allowing the modeler to control the output of the kernel machine. For example, a simple kernel is the linear kernel:

$$k(x_1, x_2) = x_1^T \cdot x_2 \tag{14.2}$$

where the basis function is $f(x) = x$.

Examples of kernel machines are the widely used models of Gaussian processes, support vector machines and kernel regression [17].

14.2.2 Kernel Modeled Gaussian Processes

Any group of random variables whose joint distribution follows a normal distribution is known as a Gaussian process (GP). Though this definition is true in statistical science, in machine learning realm GPs are characterized as member of the kernel machines group. Thus, a GP may be expressed as a function of a kernel as we derive below. The use of GP for regression problems takes the form of Gaussian process regression abbreviated as GPR and is the focal point of this section [31].

To derive the GPR framework as a kernel machine, we start from the simple linear regression model:

$$y(x, w) = w_0 + w_1\varphi_1(x) + \cdots + w_N\varphi_N(x) \tag{14.3}$$

where w_i are the regression coefficients, w_0 is the intercept and N is the number of regressors. Equation (14.2) can be consolidated into a vector form as given below:

$$\mathbf{y} = \boldsymbol{\Phi}\mathbf{w} \tag{14.4}$$

where $\boldsymbol{\Phi}$ and $\mathbf{w}$ contain the basis functions and the weights respectively. In the next step, the weights $\mathbf{w}$ follow a normal distribution with a mean value equal to zero and standard deviation taken as σ_w. Thus, it is obtained:

$$P(\mathbf{w}) = N(0, \sigma_w^2\mathbf{I}) \tag{14.5}$$

with $\mathbf{I}$ being the identity matrix. It should be noted that the selection of mean to be equal to zero is a convenient choice without affecting the derivation of the GPR framework [31].

Driven by Eqs. (14.3) and (14.4), a Gaussian process is obtained whose parameters, i.e., mean and covariance values, are taken by:

$$E[\mathbf{y}] = E[\boldsymbol{\Phi}\mathbf{w}] = \boldsymbol{\Phi}E[\mathbf{w}] = 0 \tag{14.6}$$

$$Cov[\mathbf{y}] = E[\boldsymbol{\Phi}\mathbf{w}\mathbf{w}^T\boldsymbol{\Phi}^T] = \boldsymbol{\Phi}E[\mathbf{w}\mathbf{w}^T]\boldsymbol{\Phi}^T$$
$$= \boldsymbol{\Phi}\sigma_w^2\mathbf{I}\boldsymbol{\Phi}^T = \sigma_w^2\boldsymbol{\Phi}\boldsymbol{\Phi}^T = \mathbf{K} \quad (14.7)$$

where **K** stands for the so-called *Gram matrix* with entries at position *i*, *j* is given by:

$$K_{ij} = k(x_i, x_j) = \sigma_w^2\varphi^T(x_i)\varphi(x_j) \quad (14.8)$$

and thus, the Gaussian process is expressed as:

$$P(\mathbf{y}) = N(0, \mathbf{K}) \quad (14.9)$$

However, in practice the observed values consist of the aggregation of the target value with some noise:

$$t_n = y(x) + \varepsilon_n \quad (14.10)$$

with ε_n being random noise following a normal distribution:

$$\varepsilon_n \sim N(0, \sigma_n^2) \quad (14.11)$$

where σ_n^2 denotes the variance of the noise [31]. By using Eqs. (14.9) and (14.10), we conclude that the prior distribution over targets t_n also follow a normal distribution (in vector form):

$$P(\mathbf{t}) = N(0, \mathbf{K} + \sigma_n^2\mathbf{I}) = N(0, \mathbf{C}) \quad (14.12)$$

where **C** is the covariance matrix whose entries are given by:

$$C_{ij} = k(\mathrm{x}_i,\ \mathrm{x}_j) + \sigma_n^2\delta_{ij} \quad (14.13)$$

in which δ_{km} denotes the Dirac function, and $k(x_i, x_j)$ is a valid kernel. Assuming that there exist *N* known data points, then their joint distribution with an unknown datapoint *N* + 1 denoted as $P(t_{N+1}, \mathbf{t}_N)$ is Normal [32]. Therefore, the predictive distribution of t_{N+1} at $\mathbf{x}_{N+1}$ is follows a Normal distribution [31].

Next, the covariance matrix $\mathbf{C}_{N+1}$ of the predictive distribution $P(t_{N+1}, \mathbf{t}_N)$ is subdivided into four blocks as shown below:

$$\mathbf{C}_{N+1} = \begin{bmatrix} [\mathbf{C}_N] & [\mathbf{k}] \\ [\mathbf{k}^T] & [k] \end{bmatrix} \quad (14.14)$$

where $\mathbf{C}_N$ is an *NxN* covariance matrix of the *N* known datapoints, **k** is an *Nx1* vector with entries computed by $k(\mathbf{x}_m, \mathbf{x}_{N+1})$, $m = 1,\dots, N$, and *k* is a scalar equal to $k(\mathbf{x}_{N+1}, \mathbf{x}_{N+1}) + \sigma_n^2$ [31]. By using the subdivision in 14.13 it has been shown that the predictive distribution is also a Normal distribution whose main parameters, i.e., mean and covariance functions are respectively obtained by:

$$m(\mathbf{x}_{N+1}) = \mathbf{k}^T \mathbf{C}_N^{-1} \mathbf{t}_N \quad (14.15)$$

$$\sigma^2(\mathbf{x}_{N+1}) = k - \mathbf{k}^T \mathbf{C}_N^{-1} \mathbf{k} \quad (14.16)$$

where the dependence of both the mean and covariance functions on the selected kernel is apparent [32].

Overall, the form of Eqs. (14.14) and (14.15) imply that the modeler can control the output of the predictive distribution by selecting the form of the kernel [14, 31].

14.3 Neuro-Kernel-Machine-Network

In this section the newly developed network for conducting predictive analytics is presented [30]. The developed network implements a deep learning approach [22, 26] in order to learn the historic consumption patterns of city citizens and subsequently provide a prediction of energy over a predetermined time interval [3].

The idea behind developing the NKMN is the adoption of kernel machines as the nodes of the neural network [23]. In particular, a deep architecture is adopted that is comprised of one input, L hidden (with L being larger than 3) and one output layer as shown in Fig. 14.2. Notably, the #L hidden layers are comprised of three nodes each,

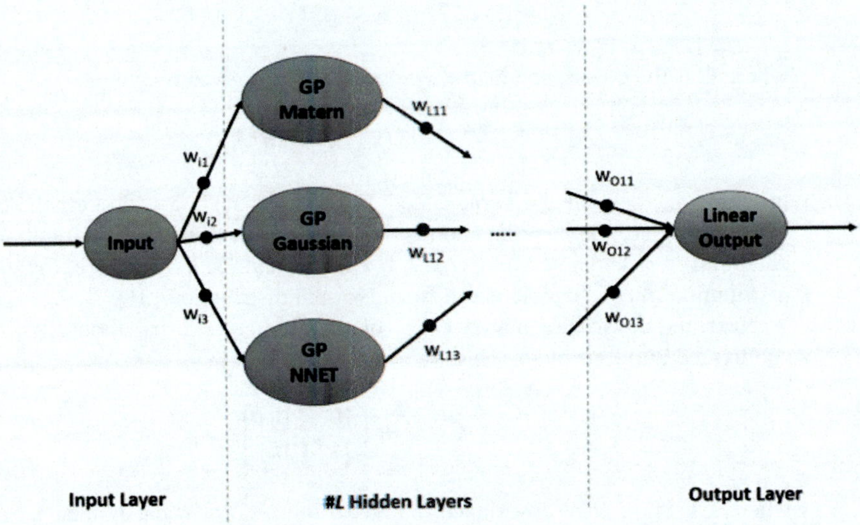

Fig. 14.2 Deep neural network architecture of NKMN

with the nodes implementing a GP equipped with a different kernel function. The input layer is not a computing layer and hence, does not perform any information processing; it only forwards the input to the hidden layers [29]. The last layer, i.e. the output, implements a linear function of the inputs coming from the last hidden layer. The presented deep network architecture is a feedforward network with a set of weights connecting the previous layer to the next one [24].

With regard to the #*L* hidden layers, it is observed that each layer has a specific structure: every hidden layer consists of three nodes (three GP as it was mentioned above). The hidden nodes are GP equipped with the i) *Matern*, ii) *Gaussian*, and iii) *Neural Net* kernel [31]. The analytical forms of those kernels are given below:

Matérn Kernel

$$k(x_1, x_2) = \left(2^{1-\theta_1}/\Gamma(\theta_1)\right)\left[\sqrt{2\theta_1}|x_1 - x_2| \Big/ \theta_2\right]^{\theta_1} K_{\theta_1}\left(\sqrt{2\theta_1}|x_1 - x_2| \Big/ \theta_2\right) \tag{14.17}$$

where θ_1, θ_2 are two positive valued parameters; in the present work, θ_1 is taken equal to 3/2 (see [31] for details), whereas $K_{\theta 1}()$ is a modified Bessel function.

Gaussian Kernel

$$k(\mathbf{x}_1, \mathbf{x}_2) = \exp\left(-\frac{\|\mathbf{x}_1 - \mathbf{x}_2\|^2}{2\sigma^2}\right) \tag{14.18}$$

where σ is an adjustable parameter evaluated during the training process [31].

Neural Net Kernel

$$k(x_1, x_2) = \theta_0 \sin^{-1}\left(\frac{2\tilde{x}_1^T \Sigma \tilde{x}_2}{\sqrt{\left(1 + 2\tilde{x}_1^T \Sigma \tilde{x}_1\right)\left(1 + 2\tilde{x}_2^T \Sigma \tilde{x}_2\right)}}\right) \tag{14.19}$$

where is an augmented input vector [31], Σ is the covariance matrix of the *N* input datapoints and θ_0 is a scale parameter [17, 31].

With regard to the output layer, there is a single node that implements a linear function as is shown in Fig. 14.3 [29]. In particular, the output layer gets as input the three values coming from the preceding hidden layer. Notably, the three inputs denoted as h1, h2 and h3, are being multiplied with the respective weights denoted as w_{o11}, w_{o12} and w_{o13} and subsequently the weighted inputs are added to form the sum **S** (as depicted in Fig. 14.3). The sum S is forwarded to the linear activation function, that provides the final output of the node that is equal to **S** [29].

At this point, a more detailed description of the structure of the hidden layers is given. It should be emphasized that the goal of the hidden layer is to model via data properties the energy consumption behavior of the smart city consumers. In order to approach that, the following idea has been adopted: *Each hidden layer represents a unique citizen* (see Fig. 14.4). To make it clearer, the nodes within the hidden layer (i.e., the three GPs models) are trained using the same training data aiming at

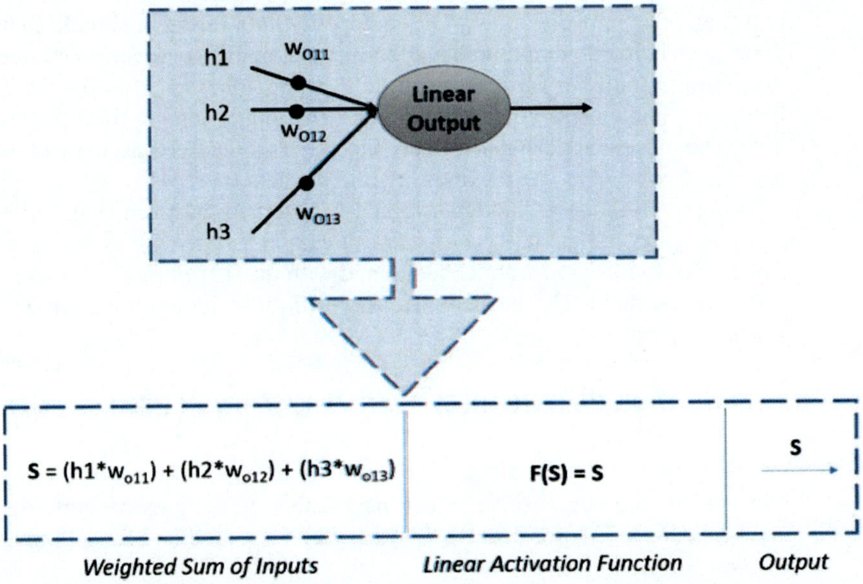

Fig. 14.3 Output layer structure of the NKMN

training three different demand behaviors for each citizen. Thus, the training data for each node contains historical demand patterns of each citizen. Overall it should be emphasized that each node is trained separately (1st stage of training in Fig. 14.5).

Then, the citizens are connected to each other via the hidden layer weights. The role of weights is to express the degree of realization of the specific behavior of the citizen in the overall city demand [15]. The underlying idea is that in smart cities the overall demand is a result of the interactive demands of the various citizens since they do have the opportunity to exchange information and morph their final demand [3, 8].

The training of the presented NKMN is performed as follows. In the first stage the training set of each citizen is put together and subsequently the nodes of each the respective hidden layer are trained. Once the node training is completed, then a training set of city demand data is put together (denoted as "city demand data" in Fig. 14.5). This newly formed training set consists of the historical demand patterns of the city (or partition of the city)—reflects the final demand and the interactions among the citizens-. The training is performed using the backpropagation algorithm.

Overall, the 2-stage process utilized for training the NKMN is comprised of two supervised learning stages: the first at the individual node level, and the second one at the overall deep neural network level. To make it clearer, the individual citizen historical data are utilized for evaluation of the GP parameters at each hidden layer while the aggregated data of the participating citizens are utilized to evaluate the parameters of the network.

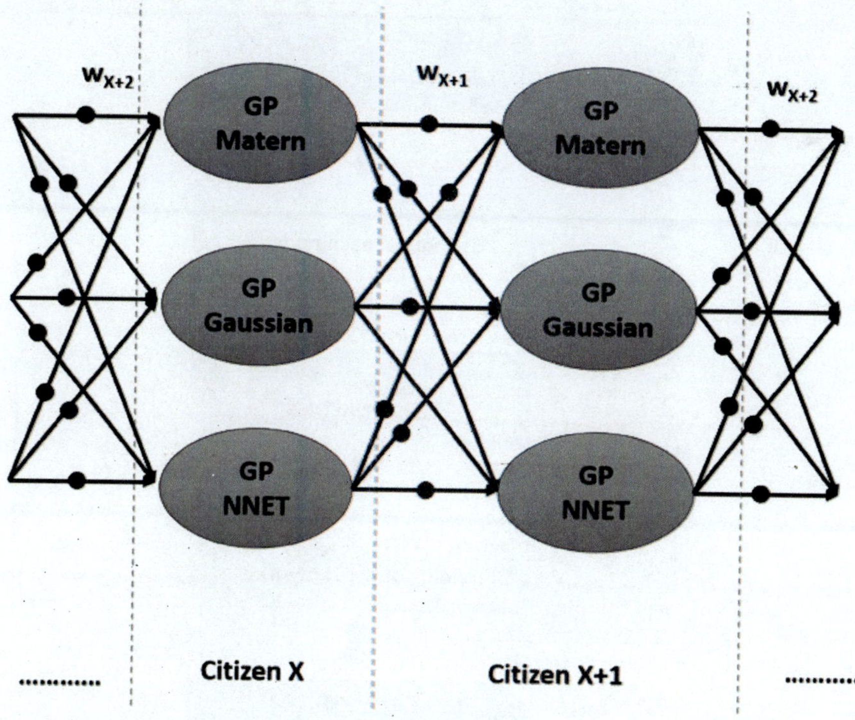

Fig. 14.4 Visualization of a hidden layer as a single citizen/consumer

Finally, once the training of the network has been completed, then the NKMN is able to make prediction over the demand of that specific group of citizens as shown at the bottom of Fig. 14.5. Notably the group might a neighborhood of 2–20 citizens or larger areas with thousands of citizens. It is anticipated in the latter case that the training process will last for long time.

14.4 Testing and Results

The presented neural-kernel-machine network for predictive analytics is applied in a set of real-world data taken from the state of the Ireland [19]. The test data contains energy demand patterns measured with smart meters for various dates. The data express the hourly electricity consumption of the respective citizens.

In order to test the presented method, a number of 10 citizens is selected (i.e., $L = 10$) and therefore the NKMN is comprised of 12 layers (1 input, 10 hidden and 1 output). The input layer is comprised of a single node and takes as input the time for which a prediction is requested, while the output is the energy demand in kW.

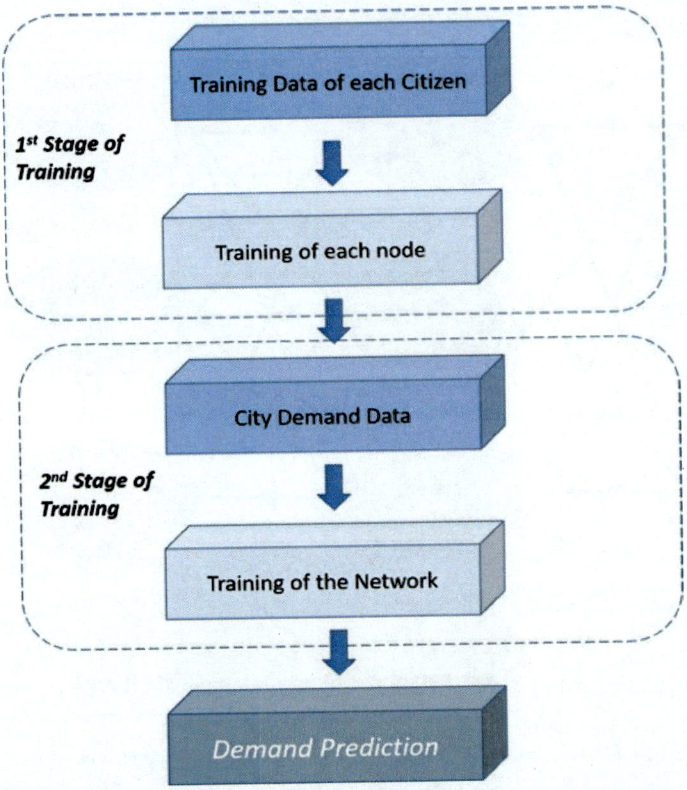

Fig. 14.5 The 2-stage training process of the deep NKMN

The training dataset for training sets for both the GP and the overall NKMN is composed as shown in Table 14.1. In particular, there are two types of training sets: one for weekdays that is comprised of all the hourly data from one day, two days, three days and the day week before the targeted day. The second set refers to weekends and comprised of hourly data from the respective day one week, two weeks and three weeks before the targeted day. The IDs of the 10 smart meters selected

Table 14.1 Composition of training sets

Hourly energy demand values	
Weekdays	Weekend
One day before	One week before
Two days before	Two weeks before
Three days before	Three weeks before
One week before	** Only for the overall NKMN training (stage 2) Morphing based on [3]

Table 14.2 Test results with respect to MAPE

Mean average percentage error	
Day	MAPE
Monday	9.96
Tuesday	8.42
Wednesday	6.78
Thursday	9.43
Friday	8.01
Saturday	10.01
Sunday	10.43

for testing were: 1392, 1625, 1783, 1310, 1005, 1561, 1451, 1196, 1623 and 1219. The days selected for testing was the week including the days 200–207 (based on the documentation of the dataset). In addition, the datasets for the weight training of the NKMN have been morphed using the method proposed in [3] given that this is a method that introduces interactions among the citizens.

The obtained results, which are recorded in terms of the Mean Average Percentage Error (MAPE), are depicted in Table 14.2. In particular, the MAPE lies within the area of 6 and 10.5. This shows that the proposed methodology is accurate in predicting the behavior of those 10 citizens. It should be noted that the accuracy of the weekdays is higher than those taken for the weekend days. This is something expected given that the training dataset contains data closer to the targeted days as opposed to weekends. Therefore, the weekday training data was able to capture the most recent dynamics of the citizen interactions, while those interactions were less successfully captured in the weekends (note: the difference is not high but it still exists.

For visualization purposes, the actual against the predicted demand for days Monday and Saturday are given in Figs. 14.6 and 14.7 respectively. Inspection of those Figs. clearly shows that the predicted curve is close to the actual one.

14.5 Conclusion

In this chapter a new deep architecture for data analytics applied to smart cities operation is presented. In particular, a deep feedforward neural network is introduced where the nodes of the network are implemented by kernel machines. Getting into more details the deep network is comprised of a single input layer, L hidden and a single output layer. The number of hidden layers is equal to the number of citizens participating in the shaping of the energy demand under study. The aim of the deep learning architecture is to model the energy (load) behavior and the interactions among the citizens that affect the overall demand shaping. In order to capture citizen behavior each hidden layer is comprised of three different nodes with each node implementing a kernel based Gaussian process with different kernel, namely, the

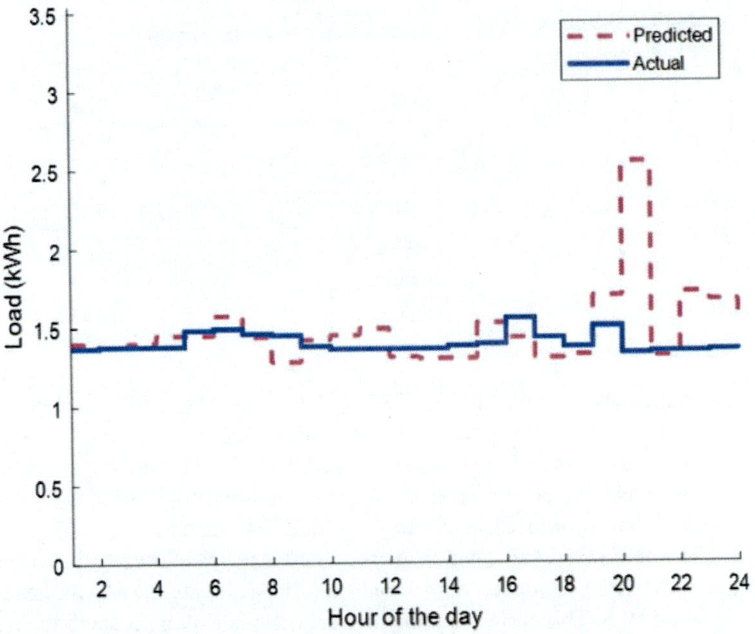

Fig. 14.6 Predicted with NKMN against actual demand for the tested day monday

Matérn, Gaussian and Neural Net kernel. The three nodes of each layer are trained on the same dataset that contains historical demand patterns of the respective citizen. The interactions among the citizens are modeled in the form of the neural network weights.

With the above deep learning architecture, we are able to capture the new dynamics in the energy demand that emerge from the introduction of smart cities technologies. Therefore, the proposed method is applicable to smart cities, and more specifically to partitions (or subgroups) within the smart city. The proposed method was tested on a set of real-world data that were morphed [3] obtained from a set of smart meters deployed in the state of Ireland. Results exhibited that the presented deep learning architecture has the potency to analyze the past behavior of the citizens and provide high accurate group demand predictions.

Future work will move into two directions. The first direction would be to test the presented method in a higher number of citizens, whereas the second direction will move toward testing various kernel machines except for GP as the network nodes.

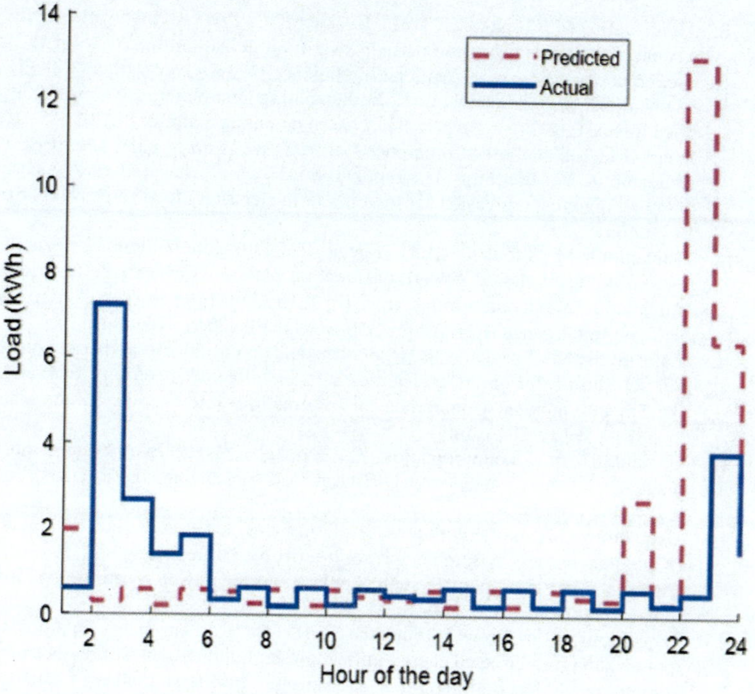

Fig. 14.7 Predicted with NKMN against actual demand for the tested day saturday

References

1. Al-Hader, M., Rodzi, A., Sharif, A.R., Ahmad, N.: Smart city components architicture. In: 2009 International Conference on Computational Intelligence, Modelling and Simulation, pp. 93–97. IEEE (2009, September)
2. Alamaniotis, M.: Multi-kernel Analysis Paradigm Implementing the Learning from Loads. Mach. Learn. Paradigms Appl. Learn. Analytics Intell. Syst. **131** (2019)
3. Alamaniotis, M., Gatsis, N.: Evolutionary multi-objective cost and privacy driven load morphing in smart electricity grid partition. Energies **12**(13), 2470 (2019)
4. Alamaniotis, M., Bourbakis, N., Tsoukalas, L.H.: Enhancing privacy of electricity consumption in smart cities through morphing of anticipated demand pattern utilizing self-elasticity and genetic algorithms. Sustain. Cities Soc. **46**, 101426 (2019)
5. Alamaniotis, M., Gatsis, N., Tsoukalas, L.H.: Virtual Budget: Integration of electricity load and price anticipation for load morphing in price-directed energy utilization. Electr. Power Syst. Res. **158**, 284–296 (2018)
6. Alamaniotis, M., Tsoukalas, L.H., Bourbakis, N.: Anticipatory driven nodal electricity load morphing in smart cities enhancing consumption privacy. In 2017 IEEE Manchester PowerTech, pp. 1–6. IEEE (2017, June)
7. Alamaniotis, M., Tsoukalas, L.H.: Multi-kernel assimilation for prediction intervals in nodal short term load forecasting. In: 2017 19th International Conference on Intelligent System Application to Power Systems (ISAP), pp. 1–6. IEEE, (2017)

8. Alamaniotis, M., Tsoukalas, L.H., Buckner, M.: Privacy-driven electricity group demand response in smart cities using particle swarm optimization. In: 2016 IEEE 28th International Conference on Tools with Artificial Intelligence (ICTAI), pp. 946–953. IEEE, (2016a)
9. Alamaniotis, M., Tsoukalas, L.H.: Implementing smart energy systems: Integrating load and price forecasting for single parameter based demand response. In: 2016 IEEE PES Innovative Smart Grid Technologies Conference Europe (ISGT-Europe), pp. 1–6. IEEE (2016, October)
10. Alamaniotis, M., Bargiotas, D., Tsoukalas, L.H.: Towards smart energy systems: application of kernel machine regression for medium term electricity load forecasting. SpringerPlus **5**(1), 58 (2016b)
11. Alamaniotis, M., Tsoukalas, L.H., Fevgas, A., Tsompanopoulou, P., Bozanis, P.: Multiobjective unfolding of shared power consumption pattern using genetic algorithm for estimating individual usage in smart cities. In: 2015 IEEE 27th International Conference on Tools with Artificial Intelligence (ICTAI), pp. 398–404. IEEE (2015, November)
12. Alamaniotis, M., Tsoukalas, L.H., Bourbakis, N.: Virtual cost approach: electricity consumption scheduling for smart grids/cities in price-directed electricity markets. In: IISA 2014, The 5th International Conference on Information, Intelligence, Systems and Applications, pp. 38–43. IEEE (2014, July)
13. Alamaniotis, M., Ikonomopoulos, A., Tsoukalas, L.H.: Evolutionary multiobjective optimization of kernel-based very-short-term load forecasting. IEEE Trans. Power Syst. **27**(3), 1477–1484 (2012)
14. Alamaniotis, M., Ikonomopoulos, A., Tsoukalas, L.H.: A Pareto optimization approach of a Gaussian process ensemble for short-term load forecasting. In: 2011 16th International Conference on Intelligent System Applications to Power Systems, pp. 1–6. IEEE, (2011, September)
15. Alamaniotis, M., Gao, R., Tsoukalas, L.H.: Towards an energy internet: a game-theoretic approach to price-directed energy utilization. In: International Conference on Energy-Efficient Computing and Networking, pp. 3–11. Springer, Berlin, Heidelberg (2010)
16. Belanche, D., Casaló, L.V., Orús, C.: City attachment and use of urban services: benefits for smart cities. Cities **50**, 75–81 (2016)
17. Bishop, C.M.: Pattern Recognition and Machine Learning. springer, (2006)
18. Bourbakis, N., Tsoukalas, L.H., Alamaniotis, M., Gao, R., Kerkman, K.: Demos: a distributed model based on autonomous, intelligent agents with monitoring and anticipatory responses for energy management in smart cities. Int. J. Monit. Surveill. Technol. Res. (IJMSTR) **2**(4), 81–99 (2014)
19. Commission for Energy Regulation (CER).: CER Smart Metering Project—Electricity Customer Behaviour Trial, 2009–2010 [dataset]. 1st (edn.) Irish Social Science Data Archive. SN: 0012-00, (2012). www.ucd.ie/issda/CER-electricity
20. Feinberg, E.A., Genethliou, D.: Load forecasting. In: Applied Mathematics for Restructured Electric Power Systems, pp. 269–285. Springer, Boston, MA (2005)
21. Kraas, F., Aggarwal, S., Coy, M., Mertins, G. (eds.): Megacities: our global urban future. Springer Science & Business Media, (2013)
22. Liu, W., Wang, Z., Liu, X., Zeng, N., Liu, Y., Alsaadi, F.E.: A survey of deep neural network architectures and their applications. Neurocomputing **234**, 11–26 (2017)
23. Mathew, J., Griffin, J., Alamaniotis, M., Kanarachos, S., Fitzpatrick, M.E.: Prediction of welding residual stresses using machine learning: comparison between neural networks and neuro-fuzzy systems. Appl. Soft Comput. **70**, 131–146 (2018)
24. Mohammadi, M., Al-Fuqaha, A.: Enabling cognitive smart cities using big data and machine learning: approaches and challenges. IEEE Commun. Mag. **56**(2), 94–101 (2018)
25. Mohanty, S.P., Choppali, U., Kougianos, E.: Everything you wanted to know about smart cities: the internet of things is the backbone. IEEE Consum. Electron. Mag. **5**(3), 60–70 (2016)
26. Najafabadi, M.M., Villanustre, F., Khoshgoftaar, T.M., Seliya, N., Wald, R., Muharemagic, E.: Deep learning applications and challenges in big data analytics. J. Big Data **2**(1), 1 (2015)
27. Nasiakou, A., Alamaniotis, M., Tsoukala, L.H.: Power distribution network partitioning in big data environment using k-means and fuzzy logic. In: proceedings of the Medpower 2016 Conference, Belgrade, Serbia, pp. 1–7, (2016)

28. Nam, T., Pardo, T.A.: Conceptualizing smart city with dimensions of technology, people, and institutions. In: Proceedings of the 12th Annual International Digital Government Research Conference: Digital Government Innovation in Challenging Times, pp. 282–291. ACM, (2011)
29. Tsoukalas, L.H., Uhrig, R.E.: Fuzzy and Neural Approaches in Engineering, p. 1997. Wiley. Inc, New York (1997)
30. Waller, M.A., Fawcett, S.E.: Data science, predictive analytics, and big data: a revolution that will transform supply chain design and management. J. Bus. Logistics **34**(2), 77–84 (2013)
31. Williams, C.K., Rasmussen, C.E.: Gaussian processes for machine learning, vol. 2(3), p. 4. Cambridge, MA, MIT press, (2006)
32. Williams, C.K., Rasmussen, C.E.: Gaussian processes for regression. In: Advances in Neural Information Processing Systems, pp. 514–520, (1996)

Chapter 15
Learning Approaches for Facial Expression Recognition in Ageing Adults: A Comparative Study

Andrea Caroppo, Alessandro Leone, and Pietro Siciliano

Abstract Average life expectancy has increased steadily in recent decades. This phenomenon, considered together with aging of the population, will inevitably produce in the next years deep social changes that lead to the need of innovative services for elderly people, focused to improve the wellbeing and the quality of life. In this context many potential applications would benefit from the ability of automatically recognize facial expression with the purpose to reflect the mood, the emotions and also mental activities of an observed subject. Although facial expression recognition (FER) is widely investigated by many recent scientific works, it still remains a challenging task for a number of important factors among which one of the most discriminating is the age. In the present work an optimized Convolutional Neural Network (CNN) architecture is proposed and evaluated on two benchmark datasets (FACES and Lifespan) containing expressions performed also by aging adults. As baseline, and with the aim of making a comparison, two traditional machine learning approaches based on handcrafted features extraction process are evaluated on the same datasets. Experimentation confirms the efficiency of the proposed CNN architecture with an average recognition rate higher than 93.6% for expressions performed by ageing adults when a proper set of CNN parameters was used. Moreover, the experimentation stage showed that the deep learning approach significantly improves the baseline approaches considered, and the most noticeable improvement was obtained when considering facial expressions of ageing adults.

C. Andrea (✉) · L. Alessandro · S. Pietro
National Research Council of Italy, Institute for Microelectronics and Microsystems, Via Monteroni c/o Campus Universitario Ecotekne-Palazzina A3, 73100 Lecce, Italy
e-mail: andrea.caroppo@cnr.it

L. Alessandro
e-mail: alessandro.leone@cnr.it

S. Pietro
e-mail: pietroaleardo.siciliano@cnr.it

G. Phillips-Wren et al. (eds.), *Advances in Data Science: Methodologies and Applications*, Intelligent Systems Reference Library 189,
https://doi.org/10.1007/978-3-030-51870-7_15

15.1 Introduction

The constant increase of the life expectancy and the consequent aging phenomenon will inevitably produce in the next 20 years deep social changes that lead to the need of innovative services for elderly people, focused to maintain independence, autonomy and, in general, improve the wellbeing and the quality of life of ageing adults [1]. It is obvious how in this context many potential applications, such as robotics, communications, security, medical and assistive technology, would benefit from the ability of automatically recognize facial expression [2–4], because different facial expressions can reflect the mood, the emotions and also mental activities of an observed subject.

Facial expression recognition (FER) is related to systems that aims to automatically analyse the facial movements and facial features changes of visual information to recognize a facial expression. It is important to mention that FER is different from emotion recognition. The emotion recognition requires a higher level of knowledge. Despite the facial expression could indicate an emotion, to the analysis of the emotion information like context, body gesture, voice, cultural factors are also necessary [5]. A classical automatic facial expression analysis usually employs three main stages: face acquisition, facial data extraction and representation (feature extraction), and classification. Ekman's initial research [6] determined that there were six basic classes in FER: anger, disgust, fear, happiness, sadness and surprise.

Proposed solutions for the classification of aforementioned facial expressions can be divided into two main categories: the first category includes the solutions that perform the classification by processing a set of consecutive images while, the second one, includes the approaches which carry out FER on each single image.

By working on image sequences much more information is available for the analysis. Usually, the neutral expression is used as a reference and some characteristics of facial traits are tracked over time in order to recognize the evolving expression. The major drawback of these approaches is the inherent assumption that the sequence content evolves from the neutral expression to another one that has to be recognized. This constrain strongly limits their use in real world applications where the evolution of facial expressions is completely unpredictable. For this reason, the most attractive solutions are those performing facial expression recognition on a single image.

For static images various types of features might be used for the design of a FER system. Generally, they are divided into the following categories: geometric-based, appearance-based and hybrid-based approaches. More specifically, geometric-based features are able to depict the shape and locations of facial components such as mouth, nose, eyes and brows using the geometric relationships between facial points to extract facial features. Three typical geometric feature-based extraction methods are active shape models (ASM) [7], active appearance models (AAM) [8] and scale-invariant feature transform (SIFT) [9]. Appearance-based descriptors aim to use the whole-face or specific regions in a face image to reflect the underlying information in a face image. There are mainly three representative appearance-based feature extraction methods, i.e. Gabor Wavelet representation [10], Local Binary

Patterns (LBP) [11] and Histogram of Oriented Gradient (HOG) [12]. Hybrid-based approaches combine the two previous features type in order to enhance the system's performance and it might be achieved either in features extraction or classification level.

Geometric-based, appearance-based and hybrid-based approaches have been widely used for the classification of facial expressions even if it is important to emphasize how all the aforementioned methodologies require a process of feature definition and extraction very daunting. Extracting geometric or appearance-based features usually requests an accurate feature point detection technique and generally this is difficult to implement in real-world complex background. In addition, this category of methodologies easily ignore the changes in skin texture such as wrinkles and furrows that are usually accentuated by the age of the subject. Moreover, the task often expects the development and subsequent analysis of complex models with a further process of fine-tuning of several parameters, which nonetheless can show large variances depending on individual characteristics of the subject that performs facial expressions. Last but not least recent studies have pointed out that classical approaches used for the classification of facial expression are not performing well when used in real contexts where face pose and lighting conditions are broadly different from the ideal ones used to capture the face images within the benchmark datasets.

Among the factor that make FER very difficult, one of the most discriminating is the age [13, 14]. In particular, expressions of older individuals appeared harder to decode, owing to age-related structural changes in the face which supports the notion that the wrinkles and folds in older faces actually resemble emotions. Consequently, state of the art approaches based on handcrafted features extraction may be inadequate for the classification of FER performed by aging adults.

It seems therefore very important to analyse automatic systems that make the recognition of facial expressions of the ageing adults more efficient, considering that facial expressions of elderly, as highlighted above, are broadly different from those of young or middle-aged for a number of reasons. For example, in [15] researchers found that the expressions of aging adults (women in this case) were more telegraphic in the sense that their expressive behaviours tended to involve fewer regions of the face, and yet more complex in that they used blended or mixed expressions when recounting emotional events. These changes, in part, account for why the facial expressions of ageing adults are more difficult to read. Another study showed that when emotional memories were prompted and subjects asked to relate their experiences, ageing adults were more facially expressive in terms of the frequency of emotional expressions than younger individuals across a range of emotions, as detected by an objective facial affect coding system [16]. One of the other changes that comes with age, making an aging facial expression difficult to recognize, involves the wrinkling of the facial skin and the sag of facial musculature. Of course, part of this is due to biologically based aspects of aging, but individual differences also appear linked to personality process, as demonstrated in [17].

To the best of our knowledge, only few works in literature address the problem of FER in aging adults. In [13] the authors perform a computational study within and across different age groups and compare the FER accuracies, founding that the recognition rate is influenced significantly by human aging. The major issue of this work is related to the feature extraction step, in fact they manually labelled the facial fiducial points and, given these points, Gabor filters are used to extract features for subsequent FER. Consequently, this process is inapplicable in the application context under consideration, where the objective is to provide new technologies able to function automatically and without human intervention.

On the other hand, the application described in [18] recognizes emotions of ageing adults using an Active Shape Model [7] for feature extraction. To train the model the authors employ three benchmark datasets that do not contain adult faces getting an average accuracy of 82.7% on the same datasets. Tests performed on older faces acquired with the webcam reached an average accuracy of 79.2%, without any verification of how the approach works for example on a benchmark dataset with older faces.

Analysing the results achieved it seems appropriate to investigate new methodologies which must make the feature extraction process less difficult, while at the same time strengthening the classification of facial expressions.

Recently, a viable alternative to the traditional feature design approaches is represented by deep learning (DL) algorithms which straightforwardly leads to automated feature learning [19]. Research using DL techniques could make better representations and create innovative models to learn these representations from unlabelled data. These approaches became computationally feasible thanks to the availability of powerful GPU processors, allowing high-performance numerical computation in graphics cards. Some of the DL techniques like Convolutional Neural Networks (CNNs), Deep Boltzmann Machine, Deep Belief Networks and Stacked Auto-Encoders are applied to practical applications like pattern analysis, audio recognition, computer vision and image recognition where they produce challenging results on various tasks [20].

It comes as no surprise that CNNs, for example, have worked very well for FER, as evidenced by their use in a number of state-of-the-art algorithms for this task [21–23], as well as winning related competitions [24], particularly previous years' EmotiW challenge [25, 26]. The problem with CNNs is that this kind of neural network has a very high number of parameters and moreover achieves better accuracy with big data. Because of that, it is prone to overfitting if the training is performed on a small sized dataset. Another not negligible problem is that there are no publicly available datasets with sufficient data for facial expression recognition with deep architectures.

In this paper, an automatic FER approach that employs a supervised machine learning technique derived from DL is introduced and compared with two traditional approaches selected among the most promising ones and effective present in the literature. Indeed, a CNN inspired from a popular architecture proposed in [27] was designed and implemented. Moreover, in order to tackle the problem of the overfitting, this work proposes also in the pre-processing step, standard methods for data generation in synthetic way (techniques indicated in the literature as "data augmentation") to cope with the limitation inherent the amount of data.

The structure of the paper is as follows. Section 15.2 reports some details about the implemented pipeline for FER in ageing adults, emphasizing theoretical details for pre-processing steps. The same section describes also the implemented CNN architecture and both traditional machine learning approaches used for comparison. Section 15.3 presents the results obtained, while discussion and conclusion are summarized in Sect. 15.4.

15.2 Methods

Figure 15.1 shows the structure of our FER system. First, the implemented pipeline performs a pre-processing task on the input images (data augmentation, face detection, cropping and down sampling, normalization). Once the images are preprocessed they can be either used to train the implemented deep network or to extract handcrafted features (both geometric and appearance-based).

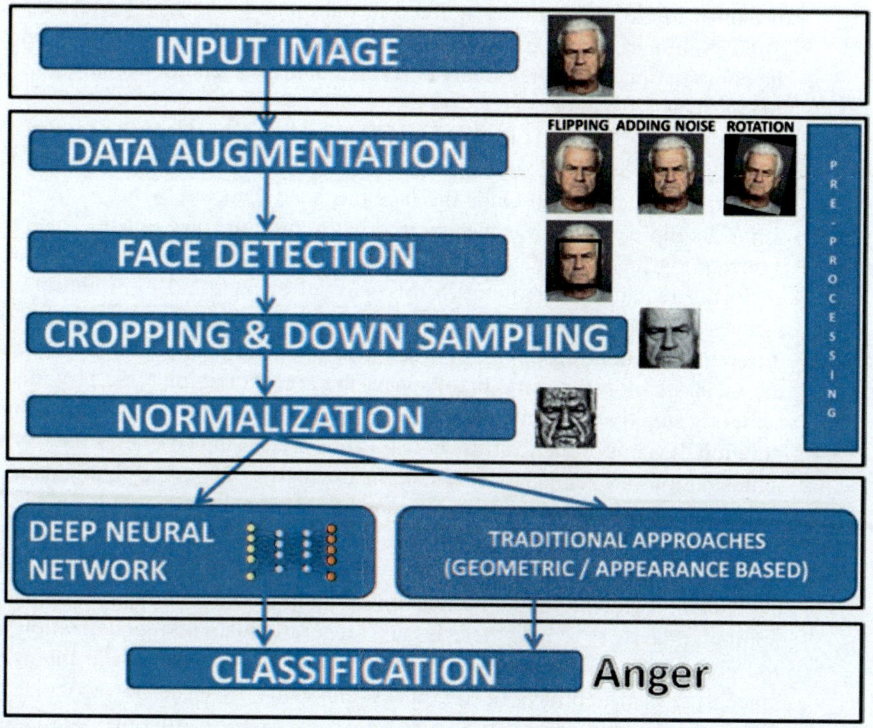

Fig. 15.1 Pipeline of the proposed system. First a pre-processing task on the input images was performed. The obtained normalized face image is used to train the deep neural network architecture. Moreover, both geometrical and appearance-based features are extracted from normalized image. Finally, each image is classified associating it with a label of most probably facial expression

15.2.1 Pre-processing

Here are some details about the blocks that perform the pre-processing algorithmic procedure, whereas the next sub-sections illustrate the theoretical details of the DL methodology and the two classical machine learning approaches used for comparison. It is well known that one of the main problems of deep learning methods is that they need a lot of data in the training phase to perform this task properly.

In the present work the problem is accentuated from having very few datasets containing images of facial expressions performed by ageing subjects. So before training the CNN model, we need to augment the data with various transformations for generate various small changes in appearances and poses.

The number of available images has been increased with three data augmentation strategies. The first strategy is to use flip augmentation, mirroring images about the *y-axis* producing two-samples from each image. The second strategy is to change the lighting condition of the images. In this work lighting condition is varied by adding Gaussian noise in the available face images. The last strategy consists in rotating the images of a specific angle. Consequently each facial image has been rotated through 7 angles randomly generated in the range $[-30°; 30°]$ with respect to the *y-axis*. Summarizing, starting from each image present in the datasets, and through the combination of the previously described data augmentation techniques, 32 facial images have been generated.

The next step consists in the automatic detection of the facial region. Here, the facial region is automatically identified on the original image by means of the Viola-Jones face detector [28]. Once the face has been detected by the Viola-Jones algorithm, a simple routine was written in order to crop the face image. This is achieved by detecting the coordinates of the top-left corner, the height and width of the face enclosing rectangle, removing in this way all background information and image patches that are not related to the expression. Since the facial region could be of different sizes after cropping, in order to remove the variation in face size and keep the facial parts in the same pixel space, the algorithmic pipeline provides a down-sampling step that generates face images with a fixed dimension using a linear interpolation. It is important to stress how this pre-processing task helps the CNN to learn which regions are related to each specific expression. Next, the obtained cropped and down-sampled RGB face image is converted into grayscale by eliminating the hue and saturation information while retaining the luminance. Finally, since the image brightness and contrast could vary even in images that represent the same facial expression performed by the same subject, an intensity normalization procedure was applied in order to reduce these issues. Generally histogram equalization is applied to enhance the contrast of the image by transforming the image intensity values since images which have been contrast enhanced are easier to recognize and classify. However, the noise can also be amplified by the histogram equalization when enhancing the contrast of the image through a transformation of its intensity value since a number of pixels fall inside the same gray level range. Therefore, instead of applying the histogram equalization, in this work the method introduced in [29] called “contrast limited adaptive histogram equalization” (CLAHE) was used. This algorithm is an improvement of the histogram equalization algorithm and essentially consists in the division of the original image into contextual regions, where histogram equalization was made on each of these sub regions. These sub regions are called tiles. The neighboring tiles are combing by using a bilinear interpolation to eliminate artificially induced boundaries. This could give much better contrast and provide accurate results.

15.2.2 *Optimized CNN Architecture*

CNN is a type of deep learning model for processing data that has a grid pattern, such as images, which is inspired by the organization of animal visual cortex [30] and designed to automatically and adaptively learn spatial hierarchies of features, from low to high-level patterns. CNN is a mathematical construct that is typically composed of three types of layers (or building blocks): convolution, pooling, and fully connected layers.

The first two, convolution and pooling layers, perform feature extraction, whereas the third, a fully connected layer, maps the extracted features into final output, such as classification. A typical implementation of CNN for FER encloses three learning stages in just one framework. The learning stages are: (1) feature learning, (2) feature selection and (3) classifier construction. Moreover, two main phases are provided: training and test. During training, the network acquires grayscale facial images (the normalized image output of pre-processing step), together with the respective expression labels, and learns a set of weights.

The process of optimizing parameters (i.e. training) is performed with the purpose to minimize the difference between outputs and ground truth labels through an optimization algorithm. Generally the order of presentation of the facial images can influence the classification performance. Consequently to avoid this problem. usually a group of images is selected and separated for a validation procedure, useful to choose the final best set of weights out of a set of trainings performed with samples presented in different orders. After, in the test step, the architecture receives a gray-scale image of a face and outputs the predicted expression by using the final network weights learned during training.

The CNN designed and implemented in the present work (Fig. 15.2) is inspired at the classical LeNet-5 architecture [27], a pioneering work used mainly for character recognition. It consists of two convolutional layers each of which followed by a subsampling layer. The resolution of the input grayscale image is 32 × 32, the outputs are numerical value which correspond with the confidence of each expression. The maximum confidence value is selected as the expression detected in the image.

The first main operation is the convolution. Each convolution operation can be represented by the following formula:

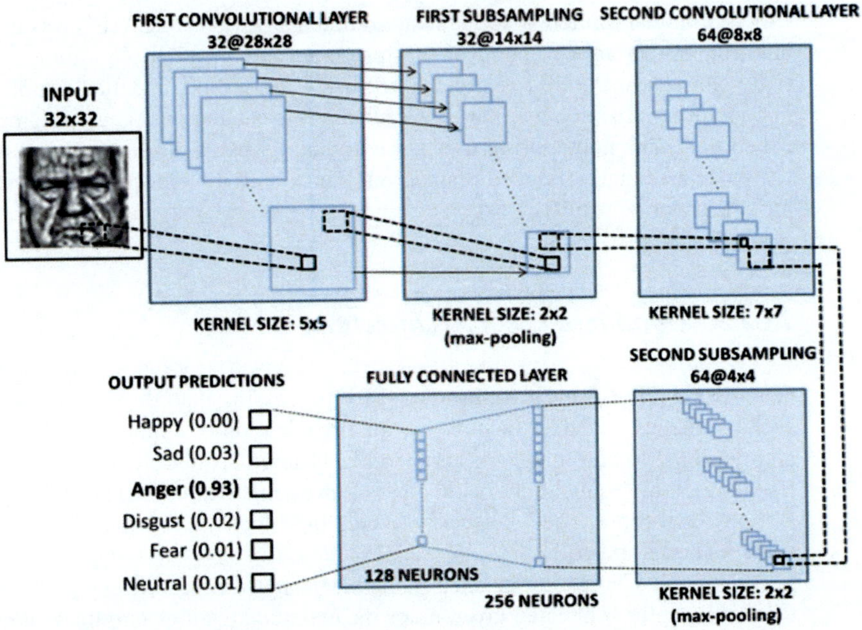

Fig. 15.2 Architecture of the proposed CNN. It comprises of seven layers: 2 convolutional layers, 2 sub-sampling layers and a classification (fully connected layer) in which the last layer has the same number of output nodes (i.e. facial expressions)

$$x'_j = f\left(\sum_{i\epsilon\omega_j} x_i^{l-1} * ck_{ij}^l + b_j^l\right)$$

where x_i^{l-1} and x'_j indicate respectively the i-th input feature map of layer $(l - 1)$ and j-th output feature map of layer l, whereas ω_j represents a series of input feature maps and ck_{ij}^l is the convolutional kernel which connects the i-th and j-th feature map. b'_j is a term called bias (an error term) and f is the activation function. In the present work the widely used Rectified Linear Unit function (ReLU) was applied because it was demonstrated that this kind of nonlinear function has better fitting abilities than hyperbolic tangent function or logistic sigmoid function [31].

The first convolution layer applies a convolution kernel of 5×5 and outputs 32 images of 28×28 pixels. It aims to extract elementary visual features, like oriented edges, end-point, corners and shapes in general. In FER problem, the features detected are mainly the shapes, corners and edges of eyes, eyebrow and lips. Once the features are detected, its exact location is not so important, just its relative position compared to the other features.

For example, the absolute position of the eyebrows is not important, but their distances from the eyes are, because a big distance may indicate, for instance, the surprise expression. This precise position is not only irrelevant but it can also pose a problem, because it can naturally vary for different subjects in the same expression.

The first convolution layer is followed by a sub-sampling (pooling) layer which is used to reduce the image to half of its size and control the overfitting. This layer takes small square blocks (2 × 2) from the convolutional layer and subsamples it to produce a single output from each block. The operation aims to reduce the precision with which the position of the features extracted by the previous layer are encoded in the new map. The most common pooling form is average pooling or max pooling. In the present paper the max-pooling strategy has been employed, which can be formulated as:

$$y_{j,k}^{i} = \max_{0 \leq m,n \leq s} \left\{ x_{j \cdot s+m,k \cdot s+n}^{i} \right\}$$

where i represents the feature map of the previous convolutional layer. The aforementioned expression takes a region (with dimension $s \times s$) and output the maximum value in that region ($y_{j,k}^{i}$). With this operation we are able to reduces an $N \times N$ input image to a $\frac{N}{s} \times \frac{N}{s}$ output image. After the first convolution layer and first subsampling/pooling layer, a new convolution layer performs 64 convolutions with a kernel of 7 × 7, followed by another subsampling/pooling layer, again with a 2 × 2 kernel. The aforementioned two layers (second convolutional layer and second sub-sampling layer) aim to do the same operations that the first ones, but handling features in a lower level, recognizing contextual elements (face elements) instead of simple shapes, edges and corners. The concatenation of sets of convolution and sub-sampling layers achieves a high degree of invariance to geometric transformation of the input.

The generated feature maps, obtained after the execution of the two different stages of features extraction, are reshaped into a one-dimensional (1D) array of numbers (or vector), and connected to a classification layer, also known as fully connected or dense layer, in which every input is connected to every output by a learnable weight. The final layer typically has the same number of output nodes as the number of classes that in the present work is set to six (the maximum number of facial expressions labeled in the analyzed benchmark datasets).

Let $\boldsymbol{x}$ denotes the output of the last hidden layer nodes, and $\boldsymbol{w}$ is the connected weights between the last hidden layer and output layer. The output is defined as $\boldsymbol{f} = \boldsymbol{w}^T \boldsymbol{x} + b$ and it is fed to a *SoftMax()* function able to generate the different probabilities corresponding to the k different facial expression (where k is the total number of expressions contained in a specific dataset), through the following formula:

$$p_n = \frac{\exp(f_n)}{\sum_{c=1}^{k} \exp(f_c)}$$

where p_n is the probability of the k-th class of facial expression and $\sum_{n=1}^{k} p_n = 1$. The proposed CNN was trained using stochastic gradient descendent method [32] with different batch sizes (the number of training examples utilized in one iteration).

After an experimental validation we set a batch size of 128 examples. The weights of the proposed CNN architecture have been update with a weight decay of 0.0005 and an adopt momentum of 0.9, following a methodology widely accepted form the scientific community and proposed in [33]. Consequently, the update rule adopted for a single weight w is:

$$v_{i+1} = 0.9 \cdot v_i - 0.0005 \cdot lr \cdot w_i - lr\frac{\delta L}{\delta w}$$

$$w_{i+1} = w_i + v_{i+1}$$

where i is the iteration index and lr is the learning rate, one of the most important hyper-parameter to tune in order to train a CNN. This value was fixed at 0.01 using the technique described in [34]. Finally, in order to reduce the overfitting during training, a "dropout" strategy was implemented. The purpose of this strategy is to drop out some units in the CNN in a random way. In general it is appropriate to set a fixed probability value p for each unit to dropped out. In the implemented architecture p was set to 0.5 only in the second convolutional layer as it was considered irrelevant to drop out the units from all the hidden layers.

15.2.3 FER Approaches Based on Handcrafted Features

In contrast to deep learning approaches, FER approaches based on handcrafted features do not provide a feature learning stage but a manual feature extraction process. The commonality of various types of conventional approaches is detecting the face region and extracting geometric features or appearance-based features. Even in this category of approaches, the behavior and relative performance of algorithms is poorly analyzed by scientific literature with images of expressions performed by ageing adults. Consequently, in this work, two of the best performing handcrafted features extraction methodologies have been implemented and tested on benchmark datasets.

Generally, geometric features methods are focused on the extraction from the shape or salient point locations of specific facial components (e.g. eyes, mouth, nose, eyebrows, etc.). From an evaluation of the recent research activity in this field, Active Shape Model (ASM) [7] turns out to be a performing method for FER. Here, the face of an ageing subject was processed with a facial landmarks extractor exploiting the Stacked Active Shape Model (STASM) approach. STASM uses Active Shape Model for locating 76 facial landmarks with a simplified form of Scale-invariant feature transform (SIFT) descriptors and it operates with Multivariate Adaptive Regression Splines (MARS) for descriptor matching [35]. After, using the obtained landmarks, a set of 32 features, useful to recognize facial expressions, has been defined. The 32 geometric features extracted are divided into the following three categories:

linear features (18), elliptical features (4) and polygonal features (10) and detailed in Table 15.1.

The last step provides a classification module that uses a Support Vector Machine (SVM) for the analysis of the obtained features vector in order to get a prediction in terms of facial expression (Fig. 15.3).

Regarding the use of appearance-based features, local binary pattern (LBP) [11] is an effective texture description operator, which can be used to measure and extract the adjacent texture information in an image. The LBP feature extraction method used in the present work contains three crucial steps. At first, the facial image is divided into several non-overlapping blocks (set to 8 × 8 after experimenting with different block

Table 15.1 Details of the 32 geometric features computed after the localization of 76 facial landmarks. For each category of features is reported the description related to the formula used for the numeric evaluation of the feature. Moreover, the last column reports details about the localization of the specific feature and the number of extracted features in the specific facial region

Category of features	Description	Details
Linear features (18)	Euclidean distance between 2 points	Mouth (6) Left eye (2) Left eyebrow (1) Right eye (2) Right eyebrow (1) Nose (3) Cheeks (3)
Elliptical features (4)	Major and minor ellipse axes ratio	Mouth (1) Nose (1) Left eye (1) Right eye (1)
Polygonal features (10)	Area of irregular polygons constructed on three or more facial landmark points	Mouth (2) Nose (2) Left eye (2) Right eye (2) Left eyebrow (1) Right eyebrow (1)

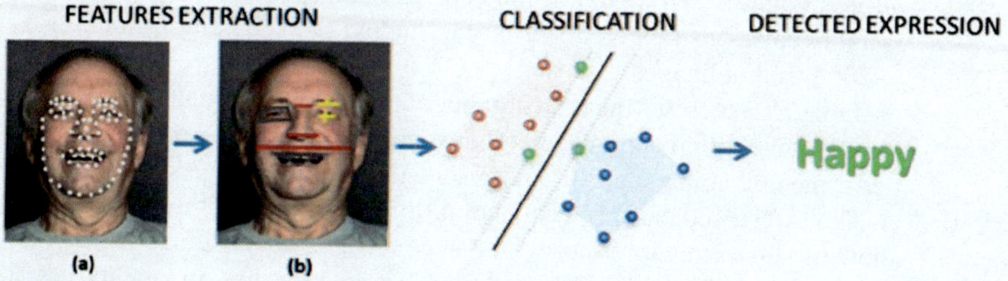

Fig. 15.3 FER based on the geometric features extraction methodology: **a** facial landmark localization, **b** extraction of 32 geometric features (linear, elliptical and polygonal) using the obtained landmarks

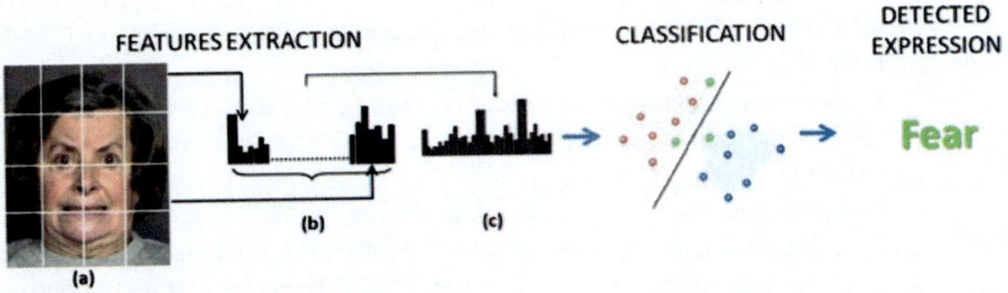

Fig. 15.4 Appearance-based approach used for FER in ageing adults: **a** facial image is divided into non-overlapping blocks of 8 × 8 pixels, **b** for each block the LBP histogram is computed and then concatenated into a single vector (**c**)

sizes). Then, LBP histograms are calculated for each block. Finally, the block LBP histograms are concatenated into a single vector. The resulting vector encodes both the appearance and the spatial relations of facial regions. In this spatially enhanced histogram, we effectively have a description of the facial image on three different levels of locality: the labels for the histogram contain information about the patterns on a pixel-level, the labels are summed over a small region to produce information on a regional level and the regional histograms are concatenated to build a global description of the face image. Finally, also in this case, a SVM classifier is used for the recognition of facial expression (Fig. 15.4).

15.3 Experimental Setup and Results

To validate our methodology a series of experiments were conducted using the age-expression datasets FACES [36] and Lifespan [37].

The FACES dataset is comprised of 58 young (age range: 19–31), 56 middle-aged (age range: 39–55), and 57 older (age range: 69–80) Caucasian women and men (in total 171 subjects). The faces are frontal with fixed illumination mounted in front and above of the faces. The age distribution is not uniform and in total there are 37 different ages. Each model in the FACES dataset is represented by two sets of six facial expressions (anger, disgust, fear, happy, sad and neutral) totaling 171 * 2 * 6 = 2052 frontal images.

Table 15.2 presents the total number of persons in the final FACES dataset, broken down by age group and gender, whereas in Fig. 15.5 some examples of expressions performed by aging adults are represented (one for each class of facial expression).

The Lifespan dataset is a collection of faces of subjects from different ethnicities showing different expressions. The ages of the subjects range from 18 to 93 years and in total there are 74 different ages. The dataset has no labeling for the subject identities. The expression subsets have the following sizes: 580, 258, 78, 64, 40, 10, 9, and 7 for neutral, happy, surprise, sad, annoyed, anger, grumpy and disgust, respectively. Although both datasets cover a wide range of facial expressions, the

Table 15.2 Total number of subjects contained in FACES dataset broken down by age and gender

Gender	Age (years)			
	(19–31)	(39–55)	(69–80)	Total (19–80)
Male	29	27	29	85
Female	29	29	28	86
Total	58	56	57	171

Fig. 15.5 Some examples of expressions performed by aging adults from the FACES database

FACES dataset is more challenging for FER as it contains all the facial expressions to test the methodology. Instead, only four facial expressions (neutral, happy, surprise and sad) can be considered for the Lifespan dataset due to the limited number of images in the other categories of facial expression. Table 15.3 presents the total number of persons in the Lifespan dataset, divided into four different age groups and further distinguished by gender, whereas in Fig. 15.6 some examples of expressions performed by ageing adults are represented (only for “happy”, “neutral”, “surprise” and “sad” expression).

The training and testing phase were performed on Intel i7 3.5 GHz workstation with 16 GB DDR3 and equipped with GPU NVidia Titan X using the Python library for machine learning Tensorflow, developed for implementing, training, testing and deploying deep learning models [38].

Table 15.3 Total number of subjects contained in Lifespan dataset broken down by age and gender

Gender	Age (years)				
	(18–29)	(30–49)	(50–69)	(70–93)	Total (18–93)
Male	114	29	28	48	219
Female	105	47	95	110	357
Total	219	76	123	158	576

Fig. 15.6 Some examples of expressions performed by aging adults from the Lifespan database

For the performance evaluation of the methodologies all the images of FACES dataset were pre-processed, whereas only the facial images of Lifespan with the four facial expressions considered in the present work were considered. Consequently, applying the data augmentation techniques previously described (see Sect. 15.2), in total 65,664 facial images of FACES (equally distributed among the facial expression classes) and 31,360 facial images of Lifespan were used, a sufficient number for using a deep learning technique.

15.3.1 Performance Evaluation

As described in Sect. 15.2.2, for each performed experiment the facial images were separated in three main sets: training set, validation set and test set. Moreover, since gradient descent method was used for training and considering that it is influenced by the order of presentation of the images, the accuracy obtained was an average of the values calculated in 20 different experiments, in each of which the images were randomly ordered. To be less affected by this accuracy variation, a training methodology that uses a validation set to choose the best network weights was implemented.

Since the proposed deep learning FER approach is mainly based on an optimized CNN architecture that is inspired from LeNet-5, it was considered appropriate to first compare the proposed CNN and LeNet-5 on the whole FACES and Lifespan datasets. The metric used in this work for evaluating the methodologies is the accuracy, whose value is calculated using the average of n-class classifier accuracy for each expression (i.e. number of hits of an expression per total number of image with the same expression):

$$Acc = \frac{\sum_1^n Acc_{expr}}{n}, \; Acc_{expr} = \frac{Hit_{expr}}{Total_{expr}}$$

where Hit_{expr} is the number of hits in the expression *expr*, $Total_{expr}$ represents the total number of samples of that expression and n is the number of expressions to be considered.

Figure 15.7 reports the average accuracy and the convergence obtained. The drawn curve emphasizes that the architecture proposed in the present work allows a faster convergence and a higher accuracy value compared to the LeNet-5 architecture, and this happens for both the analysed datasets. In particular, the proposed CNN reaches convergence after about 250 *epoch* for both datasets while LeNet-5 reaches it after 430 *epoch* for FACES dataset and 480 *epoch* for Lifespan dataset. Moreover, the accuracy obtained is considerably higher, with an improvement of around 18% for FACES dataset and 16% for Lifespan dataset.

On the other hand, the final accuracy obtained by the proposed CNN for each age group of FACES and Lifespan dataset is reported in Table 15.4 and Table 15.5. It was computed using the network weights of the best run out of 20 runs, having a validation set for accuracy measurement.

In order to make a comparison, the same tables show the accuracy values obtained using traditional machine learning techniques described in Sect. 15.2.3 (ASM + SVM and LBP + SVM).

The results reported confirm that our proposed CNN approach is superior to traditional approaches based on handcrafted features and this is true for any age group in which the datasets are partitioned. Analysing in more detail the results obtained, it is clear that the proposed CNN obtains a better improvement in the case of recognition of facial expressions performed by ageing adults. Moreover, the hypotheses concerning the difficulties of traditional algorithms in extracting features from an ageing face was confirmed from the fact that ASM and LBP get a greater accuracy with faces of young and middle-aged for each analysed dataset.

As described in Sect. 15.2.1 the implemented pipeline, designed specifically for FER in ageing adults, combines a series of pre-processing steps after data augmentation, with the purpose to remove non-expression specific features of a facial image. Therefore it is appropriate to evaluate the impact in the classification accuracy of each operation in the pre-processing step for the considered methodologies. Four different experiments, which combine the pre-processing steps, were carried

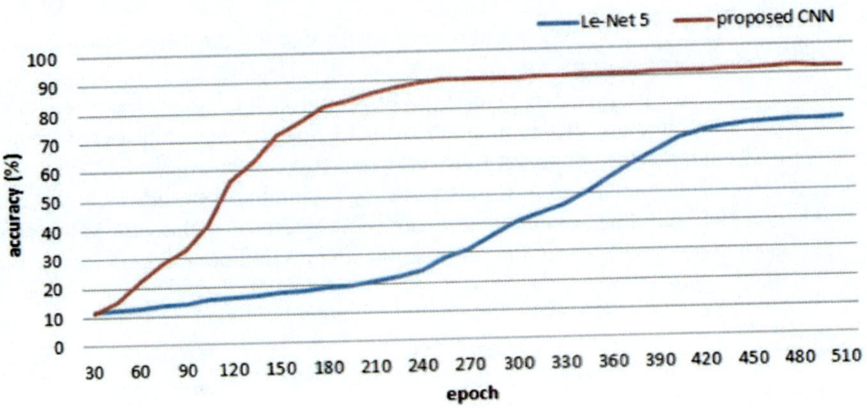

(a)

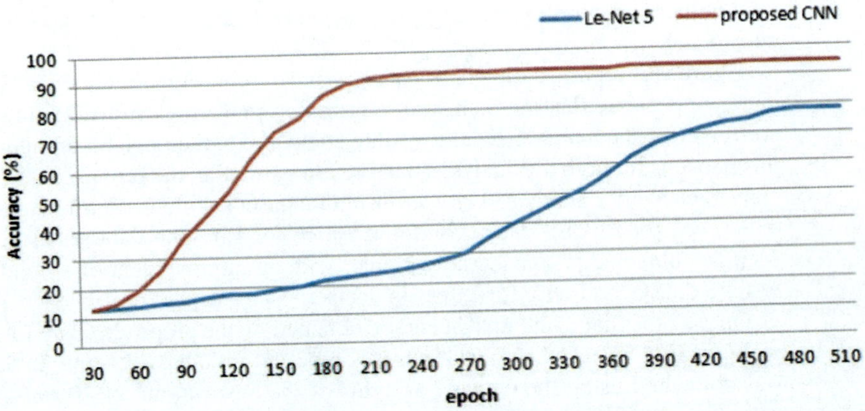

(b)

Fig. 15.7 Comparison in terms of accuracy between Le-Net 5 architecture and the proposed CNN for **a** FACES and **b** Lifespan

Table 15.4 FER accuracy on FACES dataset evaluated for different age group with proposed CNN and traditional machine learning approaches

Age group	Proposed CNN (%)	ASM + SVM (%)	LBP + SVM (%)
Young (19–31 years)	92.43	86.42	87.22
Middle-aged (39–55 years)	92.16	86.81	87.47
Older (69–80 years)	93.86	84.98	85.61
Overall accuracy	92.81	86.07	86.77

Table 15.5 FER accuracy on Lifespan dataset evaluated for different age group with proposed CNN and traditional machine learning approaches

Age group	Proposed CNN (%)	ASM + SVM (%)	LBP + SVM (%)
Young (18–29 years)	93.01	90.16	90.54
Middle-aged (30–49 years)	93.85	89.24	90.01
Older (50–69 years)	95.48	86.12	86.32
Very old (70–93 years)	95.78	85.28	86.01
Overall accuracy	94.53	87.70	88.22

out starting from the images contained in the benchmark datasets: (1) *Only Face Detection*, (2) *Face Detection + Cropping*, (3) *Face Detection + Cropping + Down Sampling*, (4) *Face Detection + Cropping + Down Sampling + Normalization* (Tables 15.6, 15.7 and 15.8).

The results reported in the previous tables show that the introduction of pre-processing steps in the pipeline allows to improve the performance of the whole system both in the case of a FER approach based on deep learning methodology that

Table 15.6 Average classification accuracy obtained for FACES and Lifespan datasets with four different combination of pre-processing steps using the proposed CNN architecture and at varying of age groups

Age range	FACES			Lifespan			
	19–31 (%)	39–55 (%)	69–80 (%)	18–29 (%)	30–49 (%)	50–69 (%)	70–93 (%)
(1)	87.46	86.56	88.31	89.44	89.04	90.32	90.44
(2)	89.44	89.34	91.45	91.13	89.67	92.18	92.15
(3)	91.82	91.88	92.67	92.08	91.99	93.21	94.87
(4)	92.43	92.16	93.86	93.01	93.85	95.48	95.78

Table 15.7 Average classification accuracy obtained for FACES and Lifespan datasets with four different combination of pre-processing steps using ASM + SVM and at varying of age groups

Age range	FACES			Lifespan			
	19–31 (%)	39–55 (%)	69–80 (%)	18–29 (%)	30–49 (%)	50–69 (%)	70–93 (%)
(1)	65.44	66.32	63.33	68.61	69.00	64.90	65.58
(2)	70.18	71.80	69.87	73.44	74.67	71.14	70.19
(3)	74.32	75.77	73.04	79.15	78.57	75.12	74.45
(4)	86.42	86.81	84.98	90.16	89.24	86.12	85.28

Table 15.8 Average classification accuracy obtained for FACES and Lifespan datasets with four different combination of pre-processing steps using LBP + SVM and at varying of age groups

Age range	FACES			Lifespan			
	19–31 (%)	39–55 (%)	69–80 (%)	18–29 (%)	30–49 (%)	50–69 (%)	70–93 (%)
(1)	67.47	68.08	65.54	70.34	71.19	68.87	67.56
(2)	71.34	70.67	69.48	77.89	76.98	71.34	70.84
(3)	76.56	76.43	74.38	82.48	83.32	78.38	77.43
(4)	87.22	87.47	85.61	90.54	90.01	86.32	86.01

in the case of a FER approach based on traditional machine learning techniques, and this is true for any age group. However it is possible to notice how the pre-processing operations improve the FER system more in the case of a methodology based on handcrafted feature extraction because, after the introduction of data augmentation techniques, the proposed CNN manages the variations in the image introduced by the pre-processing steps in an appropriate manner. A further important conclusion reached by the previous test phase is that the performance are not influenced by the age of the subject performing the facial expression, since the improvement obtained in the accuracy value remains almost constant when age changes.

Often, in real-life applications, the expression performed by an observed subject could be very different from the training samples used, in terms of uncontrolled variations such as illumination, pose, age and gender. Therefore, it is important for a FER system to have a good generalization power. As a result it turns out to be essential to design and implement a methodology for feature extraction and classification that is still able to achieve a good performance when the training and test sets are from different datasets. In this paper, we also conduct experiments to test the robustness and accuracy of the compared approaches in the scenario of cross-dataset FER.

Table 15.9 shows the results when the training and the testing sets are two different datasets (FACES and Lifespan) within which there are subjects of different ethnicity and of different ages. Furthermore, image resolution and acquisition conditions are also significantly different. From the results obtained it is evident how the recognition rates for the 3 basic emotions in common between the two datasets ("happy",

Table 15.9 Comparison of the recognition rates of the methodologies on cross-dataset FER

Training on	FACES			Lifespan		
Testing on	Lifespan			FACES		
	Proposed CNN (%)	ASM + SVM (%)	LBP + SVM (%)	Proposed CNN (%)	ASM + SVM (%)	LBP + SVM (%)
Young	51.38	42.44	44.56	53.47	41.87	41.13
Middle-aged	57.34	46.89	50.13	55.98	45.12	47.76
Older-very old	59.64	51.68	52.78	60.07	49.89	51.81

"neutral" and "sad") decrease significantly, because cross-dataset FER is a challenging task. Moreover, this difficulty in classification is greater in the case of facial expressions of young subjects who express emotions more strongly than the ageing adults.

In a multi-class recognition problem, as the FER one, the use of an average recognition rate (i.e. accuracy) among all the classes could be not exhaustive since there is no possibility to inspect what is the separation level, in terms of correct classifications, among classes (in our case, different facial expressions). To overcome this limitation, for each dataset the confusion matrices are then reported in Tables 15.10 and 15.11 (only the facial images of ageing adults were considered). The numerical results obtained makes possible a more detailed analysis of the misclassification and the interpretation of their possible causes. First of all, from the confusion matrices it is possible to observe that the pipeline based on the proposed CNN architecture achieved an average detection rate value over 93.6% for all the tested datasets and that, as expected, its FER performance decreased when the number of classes, and consequently the problem complexity, increased. In fact, in the case of the FACES dataset with 6 expressions, the obtained average accuracy was of 92.81% whereas the average accuracy obtained on Lifespan dataset (4 expressions) was 94.53%.

Going into a more detailed analysis on the results reported in Table 15.9 and related to FACES dataset, "anger" and "fear" are the facial expression better recognized,

Table 15.10 Confusion matrix of six basic expression on FACES dataset (performed by older adults) using the proposed CNN architecture

		Estimated (%)					
		Anger	Disgust	Fear	Happy	Sad	Neutral
Actual (%)	Anger	96.8	0	0	0	2.2	1.0
	Disgust	3.1	93.8	0	0.7	1.8	0.6
	Fear	0	0	95.2	1.5	3.3	0
	Happy	0.7	2.8	1.1	94.3	0	1.1
	Sad	0.6	0	4.1	0	90.2	5.1
	Neutral	2.5	2.0	2.6	0	0	92.9

Table 15.11 Confusion matrix of four basic expression on Lifespan dataset (performed by older and very old adults) using the proposed CNN architecture

	Estimated (%)				
		Happy	Neutral	Surprise	Sad
Actual (%)	Happy	97.7	0.3	1.8	0.2
	Neutral	2.1	96.4	0.6	0.9
	Surprise	4.6	0.1	93.8	1.5
	Sad	0.6	3.8	1.1	94.5

whereas "sad" and "neutral" are the facial expression confused the most. Finally, "sad" is the facial expression with the lowest accuracy.

Instead, the confusion matrix reported in Table 15.10 and related to facial expression classes of Lifespan dataset highlights that "happy" is the facial expression with the best accuracy, whereas the expression "surprise" is the worst expression recognized. "Surprise" and "happy" are the facial expression confused the most.

15.4 Discussion and Conclusions

The main objective of the present study was to compare a deep learning technique with two machine learning techniques for FER in ageing adults, considering that the majority of the works in the literature that address FER topic are based on benchmark datasets that contain facial images with a small span of lifetime (generally young and middle-aged subjects). It is important to stress that one of the biggest limitation in this research area is the availability of datasets containing facial expression of ageing adults, consequently scientific literature is lacking in publications.

Recent studies have demonstrated that human aging has significant impact on computational FER. In fact, by comparing the expression recognition accuracies across different age groups, it was found that the same classification scheme for the recognition of facial expressions cannot be used. Consequently, it was necessary first to evaluate how classical approaches perform on the faces of the elderly, and then consider more general approaches able to automatically learn what features are the most appropriate for expression classification. It is worth pointing out that hand designed feature extraction methods generally rely on manual operations with labelled data, with the limitation that they are supervised. In addition, the hand designed features are able to capture low-level information of facial images, except for high-level representation of facial images. However, deep learning, as a recently emerged machine learning theory, has shown how hierarchies of features can be directly learned from original data. Different from the traditional shallow learning approaches, deep learning is not only multi-layered, but also highlights the importance of feature learning. Motivated by very little work done on deep learning for facial expression recognition in ageing adults, we have firstly investigated an optimized CNN architecture, especially because of his ability to model complex data distributions which can be, for example, a facial expression performed by ageing adults. The basic idea of the present work was to optimize a consolidated architecture like LeNet-5 (which represents the state of the art for the recognition of characters) since revised version of the same architecture has been used in recent years also for the recognition of facial expressions.

From the results obtained it is clear how the optimized CNN architecture proposed achieves better results in terms of accuracy on both the datasets taken into consideration (FACES and Lifespan) compared to classic LeNet-5 architecture (average improvement of around 17%). Moreover, the implemented CNN converges faster than LeNet-5. By a careful analysis of the results obtained, it is possible to observe how two convolutional layers following by two sub-sampling layers are sufficient for the distinction of the facial expression, probably because the high-level features learned have the best distinctive elements for the classification of the six facial expressions contained in FACES and of four facial expressions extracted from Lifespan dataset. Experiments performed with a higher number of layers did not get better recognition percentages, on the contrary, they increased computational time, and therefore it seemed suitable to not investigate more "deeper" architecture.

Another important conclusion that has been reached in the present work is that the proposed CNN is more effective in the classification of facial expressions with respect to the two considered methodologies of machine learning, and the greatest progress in terms of accuracy was found in correspondence with the recognition of facial expressions of elderly subjects. Probably, these results are related to the deformations (wrinkles, folds, etc.) that are more present on the face of the elderly, which greatly affect the use of handcrafted features for classification purposes.

A further added value of this work lies in the implementation of pre-processing blocks. First of all it was necessary to implement "data augmentation" methodologies as the facial images available in FACES and Lifespan datasets were not sufficient for a correct use of a deep learning methodology. The implemented pipeline also provided a series of algorithmic steps which produced normalized facial images, which represented the input for the implemented FER methodologies. Consequently, in the results section, it was also considered appropriate to compare the impact of the algorithmic steps on the classification of the expressions. The results reported show that the optimized CNN architecture benefits less from the implementation of facial pre-processing techniques compared to the proposed machine learning architectures, and this consideration leads to prefer it in real contexts where for example it could be difficult to have always "optimized" images.

It is appropriate, however, a mention on the main limitations of this study. Firstly, the data available for the validation of the methodology are very few, and only thanks to FACES dataset it was possible to distinguish the six facial expressions that are considered necessary to evaluate the mood progression of elderly. Be able to distinguish between a lower number of expression (as happened for Lifespan dataset) may not be enough to extract important information about the mood of the subject being observed.

Another limitation has emerged during cross-dataset experiments. The low percentage of accuracy reached shows that FER in ageing adults is still a topic to be investigated in depth, even the difficulty in classification has been accentuated more in the case of facial expressions of young and middle-aged subjects, but that is probably due to the fact that these subjects express emotions more strongly than the ageing adults.

A final limitation of this work is found in the training of CNN with facial images available only with a frontal-view. Since an example of interesting application might be to monitor an ageing adult within their own home environment, it seems necessary to first study a methodology that automatically locates the face in the image and then extract the most appropriate features for the recognition of expressions. In this case the algorithmic pipeline should be changed, given that the original Viola-Jones face detector has limitations for multi-view face detection [39] (because it only detects frontal upright human faces with approximately up to 20 degree rotation around any axis).

Future works will deal with three main aspects. First of all, the proposed CNN architecture will be tested in the field of assistive technologies, first validating it in a smart home setup and after testing the pipeline in a real ambient assisted living environment, which is the older person's home. In particular, the idea is to develop an application that uses the webcam integrated in TV, smartphone or tablet with the purpose to recognize the facial expression of aging adults in real time and through various cost-effective commercially available devices that are generally present in the living environments of the elderly. The application to be implemented will have to be the starting point to evaluate and eventually modify the mood of the older people living alone at their homes, for example by subjecting it to external sensory stimuli, such as music and images. Secondly, a more wide analysis of how a non-frontal view of the face can affect the facial expression detection rate using the proposed CNN approach will be done, as it may be necessary to monitor the mood of the elderly by using for example a camera installed in the "smart" home for other purposes (e.g. activity recognition or fall detection), and the position of these cameras almost never allows to have a frontal face image of the monitored subject.

Finally, as noted in the introduction of the present work, since the datasets present in literature contain few images of facial expressions of elderly subjects and considering that there are a couple of techniques available to train a model efficiently on a smaller dataset ("data augmentation" and "transfer learning"), a future development will be to focus on transfer learning. Transfer learning is a common and recently strategy to train a network on a small dataset, where a network is pre-trained on an extremely large dataset, such as ImageNet [34], which contains 1.4 million images with 1000 classes, then reused and applied to the given task of interest. The underlying assumption of transfer learning is that generic features learned on a large enough dataset can be shared among seemingly disparate datasets. This portability of learned generic features is a unique advantage of deep learning that makes itself useful in various domain tasks with small datasets.

Consequently, one of the developments of this work will be to test: (1) images containing facial expression of ageing adults present within the datasets, and (2) images containing faces of elderly people acquired within their home environment (even with non-frontal pose) starting from a training derived from models pre-trained on the ImageNet challenge dataset, that are open to the public and readily accessible, along with their learned kernels and weights, such VGG [40], ResNet [41] and GoogleNet/Inception [42].

References

1. United Nations Programme on Ageing. The ageing of the world's population, December 2013. http://www.un.org/en/development/desa/population/publications/pdf/ageing/WorldPopulationAgeing2013.pdf. Accessed July 2018
2. Zeng, Z., Pantic, M., Roisman, G.I., Huang, T.S.: A survey of affect recognition methods: audio, visual, and spontaneous expressions. IEEE Trans. Pattern Anal. Mach. Intell. **31**(1), 39–58 (2009). https://doi.org/10.1109/tpami.2008.52
3. Pantic, M., Rothkrantz, L.J.M.: Automatic analysis of facial expressions: the state of the art. IEEE Trans. Pattern Anal. Mach. Intell. **22**(12), 1424–1445 (2000). https://doi.org/10.1109/34.895976
4. Fasel, B., Luettin, J.: Automatic facial expression analysis: a survey. Pattern Recogn. **36**(1), 259–275 (2003). https://doi.org/10.1016/s0031-3203(02)00052-3
5. Carroll, J.M., Russell, J.A.: Do facial expressions signal specific emotions? Judging emotion from the face in context. J. Pers. Soc. Psychol. **70**(2), 205 (1996). https://doi.org/10.1037//0022-3514.70.2.205
6. Ekman, P., Rolls, E.T., Perrett, D.I., Ellis, H.D.: Facial expressions of emotion: an old controversy and new findings [and discussion]. Philoso. Trans. R Soc. B Biolog. Sci. **335**(1273), 63–69 (1992). https://doi.org/10.1098/rstb.1992.0008
7. Shbib, R., Zhou, S.: Facial expression analysis using active shape model. Int. J. Sig. Process. Image Process. Pattern Recogn. **8**(1), 9–22 (2015). https://doi.org/10.14257/ijsip.2015.8.1.02
8. Cheon, Y., Kim, D.: Natural facial expression recognition using differential-AAM and manifold learning. Pattern Recogn. **42**(7), 1340–1350 (2009). https://doi.org/10.1016/j.patcog.2008.10.010
9. Soyel, H., Demirel, H.: Facial expression recognition based on discriminative scale invariant feature transform. Electron. Lett. **46**(5), 343–345 (2010). https://doi.org/10.1049/el.2010.0092
10. Gu, W., Xiang, C., Venkatesh, Y.V., Huang, D., Lin, H.: Facial expression recognition using radial encoding of local Gabor features and classifier synthesis. Pattern Recogn. **45**(1), 80–91 (2012). https://doi.org/10.1016/j.patcog.2011.05.006
11. Shan, C., Gong, S., McOwan, P.W.: Facial expression recognition based on local binary patterns: a comprehensive study. Image Vis. Comput. **27**(6), 803–816 (2009). https://doi.org/10.1016/j.imavis.2008.08.005
12. Chen, J., Chen, Z., Chi, Z., Fu, H.: Facial expression recognition based on facial components detection and hog features. In: International Workshops on Electrical and Computer Engineering Subfields, pp. 884–888 (2014)
13. Guo, G., Guo, R., Li, X.: Facial expression recognition influenced by human aging. IEEE Trans. Affect. Comput. **4**(3), 291–298 (2013). https://doi.org/10.1109/t-affc.2013.13
14. Wang, S., Wu, S., Gao, Z., Ji, Q.: Facial expression recognition through modeling age-related spatial patterns. Multimedia Tools Appl. **75**(7), 3937–3954 (2016). https://doi.org/10.1007/s11042-015-3107-2
15. Malatesta C.Z., Izard C.E.: The facial expression of emotion: young, middle-aged, and older adult expressions. In: Malatesta C.Z., Izard C.E. (eds.) Emotion in Adult Development, pp. 253–273. Sage Publications, London (1984)
16. Malatesta-Magai, C., Jonas, R., Shepard, B., Culver, L.C.: Type A behavior pattern and emotion expression in younger and older adults. Psychol. Aging **7**(4), 551 (1992). https://doi.org/10.1037//0882-7974.8.1.9
17. Malatesta, C.Z., Fiore, M.J., Messina, J.J.: Affect, personality, and facial expressive characteristics of older people. Psychol. Aging **2**(1), 64 (1987). https://doi.org/10.1037//0882-7974.2.1.64
18. Lozano-Monasor, E., López, M.T., Vigo-Bustos, F., Fernández-Caballero, A.: Facial expression recognition in ageing adults: from lab to ambient assisted living. J. Ambi. Intell. Human. Comput. 1–12 (2017). https://doi.org/10.1007/s12652-017-0464-x
19. LeCun, Y., Bengio, Y., Hinton, G.: Deep learning. Nature **521**(7553), 436–444 (2015). https://doi.org/10.1038/nature14539

20. Yu, D., Deng, L.: Deep learning and its applications to signal and information processing [exploratory dsp]. IEEE Signal Process. Mag. **28**(1), 145–154 (2011). https://doi.org/10.1109/msp.2010.939038
21. Xie, S., Hu, H.: Facial expression recognition with FRR-CNN. Electron. Lett. **53**(4), 235–237 (2017). https://doi.org/10.1049/el.2016.4328
22. Li, Y., Zeng, J., Shan, S., Chen, X.: Occlusion aware facial expression recognition using cnn with attention mechanism. IEEE Trans. Image Process. **28**(5), 2439–2450 (2018). https://doi.org/10.1109/TIP.2018.2886767
23. Lopes, A.T., de Aguiar, E., De Souza, A.F., Oliveira-Santos, T.: Facial expression recognition with convolutional neural networks: coping with few data and the training sample order. Pattern Recogn. **61**, 610–628 (2017). https://doi.org/10.1016/j.patcog.2016.07.026
24. Goodfellow, I.J., Erhan, D., Carrier, P.L., Courville, A., Mirza, M., Hamner, B., …, Zhou, Y.: Challenges in representation learning: a report on three machine learning contests. In: International Conference on Neural Information Processing, pp. 117–124. Springer, Berlin, Heidelberg (2013). https://doi.org/10.1016/j.neunet.2014.09.005
25. Kahou, S.E., Pal, C., Bouthillier, X., Froumenty, P., Gülçehre, Ç., Memisevic, R., …, Mirza, M.: Combining modality specific deep neural networks for emotion recognition in video. In: Proceedings of the 15th ACM on International Conference on Multimodal Interaction, pp. 543–550. ACM (2013)
26. Liu, M., Wang, R., Li, S., Shan, S., Huang, Z., Chen, X.: Combining multiple kernel methods on riemannian manifold for emotion recognition in the wild. In: Proceedings of the 16th International Conference on Multimodal Interaction, pp. 494–501. ACM (2014). https://doi.org/10.1145/2663204.2666274
27. LeCun, Y., Bottou, L., Bengio, Y., Haffner, P.: Gradient-based learning applied to document recognition. Proc. IEEE **86**(11), 2278–2324 (1998). https://doi.org/10.1109/5.726791
28. Viola, P., Jones, M.J.: Robust real-time face detection. Int. J. Comput. Vision **57**(2), 137–154 (2004). https://doi.org/10.1023/b:visi.0000013087.49260.fb
29. Zuiderveld, K.: Contrast limited adaptive histogram equalization. Graphics Gems 474–485 (1994). https://doi.org/10.1016/b978-0-12-336156-1.50061-6
30. Hubel, D.H., Wiesel, T.N.: Receptive fields and functional architecture of monkey striate cortex. J. Physiol. **195**(1), 215–243 (1968). https://doi.org/10.1113/jphysiol.1968.sp008455
31. Glorot, X., Bordes, A., Bengio, Y.: Deep sparse rectifier neural networks. In: Proceedings of the Fourteenth International Conference on Artificial Intelligence and Statistics, pp. 315–323 (2011)
32. Bottou, L.: Large-scale machine learning with stochastic gradient descent. In: Proceedings of COMPSTAT '2010, pp. 177–186. Physica-Verlag HD (2010). https://doi.org/10.1007/978-3-7908-2604-3_16
33. Krizhevsky, A., Sutskever, I., Hinton, G.E.: Imagenet classification with deep convolutional neural networks. In: Advances in Neural Information Processing Systems, pp. 1097–1105 (2012)
34. Smith, L.N.: Cyclical learning rates for training neural networks. In: 2017 IEEE Winter Conference on Applications of Computer Vision (WACV), pp. 464–472 IEEE (2017). https://doi.org/10.1109/wacv.2017.58
35. Milborrow, S., Nicolls, F.: Active shape models with SIFT descriptors and MARS. In: 2014 International Conference on Computer Vision Theory and Applications (VISAPP), vol. 2, pp. 380–387. IEEE (2014). https://doi.org/10.5220/0004680003800387
36. Ebner, N.C., Riediger, M., Lindenberger, U.: FACES—a database of facial expressions in young, middle-aged, and older women and men: development and validation. Behav. Res. Methods **42**(1), 351–362 (2010). https://doi.org/10.3758/brm.42.1.351
37. Minear, M., Park, D.C.: A lifespan database of adult facial stimuli. Behav. Res. Methods Instru. Comput. **36**(4), 630–633 (2004). https://doi.org/10.3758/bf03206543
38. Abadi, M., Barham, P., Chen, J., Chen, Z., Davis, A., Dean, J., …, Kudlur, M.: Tensorflow: a system for large-scale machine learning. In: OSDI, vol. 16, pp. 265–283 (2016)
39. Zhang, C., Zhang, Z.: A survey of recent advances in face detection (2010)

40. Simonyan, K., Zisserman, A.: Very deep convolutional networks for large-scale image recognition. arXiv:1409.1556 (2014)
41. He, K., Zhang, X., Ren, S., Sun, J.: Deep residual learning for image recognition. In: Proceedings of the IEEE Conference on Computer Vision and Pattern Recognition, pp. 770–778 (2016). https://doi.org/10.1109/cvpr.2016.90
42. Szegedy, C., Liu, W., Jia, Y., Sermanet, P., Reed, S., Anguelov, D., …, Rabinovich, A.: Going deeper with convolutions. In: Proceedings of the IEEE Conference on Computer Vision and Pattern Recognition, pp. 1–9 (2015). https://doi.org/10.1109/cvpr.2015.7298594